Sustainable Natural Hazard Management in Alpine Environments

Eric Veulliet · Johann Stötter ·
Hannelore Weck-Hannemann
Editors

Sustainable Natural Hazard Management in Alpine Environments

Editors
Dr. Eric Veulliet
alpS - Centre for Natural Hazard
 and Risk Management
Grabenweg 3
6020 Innsbruck
Austria

Prof. Dr. Johann Stötter
University of Innsbruck
Department of Geography
Innrain 52
6020 Innsbruck
Austria

Prof. Dr. Hannelore Weck-Hannemann
University of Innsbruck
Institute of Public Finance
Universitätsstrasse 15
6020 Innsbruck
Austria

ISBN 978-3-642-03228-8   e-ISBN 978-3-642-03229-5
DOI 10.1007/978-3-642-03229-5
Springer Heidelberg Dordrecht London New York

Library of Congress Control Number: 2009935467

© Springer-Verlag Berlin Heidelberg 2009
This work is subject to copyright. All rights are reserved, whether the whole or part of the material is concerned, specifically the rights of translation, reprinting, reuse of illustrations, recitation, broadcasting, reproduction on microfilm or in any other way, and storage in data banks. Duplication of this publication or parts thereof is permitted only under the provisions of the German Copyright Law of September 9, 1965, in its current version, and permission for use must always be obtained from Springer. Violations are liable to prosecution under the German Copyright Law.
The use of general descriptive names, registered names, trademarks, etc. in this publication does not imply, even in the absence of a specific statement, that such names are exempt from the relevant protective laws and regulations and therefore free for general use.

Cover design: deblik, Berlin

Printed on acid-free paper

Springer is part of Springer Science+Business Media (www.springer.com)

Preface

While natural disasters occurred with increased frequency at the end of the 1990s, especially in the Alpine regions including Austria, it became obvious that a joint research and development platform on natural hazards and risk management was in demand. The priority and necessity for such a forum was commonly perceived and finally met with the formation of the *alpS- Centre for Natural Hazard and Risk Management*. The *alpS*-platform was established in Innsbruck in the centre of the Alps, bringing together important stakeholders from Alpine regions under the umbrella of a one research centre. Thus, the vision of unified natural hazard research could actually be transformed into reality.

The seminal cooperation culture of *alpS* is focussed on project-specific, close and long-term ties with business and scientific partners. This strategy aligned perfectly with the Austrian research sponsorship programme Kplus which proofed to be an excellent framework for a research and development centre such as *alpS*.

The overall thematic proximity of most *alpS*-projects to public responsibilities inevitably led to closer cooperation with state agencies as well as other public authorities. According to the integral risk management concept applied, the participation of all relevant players was called for and could be realised within all *alpS*-projects related to natural hazard and risk management.

In the light of rapid demographic and economic changes, which aroused intensive conflict of interest and a significant increase of risk potentials, research into adaptation strategies for highly populated mountain areas needed to envisage such an approach. In this respect, interdisciplinary research and development clearly signifies a fundamental contribution.

This book intends both to describe central approaches to sustainable handling of natural hazards followed by *alpS* and to deliver insights into concrete projects conducted at the centre between 2002 and 2008. Additionally, natural, climatic and socio-economic issues of current and future challenges in Alpine areas are presented as well. Finally, the book echoes the experiences made in the *alpS* Centre throughout its first years of existence.

The projects described were selected on the basis of transdisciplinarity and integral risk management requirements. They intend to provide an insight into the possibilities of modelling and effective monitoring of natural hazard processes. Additionally, the focus rests upon the application of cutting edge methods and techniques in this realm in order to meet the requirements of meticulous data processing and collection. Common to all projects is their focus on preventive and active risk management, including the analysis, evaluation and sustainable governance of risks within a concept of integral risk management.

Besides natural scientific and technical aspects of natural hazard and risk management, their socio-economic perspectives are notably acknowledged and given high priority. Above all, this approach of *alpS* and its network of partners has successfully shown that an active and integral handling of natural hazards is both sustainable and effective.

Finally we would like to thank all project partners – especially authors and reviewers, the scientific and business partners as well as public authorities – for their cooperation and support in all respects. The success of *alpS* and as a result, this book could not have been achieved without the input and engagement of all these partners involved.

Eric Veulliet - Hans Stötter - Hannelore Weck-Hannemann
March 2009

Contents

1 Global Change and Natural Hazards: New Challenges, New Strategies 1
    J. Stötter, H. Weck-Hannemann, E. Veulliet
    1.1 Introduction 1
    1.2. Global Change Processes 3
        1.2.1 Changes in the temperature system 3
        1.2.2 Changes in the hydro-meteorological system 6
        1.2.3 Explanation - trying to find the key driver 10
        1.2.4 Consequences 11
    1.3 Global Climate Change and its Future Regional Implications ... 13
        1.3.1 The Global Frame 13
        1.3.2 Regional Implications for the 21st Century 15
        1.3.3 Globalization: Socio-economic Change and its Regional Implications 19
    1.4 Changing risks as a consequence of the overlay of Global Change processes 26
    1.5 Contributing to a sustainable natural hazard management: Our main task 28
    References 30

2 Flood Forecasting for the River Inn 35
    S. Senfter, G. Leonhardt, C. Oberparleiter, J. Asztalos, R. Kirnbauer, F.Schöberl, H. Schönlaub
    2.1 Introduction 35
        2.1.1 The River Inn in Tyrol 35
        2.1.2 Risk Management and Flooding 38
        2.1.3 An Overview of the Forecast Model 39
    2.2 Meteorological Data 41
        2.2.1 Forecast data 41
        2.2.2 Observed Data 41
    2.3 Hydrological Model HQ$_{SIM}$ 43
        2.3.1 Model Description 43
        2.3.2 Data 45

2.3.3 Calibration 45
2.3.4 Definition of the Hydrotopes... 46
2.4. Glacier Model SES 49
2.4.1 Model Description 49
2.4.2 Data 49
2.4.3 Calibration 50
2.5 The Hydrodynamic Model... 51
2.5.1 Model Description 51
2.5.2 Data 51
2.5.3 Calibration 53
2.5.4 Rule-Based Operation of Power Plants 58
2.5.5 Hydrodynamic Model for the Bavarian Inn 60
2.5.6 Potential for the Optimization of the Operational Management of Alpine Reservoirs – Example Based on the Flood in August 2005 62
2.6 Summary... 64
References 65

3 Runoff and bedload transport modelling for flood hazard assessment in small alpine catchments - the PROMAB^{GIS} model 69

M. Rinderer, S. Jenewein, S. Senfter, D. Rickenmann, F. Schöberl, J. Stötter, C. Hegg

3.1 Introduction 69
3.1.1 Simple approaches for total sediment load assessment 70
3.1.2 Simulation models for hydro-sedigraph estimation............. 71
3.2 PROMAB^{GIS} a process-based approach for massbalances in small alpine catchments... 74
3.3 Modell concept PROMAB^{GIS} 74
3.3.1 Runoff... 75
3.3.2 Bedload Transport 81
3.4 Case study Erlenbach 84
3.4.1 Overview 84
3.4.2 Geology and Processes 86
3.4.3 Hydrological setting 87
3.4.4 Modell validation 89
References 97

4 Modelling peak runoff in small Alpine catchments based on area properties and system status 103
C. Geitner, M. Mergili, J. Lammel, A. Moran, C. Oberparleiter, G. Meißl, H. Stötter
    4.1 Introduction 103
        4.1.1 Background and motivation 103
        4.1.2 Challenges and objectives 104
        4.1.3 State of the art 106
    4.2 Study areas 107
        4.2.1 Location and catchment characteristics 108
        4.2.2 Data basis and hydrological comparison 109
    4.3 Model 112
        4.3.1 General model layout 112
        4.3.2 Input data and parameters 113
        4.3.3 Parameter estimation 117
        4.3.4 System status model 120
        4.3.5 Pool model 123
        4.3.6 Runoff model 124
    4.4 Results 125
        4.4.1 Comparison of modelled and gauged discharges in the Stampfangertal catchment 125
        4.4.2 Comparison of modelled and gauged discharges in the Längental catchment 126
    4.5 Conclusions and discussion 128
    References 130

5 Process-based investigations and monitoring of deep-seated landslides 135
C. Zangerl, C. Prager, W. Chwatal, S. Mertl, D. Renk, B. Schneider-Muntau, H. Kirschner, R. Brandner, E. Brückl, W. Fellin, E. Tentschert, S. Eder, G. Poscher, H. Schönlaub
    5.1 Introduction 135
    5.2 Landslide classifications 136
    5.3 Temporal distribution of dated landslides in the East Alpine region 137
    5.4 Basic principles of deformation and failure processes of landslides 142
        5.4.1 Fracture mechanical processes 143
        5.4.2 Sliding processes 145
        5.4.3 Failure and temporal behaviour of sliding zone materials. 149
        5.4.4 Temporal deformation behaviour of landslides 153
    5.5 Geophysical investigation methods 158

5.5.1 Active seismic methods 158
　　5.5.2 Ground Penetrating Radar 163
　5.6 Monitoring of landslides 164
　　5.6.1 Deformation Monitoring 164
　　5.6.2 Seismic monitoring 168
　5.7 Summary 172
　References 173

6 Alpine tourist destinations – a safe haven in turbulent times? – Exploring travellers' perception of risks and events of damage 179
　C. Eitzinger, P.M. Wiedemann
　6.1 Introduction 179
　6.2 Theoretical approaches in risk perception research 180
　　6.2.1 Bounded rationality, heuristics und biases 180
　　6.2.2 The Psychometric Paradigm 181
　　6.2.3 The Cultural Theory 181
　　6.2.4 Context dependency of risk perception 182
　　6.2.5 The role of affective processes 182
　6.3 Risk and damage perception in alpine tourist destinations 183
　　6.3.1 Interviews: What are typical risks of alpine destinations? 184
　　6.3.2 Online questionnaire: Risk perception and trust building safety measures in alpine tourist destinations 185
　　6.3.3 Psychometric study: Qualitative dimensions of damage perception 189
　　6.3.4 Experiment on Story Effects in damage perception 194
　6.4 Summary and outlook 196
　References 197

7 Protective measures against natural hazards – are they worth their costs? 201
　A. Leiter, M. Thöni, H. Weck-Hannemann
　7.1 Introduction 201
　7.2 Alternative Decision Support Instruments 203
　　7.2.1 Cost Benefit Analysis and Cost Effectiveness Analysis ... 203
　　7.2.2 Cost Benefit Analysis - a conceptual approach 204
　7.3 The Valuation of Benefits 210
　　7.3.1 Survey design – primary data collection 213
　　7.3.2 Descriptive statistics 215
　　7.3.3 Econometric analysis 220
　　7.3.4 Results 221
　　7.3.5 Application of the results and an outlook 223
　7.4 Conclusion 224
　References 225

8 Analysing Decision Mechanisms for Natural Hazard Management 229
   C.D. Gamper, P.A. Raschky, H. Weck-Hannemann
     8.1 Introduction 229
     8.2 A process-oriented economic approach 232
        8.2.1 How can a process-oriented analysis be designed? 233
     8.3 Multi Criteria Decision Analysis – combining output- and process-orientation for an efficient decision-making process 239
        8.3.1 MCA – the procedure 241
        8.3.2 MCA's potential for natural hazard management 244
     8.4 Conclusions 245
     References 247

9 Alternative Risk Transfer and Alternative Risk Financing 251
   M. Gruber, R. Wiesner
     9.1 Introduction 251
     9.2 Self-insurance, Captives and Risk Retention Groups 251
     9.3 Securitisation 257
     9.4 Financial Reinsurance and Finite Risk 260
     9.5 Contingent Capital Structures 263
     9.6 Multi-line/Multi-year und Multi-trigger Products 266
     9.7 Weather Derivatives 269
     References 273

10 Risk management 277
   S. Ortner, J. Lammel, M. Pöckl, A.P. Moran
     10.1 Introduction 277
     10.2 Risk 277
     10.3 Risk management model 279
        10.3.1 System Framework - Definitions 281
        10.3.2 Process stages of risk management 284
     10.4 Application of risk management 288
        10.4.1 Communal level – the city of Innsbruck 288
        10.4.2 Risk assessment 294
        10.4.3 Corporate level – TILAK (The Tyrolean Provincial Hospital Company) 299
        10.4.4 Corporate level – Ski lift operator – Schlick 2000 Schizentrum AG, Bergbahn AG Kitzbühel 301
     10.5 Outlook 306
     References 307

11 Laser scanning - a paradigm change in topographic data acquisition for natural hazard management 309
T. Geist, B. Höfle, M. Rutzinger, N. Pfeifer, J. Stötter
    11.1 Introduction 309
    11.2 Description of the technology 311
    11.3 Description of data products 315
        11.3.1 Laser point cloud 316
        11.3.2 Digital Elevation Models (DEM) 318
        11.3.3 Digital Intensity Models 319
    11.4 Existing applications in natural hazard management 321
        11.4.1 Rockfall 322
        11.4.2 Landslides and debris flow deposits 322
        11.4.3 Hydrology (torrent activities and floods) 323
        11.4.4 Hazards related to glaciers and permafrost conditions 326
        11.4.5 Avalanches 326
        11.4.6 Protection forest 327
        11.4.7 Object protection 327
        11.4.8 Conclusion 327
    11.5 Laser scanning data and products – a substantial input for natural hazard management 328
        11.5.1 Data management: GIS and database embedding strategies 328
        11.5.2 Data analysis 334
        11.5.3 ALS land cover classification 335
        11.5.4 Topographic analysis 336
    11.6 Opening new dimensions - future potentials of laser scanning in natural hazard risk management 338
    References 339

12 Improving Safety in Alpine Regions through a combination of GSM/GPRS with satellite communication, GIS, and robust positioning technology 345
S. Baumann, J. Czaja, W. Lechner
    12.1 Introduction 345
    12.2 User requirements and market analysis 346
        12.2.1 User requirements 346
        12.2.2 Market overview 346
    12.3 Application scenario definition 352
        12.3.1 Background 352
        12.3.2 Mobile Mission Centre scenario 353
    12.4 System architecture with innovative mobile modules 355

    12.5 Validation of system components..358
        12.5.1 Simulation of SatCom availability 358
        12.5.2 GSM/SatCom availability in the Alps..........................360
        12.5.3 Performance analysis of "low-cost" GPS-receivers 363
        12.5.4 Multi-Sensor Box (MSB) 366
        12.5.5 Multi-Communication Box (MCB)................................368
    12.6 Validation of the PANORAMA core system 370
        12.6.1 System Overview..370
        12.6.2 Geo-data 370
        12.6.3 User terminal 371
        12.6.4 GIS with www-interface..371
        12.6.5 Test campaign..372
    12.7 Conclusions 374
    References 376

13 Pros and cons of four years experience of alpS............................379
    E. Veulliet, H. Weck-Hannemann, J. Stötter...............................379
    13.1 Introduction 379
    13.2 Main Objectives..380
    13.3 Basic Approach...382
    13.4 Structure 383
    13.5 Research Program...385
    13.6 Lessons learned...387
        13.6.1 Integrative Organisation 387
        13.6.2 Challenges, Development and Adaptation 389
    13.7 Future Challenges 392

List of reviewers 395

List of contributors 397

List of partners 401

1 Global Change and Natural Hazards: New Challenges, New Strategies

J. Stötter, H. Weck-Hannemann, E. Veulliet

1.1 Introduction

A contribution to a sustainable natural hazard management is, according to the title, the major goal of this volume. This has to follow certain demands which arise from the basic understanding of the idea of sustainability. Following the definition of former WCED secretary Brundtland that sustainable development must meet the needs of both present and future generations (World Commission on Environment and Development, 1987) this concept is future-oriented (Fig. 1.1). Furthermore, there is a common understanding of the sustainability-concepts' three basic components: economic, social and ecological development (e.g. Enquete Kommission 1998; UNDESA 2002). Both, understanding the complex processes and providing society-relevant solutions requires expertise from many different disciplines or in broader terms, complementary interrelated research activities.

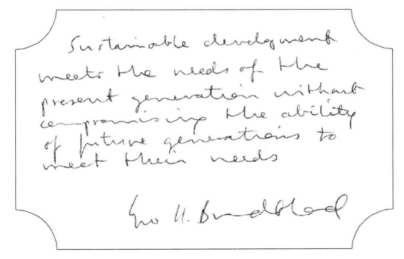

Fig. 1.1 Gro Harlem Brundtland´s definition of "Sustainability"

E. Veulliet et al. (eds.), *Sustainable Natural Hazard Management in Alpine Environments*,
DOI 10.1007/978-3-642-03229-5_1, © Springer-Verlag Berlin Heidelberg 2009

Apart from that, the idea of sustainability is strongly linked to the manifold processes of Global Change. In a systematic approach, the links between processes impacting on a global scale like global climate change and globalisation and that of a regional/local dimension of sustainability are obvious.

In such a context, it is both the conviction and the emphasis of the editors of this volume that Global Change not only involves climate change and greenhouse gas emissions, nor can it be understood in terms of a simple cause-effect paradigm. The term *Global Change* covers all natural physical, chemical, biological and societal processes, which have effects on the system earth as a whole, whereas human beings are both acting as drivers of Global Change processes and reacting as recipients of the consequences (Nationales Komitee für Global Change Forschung 2005). Consequently, research regarding these processes has to focus on the interrelations between human society and all other driving components of the earth system, both at present and in future.

Consequently, a firm disciplinary basis of the situation needs the company of integrative research approaches to fully understand the man-environment system and the complex interactions of its controlling factors and actors (see already WBGU 1993). This understanding of the interrelations between Global Change and sustainability are closely linked to the fundamental ideas of the Earth System Science Partnership (ESSP), which intends to *"bring together researchers from diverse fields, and from across the globe, to undertake an integrated study of the Earth System: its structure and functioning; the changes occurring to the System; and, the implications of those changes for global sustainability"* (stated at the Global Change Open Science Conference: Challenges of a Changing Earth at Amsterdam 2001).

Thus, many scientists across all disciplines increasingly perceive that the earth is a complex and sensitive system regulated by natural processes - and influenced, significantly and as never before, by human impacts. Moreover, it is acknowledged that bridges have to be built across disciplines in order to understand the complex interrelations within this man-environment system. This is reflected, e.g., by new theory-based scientific discussions which may end up in newly designed disciplines like "Sustainability Science" (Fig. 1.2), "Social Ecology" or even a new "Integrative Geography".

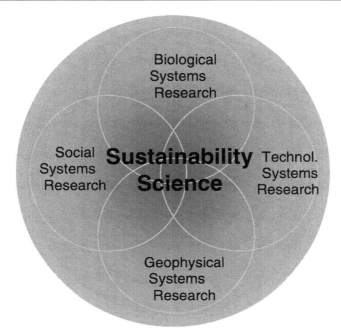

Fig. 1.2 Conceptional structure of Sustainability Research

Both Sustainability Science and Social Ecology tend to see the relation between man/society and nature as so-called "coupled systems" which are both determined by and determining each other. These disciplines aim at a deeper understanding of non-linear, complex and self-organising systems and finding of solutions for the complex problems of society which they determine to be the major research task of the 21st century (e.g. Stötter and Coy 2006). The position of these newly designed disciplines differs strongly from the situation within geography, a discipline which has understood man-environment relationship as its immanent paradigm since the very beginning of scientific geography which is, at least to some extent, now a re-discovered focus (Weichhart 2005).

1.2. Global Change Processes

1.2.1 Changes in the temperature system

The publication of the Fourth IPCC Report in early 2007 has fostered a worldwide awareness of the ongoing processes of Global Climate Change.

Although the predicted magnitude of future climate change has varied throughout the last decade (see former IPCC reports), from a present point of view all model results outlining scenarios of future potential climate conditions suggest a remarkable warming throughout the 21st century.

According to recent results of the IPCC, global average surface temperature has increased since the mid 19th century by 0.76°C ± 0.19°C (Solomon et al. 2007). This general warming trend has increased especially over the last ca. 50 years with an average of 0.13°C ± 0.03°C per decade (Fig. 1.3). The argument of an accelerating warming finds strong support and thus is highlighted by the glances taken at the developments of the last twelve years (1995 to 2006), eleven of which – the exception being 1996 – rank among the twelve warmest years on record since 1850.

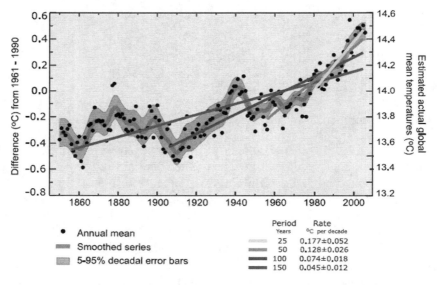

Fig. 1.3 Patterns of linear global temperature trends over the period 1850 to 2005 for the last 25 (yellow), 50 (orange), 100 (magenta) and 150 years (red). The left hand axis shows temperature anomalies relative to the 1961 to 1990 average and the right hand axis shows estimated actual temperatures, both in °C (Trenberth et al. 2007)

This global temporal development may be exemplified by two recent examples of temperature extremes on seasonal level, summer 2003 and winter 2006/2007. The summer of 2003 (June to August) was by far the warmest since the beginning of regular instrumental recording in Central Europe (e.g. Schönwiese et al. 2004). In Germany, the average temperature of summer 2003 of 19.4°C was 3.4K above the average of the period of

1961-1990, a fact that according to Schönwiese et al. (2004) occurs with a probability of $p = 0.7 \cdot 10^{-6}$ (return period > 1 million years). But when we compare the summer of 2003 with the time-period between 1994 and 2003, the occurrence probability mirrors a much more realistic picture with the values $p = 0.0022$ and an annual occurring-probability of approximately 450 years.

Following the explanations of Schär et al. (2004), the summer of 2003 constituted a comparably extreme singularity in Switzerland (Fig. 1.4) with an return period of the dimension of 10^6 years.

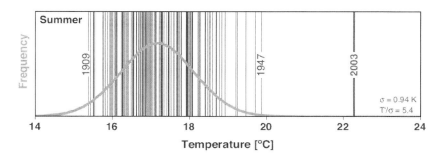

Fig. 1.4 Comparison of summer temperature 2003 to the Gaussian distribution of summer temperatures from 1864 to 2000 in Switzerland (The values in the lower right corner indicate the standard deviation (σ) and the 2003 anomaly normalised by the 1864 to 2000 standard deviation (T'/σ)) (after Schär et al. 2004)

Not less interestingly, a tentative analysis of the average temperature of the winter of 2006/2007 (September – April) leads towards similar impressions. For the meteorological station of Munich Airport, the comparison between values of "winter 2006/2007" and the average of the period 1961-1990 led to a probability of such a warm winter in $p < 10^{-6}$ (again return period in the dimension of 1 million years). But this relation changes quite strongly when we consider the recent past as a basis for our comparison. Taking into account the winters of the time-period between 1992 and 2006, the probability still appears quite extreme ($p = 0.5 \cdot 10^{-5}$; return period ca. 200.000 years). But once integrated into our comparison, winter 2006/07 seems to make the probability more understandable ($p = 0.005$; return period ca. 200 years).

1.2.2 Changes in the hydro-meteorological system

In recent years, both precipitation and runoff in Central Europe seem to have undergone a development that followed a similar pattern of exceeding values which have not been recorded since the beginning of regular measurements. Different river systems have experienced extreme floods, i.e. river Oder in July 1997, Rhine, Lech, Inn and Danube in May 1999, Weichsel in July 2001, Danube and Elbe and contributing rivers in August 2002, Lech and Inn in August 2005 (Table 1.1).

Table1.1 Extreme precipitation and runoff events in Central Europe in the last 10 years * Na (North anti-cyclonal) und NEz (North-East cyclonal), SEz (South-East cyclonal), TrW (trough above Western Europe).

Flood event	Rivers	Precipitation dimension	Runoff dimension	synoptic system*
July 1997	Oder	> 400% (Fuchs and Rudolf 2002) > 100 years (Jarabac and Chlebek 2000)	>> 100 years	TrW, Na, NEz
May 1999	Rhine, Lech, Inn, Danube	> 300% 50-100 years (Fuchs et al. 2000)	200-300 years (Vogelbacher and Kästner 1999)	SEz (= Vb)
August 2002	Danube, Elbe and contributing rivers	> 100 years	100-200 years (Gewässerkundlicher Dienst Bayern 2002)	SEz
August 2005	Inn, Lech	50-100/200 years (Habersack and Krapetz 2006)	100-200 years (> 5000 years)	SEz

By means of pure description, this data inevitably hints at a massive change in the "precipitation – runoff"-relationship. This idea is supported by first systematic analyses runoff records at different gauging stations in SW-Germany, where Caspary (2006) could explain that there is a marked change in the annual reoccurrence-rate of runoff. As shown for the river Danube, the HQ value for a flood of 100 years return period based on the recording of 45 years (1932-1976) is clearly below that similar events based on 25 years analysis of the recent past (1977-2002). This highlights that the magnitude has grown quite dramatically (Table 1.2). In the meantime, the occurrence probability of an event in the dimension which refers to former 100 years return period has increased by the factor 10.

Table 1.2 HQ-values [m^{3}s^{-1}] and confidence intervals (95% significance level) for the gauging station Beuron/Danube as function of return period based on Gumbel distribution (Caspary 2006).

Time of recording	Return period T (years)							
	2	5	10	20	25	50	100	200
1932-1976	104 ±12%	147 ±12%	175 ±16%	202 ±17%	210 ±18%	237 ±19%	263 ±20%	289 ±20%
1932-2002	125 ±12%	187 ±13%	229 ±15%	269 ±16%	281 ±16%	320 ±16%	359 ±18%	397 ±18%
1977-2002	193 ±20%	292 ±22%	358 ±25%	421 ±26%	457 ±27%	502 ±28%	563 ±29%	

Once applied to a "frequency – magnitude"-matrix, these calculations can be visualised quite nicely in their "change"-dimension (Fig. 1.5). The fundamental finding mirrors a strong increase of the hazard potential caused by the runoff/flood process.

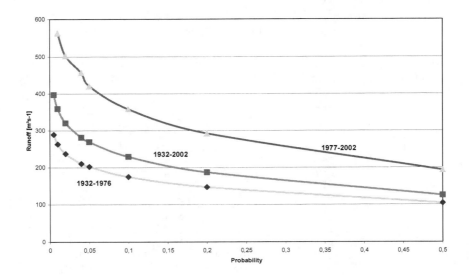

Fig. 1.5 Change of the "frequency – magnitude"-relationship at the gauging station Beuron/Danube in the period 1932-2002

The direction of this argument finds support in the analysis of two independent flood events of river Rhine in 1993 and 1995 with its ≥100 years' return period dimension ($p \leq 0.01$) by Griser and Beck (2003). The probability of the occurrence of two such extreme events follows a binominal distribution, where k = number of events, n = time between two events and p = probability:

$$P = \frac{n!}{k!(n-k)!} p^k (1-p)^{n-k}$$

(eq. 1.1)

The application of this model shows that different river systems have experienced a rather unpredictable coupling of flood events during the recent past (Table 1.3). This may also serve as an indicator for a trend towards a more intensive precipitation - runoff relationship.

1 Global Change and Natural Hazards: New Challenges, New Strategies

Table 1.3 : Probability of two independent extreme flood events (return period ≥ 100 years) within a short period of time following a binominal distribution.

River	Years of event	Probability
Rhein	1993, 1995	1 : 3360
Danube	1999, 2002	1 : 2700
Elbe	2002, 2006	1 : 1030
Lech	1999, 2005	1 : 690

These developments are even more apparent in the case of the "Bregenzer Ache"-river in Vorarlberg, Austria. Three flood events of a dimension of ≥ 100 years have occurred in the years 1999, 2002 and 2005 at the gauging station in Mellau (Passer and Partner 2006). The statistical probability of such an accumulation of extremes evidences how unlikely such an event is ($p < 0.0001$). Furthermore, it demonstrates that the changes in the hydro-meteorological subsystem are comparable to those observed in the realm of temperature.

Additionally, these findings on both temperature and precipitation-runoff system changes evidence that there exists a severe, ongoing change in the natural system which inevitably results in a new relationship between frequency and magnitude that has not been experienced within the last Millennium (for temperature see Fig. 1.6).

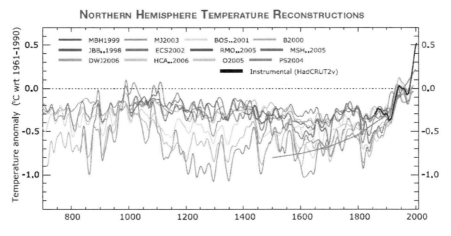

Fig. 1.6 Records of Northern Hemisphere temperature variation during the last 1300 years, with 12 reconstructions using multiple climate proxy records shown in colour and instrumental records shown in black (Solomon et al. 2007)

1.2.3 Explanation - trying to find the key driver

The explanation to all these events appears inextricably linked with changing climate conditions. Following the investigations of Fricke (2002), this development may be explained as a consequence of a significant and ongoing increase of synoptic situations (*"Großwetterlagen"*) causing rainfall with high intensity since the beginning of the 20th century. Although the record shows some drawbacks (interruption), there is a general trend (Fig. 1.7) that includes the Vb-situation apparent when a low pressure system moves from the Mediterranean Sea anti-clockwise around the Alps thus causing quite effective rainfall at the northern fringe of the Alps (e.g. the flood events in August 2002 and in August 2005).

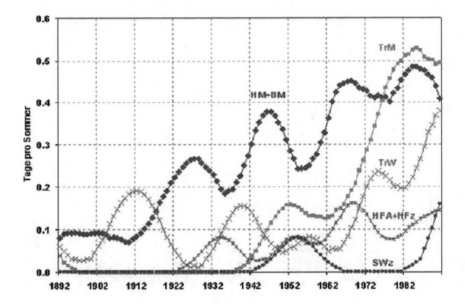

Fig. 1.7 Synoptic situations with a significant increase of summer days (June-August) with daily precipitation > 30 mm (Fricke 2002)

However, all events described above have shown that infrequent extreme temperatures and extreme precipitation intensities follow a simple temporal distribution - a pattern that appears in similar shape to a 'Normal' or 'Gaussian' curve with the extremes at the high and low ends of the range of values (Fig. 1.8). Hence the probability of occurrence of values for some variables in this range is called a "probability distribution function" (PDF). As outlined in the IPCC Report, simple statistical reasoning indicates that substantial changes in the frequency of extreme events can result from a relatively small shift of the distribution of a climate variable (Solomon et al. 2007). As shown in Fig. 1.8, a small shift (corresponding to a small change on average) can have an effect on the frequency of extremes at either end of the distribution. An increase in the frequency of one extreme (e.g., seasonal temperature or rainfall intensity) may be accompanied by a decline in the opposite extreme. Nevertheless, changes in the variability or shape of the distribution can complicate this simple picture.

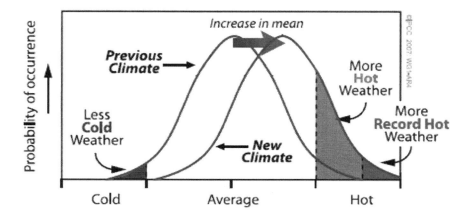

Fig. 1.8 Scheme showing the effect on extreme temperatures when the mean temperature increases, for a normal temperature distribution (Solomon et al. 2007)

Although it has to be considered that due to both the ratio type and the non-Gaussian distribution of data, linear trend functions are not appropriate means to describe the frequency - magnitude of precipitation, the shifting of the precipitation events doubtlessly can be understood as following a similar systematic background.

1.2.4 Consequences

While the regional downscaling for Southern Germany shows comparable trends towards an increasing precipitation until the mid-21st century, results from Alpine areas appear far from uniform in this regard (e.g. Arbeitskreis KLIWA 2006). This also reflects in the general problems of regional scale modelling in mountain areas. Although there is an urgent need for further regional differentiation by improved modelling, wide consensus exists on the contention that e.g. *"in all parts of Germany current vulnerability is high in the water sector, due to increasing flood risk and high potential for damage"* (Schröter et al. 2006). Consequently, first official reactions took a changing precipitation clearly into account, at the same time focusing on runoff interrelations for the planning of protection measures against flooding. In Bavaria, for instance, a so called "climate change factor" has been introduced for the dimensioning of design events for the runoff process.

The climate change factor was set to 15% up to HQ_{100} (p > 0.01) and 7.5% up to HQ_{500} (0.01 ≥ p > 0.002) as a contribution to improved future protection (StMUGV 2004). Though this first step can be identified as a future-oriented way of thinking and acting, there is a far wider need for activities. Precipitation and runoff scenarios have to be modelled on a detailed regional scale assuming different defined global climate impacts as LfU (2005) showed that there is a disperse regional pattern (Table 1.4 and Fig. 1.9).

Table 1.4 Regional climate change factors for different annualities for five hydrological regions of Baden Württemberg (see Fig. 1.9) (Arbeitskreis KLIWA 2006).

Return period [years]	Region 1	Region 2	Region 3	Region 4	Region 5
2	1.25	1.50	1.75	1.50	1.75
5	1.24	1.45	1.65	1.45	1.67
10	1.23	1.40	1.55	1.43	1.60
20	1.21	1.33	1.42	1.40	1.50
50	1.18	1.23	1.25	1.31	1.35
100	1.15	1.15	1.15	1.25	1.25
200	1.12	1.08	1.07	1.18	1.15
500	1.06	1.03	1.00	1.06	1.05
≥1000	1.00	1.00	1.00	1.00	1.00

1 Global Change and Natural Hazards: New Challenges, New Strategies   13

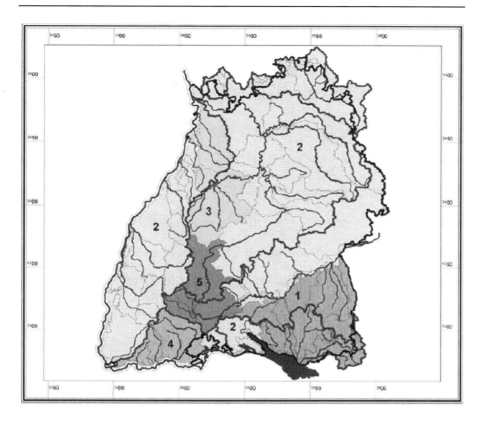

Fig. 1.9 Hydrological regions of Baden Württemberg (Arbeitskreis KLIWA 2006)

1.3 Global Climate Change and its Future Regional Implications

1.3.1 The Global Frame

As visible from the Fourth IPCC-report, climate change will go on while most-possibly causing further damages within the natural environment system that are related to natural hazard processes. No matter which decisions are made by the global society to reduce greenhouse gas emission, processes of global warming are unlikely to stop within the 21st century. Following the terminology of the IPCC working group, it is obvious (i.e. probability of occurrence > 99%) that most land areas will discover more warmer and fewer cold days and nights, warmer and more frequent hot days and nights as well as a frequency increase of warm spells/heat waves (Fig. 1.10).

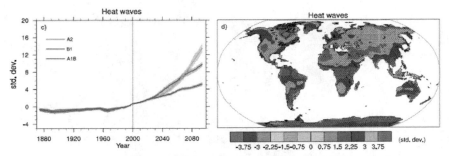

Fig. 1.10 (a) Globally averaged changes in heat waves (defined as the longest period in the year of at least five consecutive days with maximum temperature at least 5°C higher than the climatology of the same calendar day). (b) Changes in spatial patterns of simulated heat waves between two 20-year means (2080–2099 minus 1980–1999) for the A1B scenario (Meehl et al. 2007)

The dimension of the modelled surface warming differs on the basis of three different scenarios (IPCC Special report on emission scenarios). While for scenario B1 the warming might be termed rather moderate, the scenarios A1B and A2 show drastic increases in temperature (Fig. 1.11). Thus, a warming up to 7°C until the end of the 21st century needs to be taken into consideration for some regions of the northern hemisphere.

1 Global Change and Natural Hazards: New Challenges, New Strategies 15

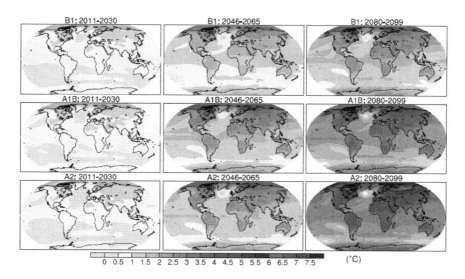

Fig. 1.11 Multi-model mean of annual mean air temperature change (°C) for the scenarios B1 (top), A1B (middle) and A2 (bottom), and three time periods, 2011 to 2030 (left), 2046 to 2065 (middle) and 2080 to 2099 (right) in comparison to the average of the period 1980 to 1999 (Meehl et al. 2007)

It is considered likely (i.e. probability > 95%) for the 21st century that the frequency of heavy precipitation events or the proportion of total rainfall from heavy falls will increase in most of the areas.

1.3.2 Regional Implications for the 21st Century

A downscaling of the global climate change trends onto European scale shows a regional differentiation between Northern and Southern Europe. Following the explanations of Christensen et al. (2007), the average annual warming in Central and Southern Europe (south of 48°, Fig. 1.12) resides within a range from 2.2°C to 5.1°C will be ca. 3.5°C in the 21st century, accompanied by extremer warming during summer seasons (Table 1.5, Fig. 1.13). Furthermore, the report puts forward the contention that all summers, autumns and years will be warmer compared to the time-period between 1980 and 1999, at the same time claiming that there is a 95%-chance to evidence such changes in winter and spring too.

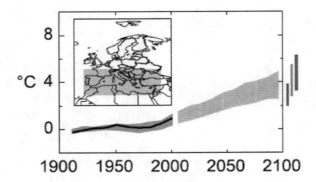

Fig. 1.12 Temperature anomalies with respect to 1901 to 1950 for Europe (south of 48°), land regions for 1906 to 2005 (black line) and as simulated (red envelope) by MMD models incorporating known forcings; and as projected for 2001 to 2100 by MMD models for the A1B scenario (orange envelope). The bars at the end of the orange envelope represent the range of projected changes for 2091 to 2100 for the B1 scenario (blue), the A1B scenario (orange) and the A2 scenario (red) (Christensen et al. 2007).

Table 1.5 Regional averages of temperature and precipitation projections from a set of 21 global models in the MMD for the A1B scenario. The mean temperature and precipitation responses are first averaged for each model over all available realisations of the 1980 to 1999 period from the 20th Century Climate in Coupled Models (20C3M) simulations and the 2080 to 2099 period of A1B. Computing the difference between these two periods, the table shows the minimum, maximum, median values among the 21 models, for temperature (°C) and precipitation (%) change (after Christensen et al. 2007).

Season	Temperature [°C] Median	Minimum	Maximum	Precipitation [%] Median	Minimum	Maximum
DJF	2.6	1.7	4.6	-6	-16	6
MAM	3.2	2.0	4.5	-16	-24	-2
JJA	4.1	2.7	6.5	-24	-53	-3
SON	3.3	2.8	5.2	-9	-29	-9
Year	3.5	2.2	5.1	-9	-27	-9

Admittedly, scenarios of both average and extreme precipitation face higher degrees of quantitative uncertainty than one would hope for – a situation that impacts on the prognosis of temperature changes too. Nevertheless, some interesting patterns can be highlighted: Once average precipitation is simulated to show only limited change, high extremes of precipitation are very likely to increase in magnitude and frequency in Central and Southern Europe at winter seasons. Furthermore, an extreme short-term precipitation may increase in summer too as an increased water vapour content of a warmer atmosphere enters despite of a general decrease in average precipitation (e.g. Frei et al. 2006, Beniston et al. 2007). With respect to the frequency – magnitude relationship, however, much deeper changes are expectable in the realm of reoccurrence frequency of precipitation extremes than in the magnitude of extremes (e.g. Huntingford et al. 2003; Barnett et al. 2006).

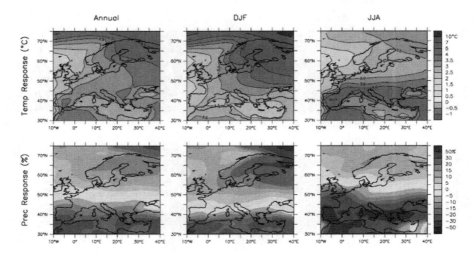

Fig. 1.13 Temperature and precipitation changes over Europe from the MMD-A1B simulations. Top row: Annual mean, DJF and JJA temperature change between 1980 to 1999 and 2080 to 2099, averaged over 21 models. Middle row: same as top, but for fractional change in precipitation. Bottom row: number of models out of 21 that project increases in precipitation

As some of the hazard processes are strongly related to precipitation, a change towards single precipitation events of higher intensity will produce, for example, floods and/or debris flow processes with a higher energy potential.

In addition to such direct impacts on the natural hazard processes, there are also indirect effects related to warmer conditions, e.g. through a melting of the alpine cryosphere. While snow immediately reacts to warmer conditions, glaciers and alpine permafrost bodies only scarcely show reactions after a time lag from years to decades. The main effects this loss of ice takes on hazard processes are: a) large areas of unstable rock material are freshly exposed to erosion processes, and b) release zones of these mass movement processes are situated at higher altitudes causing higher energy-potentials and process-dynamics. Notwithstanding positive impacts on vegetation resulting from the protection against specific erosion processes (due to the rising tree-line and vegetation boundaries), all natural hazard processes except avalanches show an increase in process magnitude and/or frequency (Bader and Kunz 1998).

1.3.3 Globalization: Socio-economic Change and its Regional Implications

Which impacts do processes of global climate change and more pronounced reoccurrences of natural hazards such as floods, storms and extreme temperatures more specifically take on society? How are physical environment, economic activities, social institutions and, more generally, living conditions and well-being of individuals affected by processes of global change? Although human activity inevitably impacts on natural hazard processes at all levels, it is of prime relevance in the discussion of global change to highlight increasing damage potential in the areas affected and provoked by these hazards. At the same time, these developments on the part of natural hazards processes are accompanied by significant socio-economic changes that are crucial for the overall development of damages. To these ends, not only an increase of risk potentials, but also a significant increase of vulnerability appears clearly identifiable.

The Global Frame

On a global scale potential risk increases links with other factors such as increase rates of the world population or increases of gross national products in the countries concerned. While the world population counted some 2.5 billion inhabitants back in 1950, it increased to a total of 6.5 billion until 2005 and will further grow to 9 billion by 2050 (Statistics Austria 2008). Additionally, rates of growth appear to display very differently: While the annual population growth rate was at a low level at dawn of the 21st century in the industrialised world (with a value of 0.25), it heralded to some 1.46 in developing countries. The increased concentration of the population in metropolitan areas and mega-cities further intensifies this situation with the fact that a natural disaster in these areas would be even more devastating. This tendency appears easily identifiable in most cities of the world's developing countries because the population sizes of Sao Paolo, Dehli, Jakarta, Dhaka or Lagos has increased dramatically there (Munich Re 2005).

Not only the citizens affected are important in this context, but also the intensity of damage potentials measured with accumulated economic data has increased dramatically. Whilst the global GDP mirrored growth-rates of 3-5% annually, regional differences seem to prevail accordingly. This is true for GDP growth rates which are relatively high in emerging economies on the one hand and for the absolute level of the GDP on the other, which is estimated on average at 40.000 $US per capita in richer countries and at 200 $US per capita in poorer ones.

Especially in poorer countries issues like confronting natural hazards are crucial due to merely revolving around pure survival. Due to their economic strength, however, developed countries are generally better equipped to confront the challenges natural hazards enshrine, because they have the possibility to implement individually or collectively financed protection measures that support prevention of damages. On the other hand, the accumulation of consumer goods subsequently alters the number of potential damages from hazard-processes too.

Developments in the Alpine Region

Within industrialised countries, people living in mountain regions are among those affected earliest by global climate change. On the one hand, indications of global climate change grasp earlier in mountain environments than in areas of moderate climate, and on the other hand, the consequences of such changes are more significant within the populations of mountain regions as there are not many options available when faced with the need to adapt to global climate change. Only a low proportion of land is suitable for permanent housing. In addition, the population depends on relatively few economic sectors, foremost on tourism and to a lesser extent on agriculture, each of which is highly dependent on the 'state of the environment' (Steininger and Weck-Hannemann 2002).

In the course of the 20th century, population growth has been accompanied by a considerable increase in individual demands for space in almost all communities. Especially small villages which are often lying in sheltered positions have been turned into accumulations of buildings greater in number and bigger in size and covering a much larger area than ever before. Furthermore, the socio-economic aspect of a change from a society predominantly determined by agriculture to a society of mobility and leisure inevitably constitutes a decisive factor in the development of the overall situation. This process inextricably links with an additional presence of externals (the "tourists") at specific periods (holidays seasons, weekends), which multiplies the original Alpine population. In addition, it is closely connected with an almost complete loss of the awareness of natural hazards and the associated risks and a related decrease in risk acceptance. All these aspects connected to a change in the man-made environment have led to a higher probability of the presence of human beings and property and thus to an increase in the value of items becoming exposed to hazard processes. For reasons of simplicity, this will be illustrated selectively by indicators referring to the development of the resident population as well as of temporary population (tourism) and the inventory of buildings in the alpine region of Tyrol (Table 1.6).

The total population of Tyrol, for instance, increased in the period between 1951 and 2001 by 73% (= factor 1.73), a trend that mirrored in 96% of the communities. This development has two reasons: First of all, communities close to Innsbruck and alongside the developing zone situated in the Inntal-valley could significantly gain in population size due to suburbanisation processes. Secondly, the total population of alpine valleys grew because tourism became the dominant economic sector and opened up for new job opportunities and wealth (Stötter et al. 2002, Stötter 2007). This development simultaneously increased the probability of presence in the communities and with it the total damage potentials.

As easily assumable, changes in population development did not occur without leaving their stamp on the basic structures of buildings: They rapidly changed in terms of numbers, quality and functions. These changes readily served as indicators for developments both within society and within the social and economic structures and processes of Alpine regions. However, this increase of the basic fabric of buildings was accompanied by an enhanced demand for surface areas in the regions concerned – a development which prominently mirrored in changes of the overall picture of cultivated landscapes, too. On average, though, the overall number of residential buildings has quintupled - a significant increase when put in comparison to the 70%-increase of population in Alpine areas. Furthermore, this development also mirrores changes in ways of life: more small families, single-households and an enhanced individual demand of living space (Kanitscheider 2008a, Kanitscheider 2008b).

Table 1.6 Factor of changes and statistic classification figures on the communal level: Resident population, residential buildings, overnight stays (35 communities failed to provide information on overnight stays in 1954), guest beds (36 communities failed to provide information on the number of guest beds), accommodation buildings (in 35 communities accommodation buildings had not existed in 1945), and the potential population.

Classification figure	Time period	Quantity	Arithmetic mean	Standard deviation	Maximum	Minimum
Resident population	1951-2001	278	1,73	0,62	6,44	0,50
Residential buildings	1945-2001	278	5,10	2,27	15,68	1,03
Overnight stays	1954-2001	243	43,43	128,95	1391,73	0,16
Guest beds	1954-2001	242	7,20	8,66	93,81	0,06
Accomodation buildings	1945-2001	243	6,12	8,41	90,00	1,00
Potential population	1951/54-2001	278	4,76	1,96	15,14	1,06

The significant increase of damage potential through an enhanced likelihood of presence of persons in their homes becomes even more serious when tourism-aspects are taken into account. Put more precisely, the factor "likelihood of presence" appears even more virulent once tourists are present in the communities. If we examine the number of overnight stays within the time-period 1954-2001 more closely, we soon discover that the factor 43 in the realm of the average increase mirrors the rapid development in the tourism sector of Tyrol quite nicely (99% of Tyrol's communities can proudly look back on a significant gain in overnight stays throughout the last years).

Additionally, the spatial spread has to be subjected to a more differential view due to the fact that there are some communities which already had a well-developed supply of guest-beds and did not gain too much in this respect during boom-times of alpine tourism. Furthermore, through the significant increase of overnight-stays in the region, a revealing process of change influenced the Federal state of Tyrol thereby managing to put tourism into the position of a leading force of economic change.

However, the total number of guest-beds serves as another important indicator of an increase of damage potentials. This number indicates how many persons are present in a community when all beds are booked and thus how many people are at risk in the case of a natural hazard event. When we assume a full-booking of all guest-beds, we can also estimate – at least for high-season times – the potential damage from the presence probability rates, thereby using the number of guest-beds as an important indicator.

An increase of damage potentials can also be measured by the increase of accommodation buildings in a community (e.g. Hotels, Bed & Breakfast etc.). Besides the changes in total numbers, the development at a smaller scale are of significant importance in this regard because a general trend towards bigger and better-equipped facilities was clearly identifiable, a fact that also increased potential damage. Due to the fact that no standardised, reproducible statistical data exist on this indicator in the Tyrol, many smaller analyses heavily depended upon case studies (e.g. Keiler 2004). These altogether indicate an inflation-adjusted increase by a factor 3, which means that damage potentials measured by the product of numbers and values of the buildings concerned have risen 10 times throughout the last years.

24 Stötter et al.

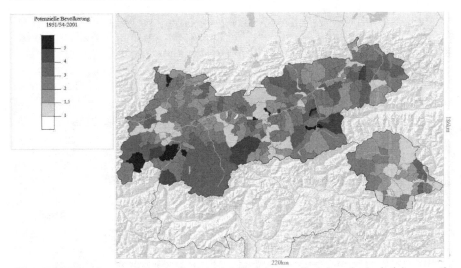

Fig. 1.14 The factor change of potential population (local and touristic) over the time-period between 1951/1954 and 2001, with a special view on the communal level (factor 2 = doubling)

Regarding the likelihood of presence in the communities, the introduction of a factor called "potential population" appears to make sense in regions that attract big numbers of travellers and tourists. This "potential population" is made up by the number of permanent residents plus that of guest beds, and therewith constitutes another important indicator for the development of damage potential. From this point of view, all communities in Tyrol evidenced a significant increase within the time-period measured whilst the margin ranges from a factor not highly above one to a rate of 15. The pattern of spatial distribution is shaped by the development of population and by that in the realm of tourism because of two different, partially independent components that drive these processes. And with it the indicator changes within a given period of time - a perception that especially mirrors the situation in the Tyrol where tourism population influences development more intensely than the population local to these places. Consequently, tourism centres have to confront higher increases in this realm (Fig. 1.14).

The developments outlined above cover the situation in the province of Tyrol, thereby making clear that damage potentials changed in times of Global Change significantly, especially in the second half of the 20th century. They most possibly reaching higher values than the changes often discussed under the headings "impacts on nature" and "increases of risk potentials". Therefore, when we talk of a supra-regionally as well as Tyrol-widely increasing risk with respect to natural hazards, we have to differentiate in making judgements with respect to the overall socio-economic situation. In the meantime, increases in values as well as in the likelihood of presence of both objects and persons acquire central significance.

Future scenarios

How does the future development of the processes described above look like? Current perspectives appear to support the perception that a change in importance of risk and damage potentials has almost already become a reality.

To the extent possible, prognosis based on statistics on the developments of the past 50 years heavily supports the contention that the number of populations and consumer goods endangered by natural hazard events has grown significantly in almost all places on our planet. In the region of Tyrol, however, a further increase in the realm of damage potentials is clearly at hands, whilst these growth-rates might be lower compared to those of the past. For instance, the growth of native permanently resident population of about10% is estimated for the Tyrol (Statistics Austria 2008). About the same is true for the development of tourism numbers where higher numbers are unlikely. This is also due to a warming-driven reduction of the winter season in lower-located areas, a process that could develop to a situation in which Alpine skiing is no longer possible in areas such as the "Kitzbühel"-region or the northern limestone Alps (Träwöger 2003; Steiger 2004). Hence tourism will not face big increases as long as no appealing alternatives offers are accessible for newly attracted and prosperous demand-groups. Due to the fact that population-increases merely derives from migration-gains, it can be assumed on the basis of current socio-economic figures that buildings will grow both in numbers and value within the coming years.

Inextricably linked with these processes, however, a further increase of damage potentials will confront societies in Alpine regions and especially in Tyrol with a set of new challenges. Hence new ways of dealing with increasing risk scenarios are highly necessary if a sustainable conservation of the Alpine environment is a main aim.

1.4 Changing risks as a consequence of the overlay of Global Change processes

Up-to-date natural hazard management inevitably rests upon the grasping and concise assessment of risks and thus contains elements of the environment as well as those of cultural, economic and social life. In the light of this broader approach, though, potentially hazardous processes are no longer regarded as existing completely isolated from other corresponding factors. Instead, they are coexisting with damage potentials that are more clearly exposed. This opening of the approach directly leads to the risk concept.

Risk defined as the likelihood of reoccurrence of processes of defined magnitudes and corresponding extents of damages inevitably constitute a hazardous event (equation 1.2). Regarded quantifiable, different risks can be compared and contrasted with each other. The basic idea of the risk concept lies in its future-oriented perspective that – through an ex ante analysis of potential effects of natural and socio-economic processes – inevitably allows for the implementation of adequate solutions and coping-measures.

$$R_{i,j} = p_{Si} \cdot A_{Oj} \cdot p_{Oj,Si} \cdot v_{Oj,Si}$$

(eq. 1.2)

$R_{i,j}$ = Risk regarding in scenario i and object j
p_{Si} = Likelihood of occurrence of scenario i
A_{Oj} = Value of the object j
$p_{Oj,Si}$ = Likelihood of presence of object j compared to scenario i
$v_{Oj,Si}$ = Resilience of object j in dependency to scenario i

By and large, risks can be viewed from two basic perspectives. Analysis from the side of emissions focuses narrowly on an event that has certain impacts on the current situation. In the frame of classic, process-oriented natural hazard appraisal this focus on emissions is widely applied in many studies. This was particularly true for scientific practice before the International Decade for Natural Disaster Reduction (IDNDR) directed world-wide attention onto dealing with natural hazards. On the other side, emission-examinations that focus on objects of value (human life, material and immaterial assets …) ground their analysis on external impacts that can lead to specific changes (results or damages). Put differently, the likelihood of occurrence of potentially dangerous natural processes as one of the two central factors of the risk function is examined thoroughly without leaving damage potentials out of account. This approach finds broader application in recent scientific projects (e.g. Keiler et al. 2006) and in the praxis of planning that increasingly mirror this tendency (e.g. in South Tyrol; Stötter and Zischg 2007). Additionally, the literature on the economic costs of global climate change (e.g. Nordhaus and Boyer 2000, Stern 2006, Cline 2007) and on the efficiency of alternative policy measures (e.g. Kemfert and Praetorius 2005, Weck-Hannemann 2006, Agrawala and Fankhauser 2008, and Raschky and Weck-Hannemann 2008) is growing rapidly.

Risks that derive from natural hazards can be analysed on very different examination levels, ranging from highly aggregated analyses on the level of national accounts to the disaggregated level of individual objects and human beings. The term object risk describes the size of a risk for a certain, pre-defined object. Under this aspect, the object signifies the smallest unit examined from the risk perspective (for instance a building or an endangered part of a traffic route). The individual risk derives from the object risk and the number of persons present in this object. The collective risk describes the overall-damages expected for society or certain pre-defined societal groups through the sum of all object and individual risks within this collective unit (Merz et al. 1995).

Equation 1.2 is always used when the likelihood of occurrence of an event can be estimated clearly and its impacts are easily identifiable on a quantitative scale. Whilst literature on natural hazards focused more on a quantitative estimation of damage potentials and its impacts on the number of persons and/or buildings potentially endangered, economic perspectives on this issue, applied e.g. in the frame of a cost-benefit analysis of different risks, merely focused on a monetary assessment of damage potentials. According to its underlying welfare-economics, a potential cease of benefits after a natural hazard event needs to be assessed monetarily and in welfare terms (e.g. Gamper et al. 2006). As long as this undertaking is usually pursued by the grasping of instauration-costs or insurance-values of buildings it only pinpoints at instead of fully capturing the true situation in this realm. In addition, such a procedure that vigorously focuses on a benefit-oriented examination of the individuals involved subsequently makes the inclusion of a risk version function considerably obsolete (Borsky and Weck-Hannemann 2009).

1.5 Contributing to a sustainable natural hazard management: Our main task

It appears irreversible that natural hazard processes strongly affected by the consequences of Climate Change will change regarding the "frequency – magnitude"-relationship too. Besides others, avalanches, debris flows, flooding, and rock falls are examples of natural, geo-morphological processes in high mountain areas - most of which assume an "occurrence-probability" under stable climatic conditions – that follow a more or less pre-defined relationship between magnitude and frequency. Once these processes interfere with human interests, they are perceived as a direct or indirect threat, or as a natural hazard.

And the human environment has inevitably been affected by theses processes ever since human beings began to settle and utilize the Alps or other high mountain regions of the world, at the same time causing damage to persons, buildings, property, and infrastructure. For centuries, both the hazard and the damage potential seem to have rested in a long-term equilibrium which was only interrupted by extreme events (e.g. Stötter and Keiler 2003). Recently, however, the increase of damages and costs of recovery arising from natural hazards experiences a recognisable trend on the regional as well as on the global scale (e.g. Munich Re 2008). Year after year the Munich Re-Insurance Group lines out in its examinations on natural hazards that an exponential increase of natural hazard events grasps ground since the middle of the 20th century. The long-term trend discovers intensified forms in its recordings on catastrophes and the economic and insured damages resulting from them – a fact that pinpoints at developments in the realm of damage potentials that subsequently alter the dangers risk potentials pose.

1 Global Change and Natural Hazards: New Challenges, New Strategies

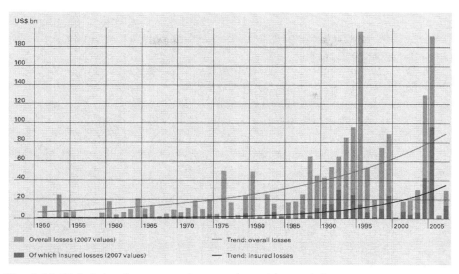

Fig. 1.15 Global developments of economic and insured damages resulting from natural disasters within the timeperiod between 1950 and 2007 (Munich Re 2008)

Though a similar accurate damage statistic does not exist for the national or regional level, frequently reoccurring natural hazard events – through the examination of compensation costs – subsequently allow for the contention that the total number of damages is clearly increasing. The devastating floods of 2002 and 2005 are mentioned here in place of many others to show that damage compensation required an increase in the disaster-fund with additional 500 million € in 2002 and with 251 million € in 2005.

In the light of these rapid changes in character of natural hazards, the socio-economic situation and the risks associated with it desperately call for a new and more active approach to corresponding coping-measures. The growing reluctance of a public that seems to willingly accept risks paired with a rising shortage of public funds subsequently indicate a pressing need to rethink all individual steps of appropriate risk management. Without doubt, a set of crucial questions arise from coping-strategies to natural hazard related risks, such as "What can happen?", "What is acceptable?", and "What needs to be done?" Therefore, analysis as well as evaluation and management strategies of natural hazards processes have to be adapted to potential future developments that cover all relevant aspects of sustainable risk management.

In this respect, the alpS-Centre for Natural Hazard Management was founded in 2002 in order to develop new strategies that would contribute to a sustainable safe *lebensraum* in the Alps in particular and in any mountain region in general. alpS tries to meet these challenges by inter- and transdisciplinary cooperation with competent commercial, scientific and administrative partners.

It is therefore both the vision and the mission of the alpS-Centre for Natural Hazard Management to improve the preparation of society for the impact of disasters and to proceed from the reaction to and protection against hazards to pro-active integrative risk management strategies as a contribution to a sustainable safety in the alpine *lebensraum*. The knowledge gained from scientific projects on a local or regional scale in the Alps is transferable to the global scale, which subsequently increases the safety of individuals as well as whole societies in the mountain areas of our planet.

References

Agrawala S, Fankhauser S (eds.) (2008) Economic Aspects of Adaptation to Climate Change: Costs, Benefits and Policy Instruments, Paris: OECD
Arbeitskreis Klimaveränderung und Wasserwirtschaft (KLIWA) (2006) *Regionale Klimaszenarien für Süddeutschland. Abschätzung der Auswirkungen auf den Wasserhaushalt*, KLIWA-Berichte 9, Mannheim
Bader St, Kunz P (1998) *Klimarisiken: Herausforderung für die Schweiz*, Zürich: vdf Hochschulverlag
Barnett DN et al. (2006) Quantifying uncertainty in changes in extreme event frequency in response to doubled CO2 using a large ensemble of GCM simulations, *Climate Dynamics*, 26, pp 489–511
Bayrisches Staatsministerium für Umwelt, Gesundheit und Verbraucherschutz (2004) *Berücksichtigung von möglichen Klimaänderungen – Interner Erlass Bayerisches Staatsministerium für Umwelt, Gesundheit und Verbraucherschutz*, unpubl. manuscript, München: BStMUGV
Becker E (2003) Soziale Ökologie: Konturen und Konzepte einer neuen Wissenschaft, in: Matschonat, G, Gerber A(eds.): *Wissenschaftstheoretische Perspektiven für die Umweltwissenschaften*, Weikersheim: Margraf Publishers, pp 165-195
Becker E, Jahn T (2003) Umrisse einer kritischen Theorie gesellschaftlicher Naturverhältnisse, in: Böhme G, Manzei A (eds.): *Kritische Theorie der Technik und der Natur*, München: Wilhelm Fink, pp 91-112
Becker E, Jahn T (eds.) (2006) *Soziale Ökologie - Grundzüge einer Wissenschaft von den gesellschaftlichen Naturverhältnissen*, Frankfurt/New York, Campus
Beniston M et al (2007) Future extreme events in European climate: An exploration of regional climate model projections, *Climate Change*, doi:101007/s10584-006-9226-z
Borsky S, Weck-Hannemann H (2009) Sozio-ökonomische Bewertung der Schutzleistung des Waldes vor Lawinen, alpS Projektbericht, January 2009

Caspary HJ (2006) *Zunahme kritischer Wetterlagen als Ursache für die Entstehung extremer Hochwasser in Südwestdeutschland*, KLIWA-Berichte 9, Mannheim, pp 135-151

Christensen JH et al. (2007) Regional Climate Projections, in: Solomon, S et al. (eds.): *Climate Change 2007: The Physical Science Basis. Contribution of Working Group I to the Fourth Assessment Report of the Intergovernmental Panel on Climate Change*, Cambridge: Cambridge University Press

Cline W (2007) Global Warming and Agriculture: Impact Estimates by Country, The Peterson Institute

Deutscher Bundestag (ed.) (1998) *Abschlussbericht der Enquete-Kommission "Schutz des Menschen und der Umwelt – Ziele und Rahmenbedingungen einer nachhaltig zukunftsverträglichen Entwicklung" des 13. Deutschen Bundestags: Konzept Nachhaltigkeit. Vom Leitbild zur Umsetzung*, Bonn: Deutscher Bundestag

Frei C, Schöll J, Schmidli S, Fukutome, Vidale PL (2006) Future change of precipitatin extremes in Europe: Intercomparison of scenarios from regional climate models, *Journal of Geophysical Research*, 111, D06105, doi:101029/2005JD005965

Fricke W (2003) Hängen vermehrte Starkniederschläge am Hohenpeißenberg mit veränderten Wetterlagen zusammen?, in: Deutscher Wetterdienst (ed.): *Klimastatusbericht 2002*, Offenbach: DWD, pp 165-171

Fuchs T., Rudolf B (2002) Niederschlagsanalyse zum Weichselhochwasser im Juli 2001 mit Vergleich zum Oderhochwasser 1997, in: Deutscher Wetterdienst (ed.): *Klimastatusbericht 2001*, Offenbach: DWD, pp 268-271

Fuchs T, Rapp J, Rudolf B (2000) Niederschlagsanalyse zum Pfingsthochwasser 1999 im Einzugsgebiet von Donau und Bodensee, in: Deutscher Wetterdienst (Hrsg.): *Klimastatusbericht 1999*, Offenbach: DWD, pp 26-34

Gamper C, Thöni M, Weck-Hannemann H (2006) A Conceptual Approach to the Use of Cost Benefit and Multi Criteria Analysis in Natural Hazard Management, Natural Hazards and Earth System Sciences, 6, pp 293-302

Gewässerkundlicher Dienst Bayern (2002) Hochwasser im August 2002. – internal report

Grieser J, Beck C (2003) Extremniederschläge in Deutschland Zufall oder Zeichen?, in: Deutscher Wetterdienst (Hrsg.): *Klimastatusbericht 2002*, Offenbach: DWD, pp 141-150

Habersack H, Krapesch G (2006) *Hochwasser 2005: Ereignisdokumentation der Bundeswasserbauverwaltung, des Forsttechnischen Dienstes für Wildbach- und Lawinenverbauung und des Hydrographischen Dienstes*, Wien

Huntingford C et al. (2003) Regional climate-model predictions of extreme rainfall for a changing climate, *Quarterly Journal of the Royal Meteorological Society*, 129, pp 1607–1621

Ihringer J (2004) Ergebnisse von Klimaszenarien und Hochwasser-Statistik, *KLIWA-Berichte* 4, Mannheim, pp 153-167

Jarabac M, Chlebek A (2000) Einfluss des Waldes des Waldes auf das Hochwasser 1997 im Nordmährischen Gebiet, *Interpraevent*, 1, pp 129-136

Kanitscheider S (2008a) Single-person Households, in: Tappeiner U, Borsdorf A, Tasser E (eds.) *Mapping the Alps: Society – Economy – Environment*, Heidelberg: Spektrum, pp 118-119

Kanitscheider S (2008b): Average Household Size, in: Tappeiner U, Borsdorf A, Tasser E (eds.): *Mapping the Alps: Society – Economy – Environment*, Heidelberg: Spektrum, pp 124-125

Keiler M, Fuchs S, Zischg A (2006) Methoden zur GIS-basierten Erhebung des Schadenpotenzials für naturgefahreninduzierte Risiken, in: Strobl, J, C Roth (eds.): *GIS und Sicherheitsmanagement*, Heidelberg: Verlag Wichmann, pp 118-128

Keiler M (2004) Development of the damage potential resulting from avalanche risk in the period 1950-2000: Case-study Galtür, *Natural Hazards and Earth System Sciences*, 4(2), pp 249-56

Kemfert C, Praetorius B (eds.) (2005) Die ökonomischen Kosten des Klimawandels und der Klimapolitik, Vierteljahreshefte des DIW, 2

Landesanstalt für Umweltschutz Baden-Württemberg (2005) *Festlegung des Bemessungshochwassers für Anlagen des technischen Hochwasserschutzes. Leitfaden Oberirdische Gewässer*, Gewässerökologie Nr. 92, Stuttgart

Landesanstalt für Umweltschutz Bayern (2006) *Endbericht August-Hochwasser 2005 in Südbayern*, München: unpubl. manuscript.

Meehl GA et al. (eds.) (2007) *Climate Change 2007: The Physical Science Basis. Contribution of Working Group I to the Fourth Assessment Report of the Intergovernmental Panel on Climate Change*, Cambridge/New York: Cambridge University Press

Merz H.A, Schneider Th, Bohnenblust H (1995) *Bewertung von technischen Risiken: Beiträge zur Strukturierung und zum Stand der Kenntnisse, Modelle zur Bewertung von Todesfallrisiken*, Polyprojekt Risiko und Sicherheit, vol. 3, Zürich: vdf Hochschulverlag

Munich Re (2005) *Mega-cities – Mega-risks: Trends and challenges for insurance and risk management*, Munich

Munich Re (2008) *Topics Geo Natural catastrophes 2007*, Munich

Nationales Komitee für Global Change Forschung (2005) *Positionspapier für eine kohärente deutsche Forschungsstrategie zum Globalen Wandel*, Bonn

Nordhaus WD, Boyer J (2000) Warming the World: Economic Models of Global Warming, Cambridge/Mass.: MIT Press

Raschky P, Weck-Hannemann H (2008) Vor- oder Nachsorge? Ökonomische Perspektiven, in: Felgentreff, C, T Glade (eds.) *Naturrisiken und Sozialkatastrophen*, Heidelberg: Spektrum, pp 269-279

Rudolf B. et al (2006) *Hydrometeorologische Aspekte des Hochwassers in Südbayern im August 2005*, DWD Hydrometeorologie, Offenbach: DWD.

Schär C et al. (2004) The role of increasing temperature variability in European summer heat waves, *Nature*, 427, pp 332–336

Schönwiese C.D, Staeger T, Trömel S, Jonas M (2004) Statistisch-klimatologische Analyse des Hitzesommers 2003 in Deutschland, in: Deutscher Wetterdienst (ed.): *Klimastatusbericht 2003*, Offenbach: DWD, pp 123-132

Schönwiese C-D, Staeger T, Trömel S (2004): The hot summer 2003 in Germany: Some preliminary results of a statistical time series analysis, *Meteorologische Zeitschrift*, 13(4), pp 323-327

Schröter D, Zebisch M, Grothmann T (2006) Climate Change in Germany - Vulnerability and Adaptation of Climate-Sensitive Sectors, in: Deutscher Wetterdienst (Hrsg.): *Klimastatusbericht 2005*, Offenbach: DWD, pp 44-56

Solomon, S et al. (eds.) (2007) *Climate Change The Physical Science Basis. Contribution of Working Group I to the Fourth Assessment Report of the Intergovernmental Panel on Climate Change*, Cambridge/New York: Cambridge University Press

Steiger R (2004) *Klimaänderung und Skigebiete im bayerischen Alpenraum*, Innsbruck: unpubl. MA-Thesis

Steininger KW, Weck-Hannemann H (eds.) (2002) *Global Environmental Change in Alpine Regions. Recognition, Impact, Adaptation and Mitigation*, Cheltenham/Northampton: Edward Elgar

Stern N (2006) The Economics of Climate Change: The Stern Review, H.M. Treasury.

Stötter J (2007) Zunahme des Schadenspotenials und Risikos in Tirol als Ausdruck der Kulturlandschaftsentwicklung seit den 1950er Jahren, in: Innsbrucker Geographische Gesellschaft (ed.): *Alpine Kulturlandschaft im Wandel: Hugo Penz zum 65 Geburtstag*, Innsbruck: Innsbrucker Geographische Gesellschaft, pp 164-178

Stötter J, Keiler M (2003) Die Rolle des Menschen bei der Sicherung des alpinen Lebensraums, in: Varotto, M, R Psenner (Hrsg): *Entvölkerung im Berggebiet: Ursachen und Auswirkungen*, Belluno: Reto Montagna, pp 217-230

Stötter J, Zischg A (2007) Alpines Risikomanagement – theoretische Ansätze, erste Umsetzungen, in: Felgentreff, C, T Glade (Hrsg.): *Naturrisiken und Sozialkatastrophen*, Heidelberg: Spektrum, pp 297-310

Stötter J, Meissl G Ploner A, Soenser T (2002) Developments in Natural Hazard Management in Alpine Countries Facing Global Environmental Change, in: Steininger, KW, Weck-Hannemann H (eds.): *Global Environmental Change in Alpine Regions. Recognition, Impact, Adaptation and Mitigation*, Cheltenham/Northampton: Edward Elgar, pp 113-130

Stötter J, Coy M (2006) Globaler Wandel - regionale Nachhaltigkeit, in: Grumiller, M, TD Märk (Hrsg.): *Zukunftsplattform Obergurgl 2006: Forschungskooperationen innerhalb der Leopold-Franzens-Universität Innsbruck*, Innsbruck: IUP, pp 169-72

Trawöger E (2003) *Mögliche Folgen eines Klimawandels für den Skitourismus*, Innsbruck: unpubl. MA-Thesis

Trenberth KE et al. (2007) Observations: Surface and Atmospheric Climate Change, in: Solomon, S et al. (eds.): *Climate Change 2007: The Physical Science Basis. Contribution of Working Group I to the Fourth Assessment Report of the Intergovernmental Panel on Climate Change*, Cambridge/New York: Cambridge University Press

United Nations Department of Economic and Social Affairs (2002) *Report of the International Forum on National Sustainable Development Strategies*, Accra: UN DESA

Vogelbacher A, Kästner W (1999a) Pfingsthochwasser 1999 - ein Jahrhundertereignis an Iller, Ammer und Donau, auf: Hochwassernachrichtendienst des Bayerischen Landesamtes für Wasserwirtschaft: <http://www.bayern.de/lfw/hnd/ereignisse.htm>, (05.01.2008)

Wald J (2004) Auswirkungen der Klimaveränderungen auf Planungen – Praxisbeispiele, *KLIWA-Berichte* 4, Würzburg, pp 169 - 185

Weck-Hannemann H (2006) Efficiency of Protection Measures, in: Ammann, WJ et al. (eds.): *Risk 21 - Coping with Risks Due to Natural Hazards in the 21st Century"*. London: Taylor, Francis, pp 147-154

Weichhart P. (2005) Auf der Suche nach der „dritten Säule. Gibt es Wege von der Rhetorik zur Pragmatik?, in: Müller-Mahn, D, Wardenga U (eds.): *Möglichkeiten und Grenzen integrativer Forschungsansätze in Physischer Geographie und Humangeographie*, Leipzig: Leibniz-Institut für Länderkunde, pp 109-136

World Commission on Environment and Development (1987) *Our Common Future*, Oxford: Oxford University Press

2 Flood Forecasting for the River Inn

S. Senfter, G. Leonhardt, C. Oberparleiter, J. Asztalos, R. Kirnbauer, F.Schöberl, H. Schönlaub

2.1 Introduction

2.1.1 The River Inn in Tyrol

The river Inn as the main river in Tyrol moulds the settlement and economic area in Northern Tyrol in a considerable way. 66 % of the area drains into the Inn, whereas the remaining 34% drain into the Lech, the Grossache and the Drau in East Tyrol. The Inn flows through Tyrol for about 200km, from the Swiss border at Martinsbruck to Kufstein, where it leaves Tyrol and flows into Bavaria/Germany (Fig. 2.1).

The 100-year flood (HQ_{100}) discharge is 512m^3/s in Martinsbruck and increases to 1,370m^3/s at Innsbruck and rises further to 2,250m^3/s at Oberaudorf after crossing the border at Kufstein (Table 2.2). Until now the highest level of discharge in the river Inn at Innsbruck (1,511m^3/s) was measured on August 23, 2005 (Fig. 2.2). It was determined for this particular flood event in Innsbruck that it took 20 hours from the first response of the gauge until the peak discharge was reached (Gattermayr 2005).

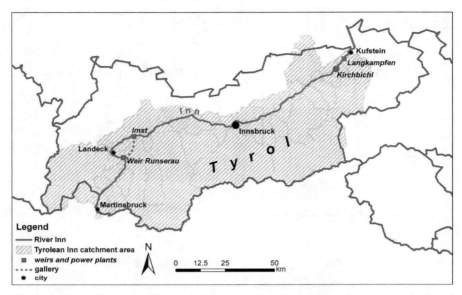

Fig. 2.1 Map of Northern Tyrol showing the river Inn, the catchment area in Tyrol and run-of-river/diversion power plants

Table 2.1 Characteristic discharges for selected gauges on the Inn

	MQ	HQ_1	HQ_{100}
Gauge Martinsbruck	-	275[m^3/s]	512[m^3/s]
Gauge Innsbruck	165[m^3/s]	641[m^3/s]	1,449[m^3/s]
Gauge Oberaudorf	307[m^3/s]	1,220[m^3/s]	2,250[m^3/s]

MQ (mean water discharge), HQ_1 and HQ_{100} (flood discharges with a statistical recurrence interval of 1 and 100 years respectively) for selected gauges, determined from a statistical series covering at least 30 years (BAFU 2006, HZB 2006, Blöschl et al. 2006, LFU 2006)

Fig. 2.2 The flood in Innsbruck on August 23, 2005. The discharge at the time (4.48 p.m.) the picture was taken was at around 1,395m^3/s. Only a few centimetres of freeboard are protecting the city of Innsbruck and the university (picture) from catastrophic damage

The discharge is influenced by several reservoirs and power stations in the Swiss catchment area of the Inn. Along the Tyrolean length of the Inn three run-of-river power stations and diversion hydropower facilities are used to produce electricity: the weir at Runserau with a power station in Imst, the power plant in Kirchbichl and in Langkampfen (Fig. 2.1). Several peak-load power plant reservoirs in the catchment areas (reservoir Gepatsch for the Kaunertal power station, reservoir Finstertal and Längental for the group of Sellrain-Silz power stations, the Achensee power station) influence the runoff situation on the Inn through demand-oriented operations as well as being able to store large volumes of water in the case of an impending flood (Widmann 1989).

The hydrological situation reflects the alpine character of the Inn catchment area. Large differences in altitude result in a high average slope of the subcatchments, which in turn leads to a heightened flow velocity and a more rapid response of the hydrograph to rainfall (Baumgartner and Liebscher 1996).

Consequently, the model is highly sensitive in regard to rainfall input and its spatial and temporal distribution.

2.1.2 Risk Management and Flooding

Risk as a product of occurrence probability and effect depends substantially on the potential damage in the areas affected by flooding. The Inn Valley as the most important settlement and economic area in Tyrol has been opened up to more intense agricultural, construction and industrial use in the last decades and centuries through numerous protection measures and river training. The concentration of important infrastructure in the valley region enhances the vulnerability of the area in the case of a catastrophic event, which increases the design discharge of protection measures. An important feature of integral risk management is to deal with exactly these events.

Thus, the central question is how to prepare for these rare events to minimize the damage using the restricted space to its optimum, while at the same time taking the sustainability of protection measures into consideration. Protection measures, in particular levees along the river-side which reduce the area around the river, have led to the regions downriver being more greatly affected by flooding. In regard to a sustainable and cross-national approach towards river basins, increasing the level of protection with new protective measures along the river, should be critically questioned.

In the alpS Centre for Natural Hazards Management, two different strategies concerning flooding are being pursued: the detailed scenario-based analysis and the provision of early warning with the help of forecasting.

Detailed Scenario-Based Analysis

Well-defined areas, for which a detailed appraisal is advisable, are dealt with more specifically in the detailed scenario-based analysis. This is because these districts are either of great importance, due to their enhanced risk potential, or due to an especially protection-worthy infrastructure and financial values. The scenarios can be based on previous events and scaled up to extreme ones. On this foundation the effects of extreme events can be investigated in advance and an analysis in regard to preventive risk management can be made. The interpretation of these scenarios is the basis for risk regulation such as the design of protection measures or the conception of a plan of action for emergency organisations. Figure 2.3 shows an example of a detailed hydrodynamic flood analysis.

Fig. 2.3 Detailed scenario-based analysis Innsbruck: flood modelling based on detailed terrain data.

Early warning with the help of forecasting

The aim of forecasting is to offer early and reliable warning for the affected areas and to be able to estimate the expected intensity of an event. Based on a forecast it is possible, with the help of the prepared scenarios from the detailed analyses, to develop an optimized strategy to deal with the threat. Thereby it is essential to maintain a temporal advantage as well as the enhanced reliability and objectivity of the statements particularly in comparison with the usual observation of the gauges and the rainfall situation, which demands a high degree of experience with the local conditions. A combination of the models and the experience supported appraisal of the complex information can in future lead to an elevated quality of flood forecasting.

2.1.3 An Overview of the Forecast Model

The flood forecast model for the Tyrolean Inn is a modular-based, hybrid system conceived and built with three main components (Fig. 2.4). The first part is the meteorological data from observation stations in the whole Tyrolean Inn catchment area as well as meteorological forecast data from the numerical weather models of the **Z**entral**A**nstalt für **M**eteorologie und **G**eodynamik (ZAMG) in Vienna (Central Institute for Meteorology and Geodynamics). These data are used in the second component, which consists of the hydrological model HQ$_{SIM}$ (Kleindienst 1996) and the glacier model SES (Asztalos 2004). The models are calibrated with historically observed data and use the current data to calculate the discharge from the tributary catchments. In the third component, the flow along the river Inn is represented by a hydrodynamic, one-dimensional model, which integrates the inflows from all the tributaries. The hydrodynamic model represents the wave propagation on the Inn whilst including the influence of the run-of-river power stations and diversion hydropower facilities. A rule-based operation of the reservoirs of the Inn's power stations can be represented with this model.

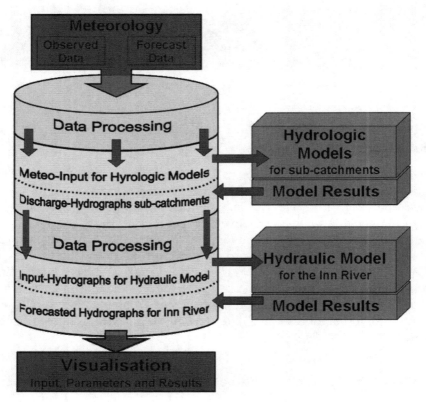

Fig. 2.4 Concept of the flood forecasting model

2.2 Meteorological Data

Meteorological data that are fed into the flood forecast are based on numerical weather models as well as data from observation stations in the Tyrolean Inn catchment area. A discharge prediction for the river Inn would also be possible with the use of only the latter; however, the inclusion of forecasts extends the lead time of the forecast significantly.

2.2.1 Forecast data

Nowcasting and NWP (Numerical Weather Prediction)-Models

Meteorological forecasts can be divided into two groups depending on their prediction period. If a prediction is made of up to 6 hours into the future one can speak of *Nowcasting*, but when faced with a longer prediction period one speaks of *Numerical Weather Prediction (NWP)-Models*.

For the Tyrolean Inn flood forecast system *Nowcasting* data from the tool INCA (**I**ntegrated **N**owcasting through **C**omprehensive **A**nalysis) (Csekits et al. 2001) operated by ZAMG is used. To operate INCA it is necessary to access meteorological data from the catchment area in short intervals (10-60 min). Thereby, nowcasting based on measured data (observation stations, satellite data, and radar data) and NWP forecasting data is combined as a time-weighted mean. INCA also provides forecast data for the time period of +6 till +48 hours (refers to prediction point in time), which is identical with the output of the NWP-Model ALADIN-AUSTRIA also operated by ZAMG. INCA supplies analyses and predictions of rainfall, temperature, atmospheric humidity, wind and wind speed, global radiance as well as cloud coverage. The horizontal resolution of ALADIN-AUSTRIA is 9,6km, while the vertical discretisation comprises 45 layers (Wang et al. 2006).

By including additional measured data as well as data from a weather radar the quality of *Nowcasting* can be further increased. Through the planned installation of a second radar in the Inn catchments a further improvement of *Nowcasting* for Tyrol can be expected.

2.2.2 Observed Data

In Tyrol several different institutions operate meteorological observation stations which form an exceptionally compact network. A total of 80 stations run by the following operators can be accessed to obtain flood prediction data:

- Hydrographischer Dienst Tirol (Hydrographic Service Tyrol) - 30 stations
- Lawinenwarndienst Tirol (Avalanche Warning Service Tyrol) - 7 stations
- **Z**entral**A**nstalt für **M**eteorologie und **G**eodynamik (Central Institute for Meteorology and Geodynamics ZAMG) - 26 stations
- **Ti**roler **W**asserkraft **AG** (Tyrolean Hydropower Company – TIWAG) - 12 stations
- VERBUND (Österreichische Elektrizitätswirtschafts-AG) (Austrian Hydro Power AG) - 5 stations

All stations conduct rainfall and temperature measurements. The data are retrieved automatically numerous times daily.

Long-term meteorological data from 1994 to 2001 were used for the calibration of the hydrological model HQ$_{SIM}$ as well as the snow and ice melt model SES (**S**chnee- und **E**is**S**chmelzmodell). These data were also supplied by the previously mentioned station operators, whereby only stations were used which in the next years will also supply real time measurements and will therefore also be incorporated in the flood forecast. On account of the large amount of data it was necessary to incorporate the time series in a database for easier management. Because most of the data had not been corrected, they had to be put through a quality inspection. The thereby used algorithms, such as plausible value check could also be used in the automatic quality control of the real-time data in future.

The calibration time period (1994-2001) was chosen because of the data availability. For each part of the catchment area a dataset with hourly rainfall distribution and vertical temperature profile was calculated. In addition, hourly information about the relative humidity, the wind speed and the global radiance was needed for the calibration of the snow and ice melt model. The rainfall data was interpolated on a 5x5 km grid using an IDW (**I**nverse **D**istance **W**eighting) algorithm. The vertical temperature profile of the surrounding stations was derived with a resolution of 100m using linear interpolation.

2.3 Hydrological Model HQ$_{SIM}$

2.3.1 Model Description

The hourly hydrological modelling of the Inn tributaries is carried out with the continuous, sub-area based precipitation-discharge model HQ$_{SIM}$ (Kleindienst 1996), an enhancement of the water balance model BROOK (Federer und Lash 1978).

Through the use of HQ$_{SIM}$ it is possible to represent the heterogeneous soil composition and the equally heterogeneous geological composition of the Tyrolean Inn catchment area in the hydrological modelling. Based on specifically chosen reference subcatchments (Table 2.2) in the main geological units the most important parameters for the model were determined and then transferred to other subcatchments in the same geological units (Kirnbauer and Schönlaub 2006).

Table 2.2 Description of the reference subcatchments. The location of the areas can be seen in Fig. 2.6

Subcatchment area	Geology	Glaciated	Influenced by hydropower plants
Brixentaler Ache	Greywacke Zone		
Brandenberger Ache	Northern Calcareous Alps		
Fagge	Metamorphic Basement Units	X	X
Ötztaler Ache	Metamorphic Basement Units	X	
Ziller	Metamorphic Basement Units	X	X

HQ$_{SIM}$ represents catchment areas in the form of homogenous sub-areas (hydrotopes). The runoff formation in every hydrotope is described by a combination of storages (Kleindienst 1996). These are the intercepted precipitation, snow cover and soil storage. The latter is divided into an upper soil zone, an unsaturated zone and a ground water zone.

All the partial processes and storages included in HQ$_{SIM}$ are depicted in Fig. 2.5, except for snow melt, which is determined with a modified degree day factor technique.

The vegetation is described with the leaf area index (LAI) and trunk area index. To determine evaporation Hamon's calculated potential evapotranspiration (Federer and Lash 1978) in dependency of the available water is used (Kleindienst 1996).

The throughfall precipitation or the snow melt is separated into surface runoff and infiltration based on the degree of soil saturation of the individual sub-area. Water can evaporate from the upper soil-layer, and the discharge from this layer reaches down into the unsaturated zone.

The outflow from the unsaturated zone depends on soil moisture and can be determined with the help of the Mualem-Van Genuchten approach (non linear storage). With the help of a coefficient the outflow is split into interflow to channel and inflow into the saturated zone. The groundwater storage is conceived as a linear storage, which means that the discharge is proportional to the storage space (dependency is also applicable in a nonlinear storage). A further coefficient divides the outflow from the saturated zone into base flow and deep percolation, which is not taken into consideration in this model.

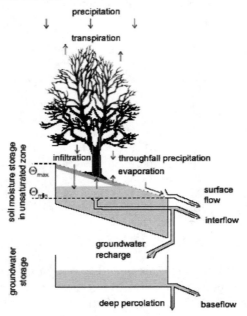

Fig. 2.5 Runoff formation in HQ$_{SIM}$: processes and storages, from Kleindienst (2001)

To represent runoff concentration, the three discharge components (surface-, inter-, and base flow) of each hydrotope are routed by a translation into the channel. Each hydrotope is divided into areas of different flow times using isochrones.

Each channel reach receives inflow from the hydrotopes and the upstream reaches. The flow velocity in the reach can be calculated according to Rickenmann (1996) or Strickler, or set to a constant value.

Intake from and return flow to a reach can be considered whereas it is not possible to model backwater.

2.3.2 Data

A Digital Elevation Model with a grid size of 250m is available for the Tyrolean Inn catchment area (data set courtesy of the Institute for Hydraulic and Water Resources Engineering, Vienna University of Technology), from which slope and aspect can be deduced. The further GIS data sets stream network, catchment area boundaries, land use and soil are taken from the Digital Hydrological Atlas of Austria (BMLFUW 2005) and are available as vector datasets (shapefiles). In addition, a more detailed stream network (data set courtesy of the Institute for Hydraulic and Water Resources Engineering, Vienna University of Technology), which includes artificial intake and return flow, is used.

The meteorological input data needed for hydrological modelling are temperature and rainfall at an hourly resolution. In addition, wind speed, cloud cover and relative humidity can be used.

2.3.3 Calibration

Due to limitations on the number of tributaries that can flow into the hydraulic model, small catchment areas must be merged. In Fig. 2.6 the combined catchment areas are depicted; the regions with red borders were modelled using the snow and ice melt model SES (Asztalos 2004, Section 2.5.1).

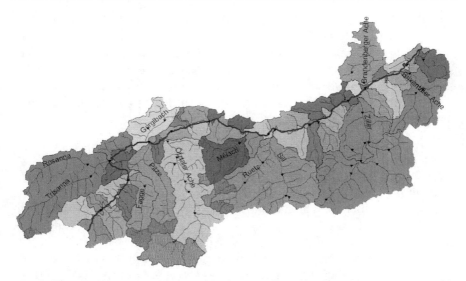

Fig. 2.6 Inn catchment area in Tyrol and merged small subcatchments. Areas with red borders: glaciated catchment areas modelled with SES

2.3.4 Definition of the Hydrotopes

The definition of the hydrotopes (also called **H**ydrological **R**esponse **U**nits HRUs) for one catchment area was achieved through the superposition of regional characteristics such as topography, slope, aspect, and soil and land use.

On the basis of the classified data the hydrotopes were manually identified (Fig. 2.7). Thereby it was ensured that not more than two classes of topographic characteristics were comprised in one hydrotope, which means that the altitude range is less than 500m, the slope range is less than 15° and the aspect range less than 90°.

Small-scale land use units were re-classified in favour of larger hydrotopes. Furthermore, the spatial connection of the hydrotopes was taken into consideration.

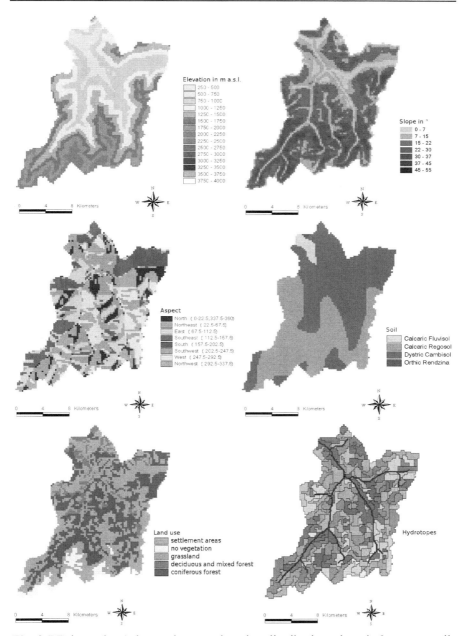

Fig. 2.7 Brixentaler Ache catchment: elevation distribution, slope in °, aspect, soil, land use and hydrotopes (from top left to right bottom)

For small catchments a simplified method was applied. Elevation-classes (max. 500m difference) and the soil type were used for the definition of the hydrotopes. This resulted in 3 to 12 hydrotopes depending on the size of the catchment. Slope, aspect and vegetation were averaged (achieved with the help of the leaf area index over one seasonal cycle).

The soil parameters and the snow melt parameters have the strongest influence on the quality of the simulation.

The initial values for the model parameters were chosen based on the results of the calibration of the Brixentaler Ache catchment using daily mean values (Drabek 2004) as well as on the values from different references for the following topics:

- Physical characteristics of soil (Leij et al. 1999; Richard and Lüscher 1983, 1987)
- Surface runoff and soil moisture (Wilson et al. 2005; Fritsche 2001)
- Degree day factor (DDF) (Hock 2003; Lundberg and Beringer 2005)
- Leaf area index (Buermann et al. 2001; Scurlock et al. 2001) Calibration Results for the Oetztaler Ache catchment can be seen in Fig. 2.8

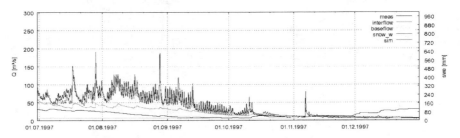

Fig. 2.8 Measured hourly mean discharge and modelled runoff and snow-water equivalent (SWE) of the snow cover with HQ_{SIM} for the Oetztaler Ache at the Brunau gauge in 1997. Black – measurement, red – channel discharge modelled with HQ_{SIM}, green – interflow, lightblue – base flow, blue – snow-water equivalent (SWE)

2.4. Glacier Model SES

2.4.1 Model Description

To include the glaciated areas of the tributary catchments in the optimal way, they are modelled using an advanced version of the snow and ice melt model SES (Asztalos 2004).

SES is a distributed energy balance model that is used for the calculation of snow, firn and ice melt at hourly intervals, which is linked to a runoff-model (consisting of parallel cascades of linear reservoirs).

Generally short wave radiation supplies the largest input of melt energy to the alpine snow cover, and therefore the albedo of the snow cover or the firn or ice surface of a glacier is of special importance. The albedo changes drastically with progressing metamorphosis of the snow cover and is influenced by possible pollution (e.g. Sahara dust), as well as exhibiting diurnal variation. Therefore, a module was developed for SES to take the temporal change of the albedo as a function of the consumed energy input into account when modelling the glacial melt (Asztalos, 2004, Kirnbauer and Schönlaub 2006). In a following step the modelled melt water amount is routed to the channel via the SES-runoff module.

The runoff module of SES consists of five parallel Nash Cascades for snow, firn, ice, non glaciated area and soil (Fig. 2.9). Each of these storage units is defined by the parameters n and k of the Nash Cascade. A portion of the melt from the glaciated (f_g) or non glaciated (f_{ng}) areas, which is determined by a coefficient, reaches the slower reacting soil storage. The sum of the output from the five reservoirs then forms the runoff.

2.4.2 Data

For high quality snowmelt modelling with SES it is necessary to include terrain data with high spatial resolution. In the glaciated areas (Table 2.2) a Digital Elevation Model with a grid size of 10m is available (supplied by the State of Tyrol). This model was lumped to a resolution of 50m.

Topographic characteristics such as slope, aspect, curvature and local horizon were determined from the digital elevation model with a resolution of 50x50m. Besides air temperature and precipitation, meteorological data such as temperature gradient, relative humidity, global radiation and wind speed are needed for the SES.

These factors were determined for the centre of each respective glaciated catchment area.

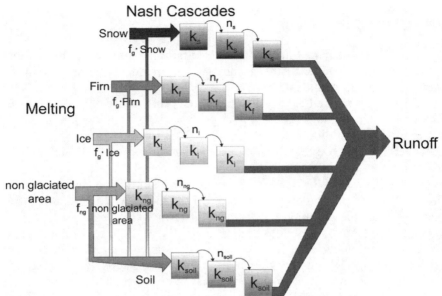

Fig. 2.9 Scheme of the runoff concentration in SES

2.4.3 Calibration

The calibration of the glaciated areas with SES was achieved in two steps. Firstly, the snowmelt module of SES was calibrated by means of the patterns of snow-free areas derived from photographs taken by an automatic camera on the Schwarzkögele above the Vernagtferner, set up by the Committee for Glaciology of the Bavarian Academy of Sciences and Humanities. These photographs were rectified to a map scale and compared to the depletion patterns simulated by SES. In addition, the hydrograph and cumulative curves of the runoff at the highly glaciated gauging station Vernagtbach were used to refine the calibration of this module. Secondly, the parameters of the runoff concentration module were calibrated against measured discharge at the gauges at Vent upstream of Niedertalbach, Vent downstream of Niedertalbach and Obergurgl.

2.5 The Hydrodynamic Model

2.5.1 Model Description

As mentioned in section 2.1.3, the scope of the hydrodynamic model is the calculation of the flood wave propagation in the river channel. Compared with hydrological flood routing methods, a lot more data about the river channel must be available to build a hydrodynamic model.

The flow in a river channel can be described mathematically by a set of partial differential equations known as the Saint-Venant-Equations, a simplification of the Navier-Stokes-Equations (Chanson 2004). These equations are solved numerically by the hydrodynamic model. The model used in this system is referred to as a one-dimensional (1D) model, which indicates that the flow velocity is assumed to be uniform over the entire cross-section of the river channel. This is of course a simplification but 1D-models have proven to be accurate enough for many applications. Of course multidimensional models (2D, 3D) are more accurate, but they require a lot more terrain data as well as computational resources and therefore computing time.

A hydrodynamic model used within a forecast system must guarantee numerically stable simulations over a wide range of discharges and simultaneously provide accurate results. Furthermore, most of these models are large as they cover a long reach of a river as a whole. As in the case of an impending flood new simulations should be performed quite frequently (up to every hour), the computation time must be kept as short as possible.

The model of the river Inn represents a section of 196.517km and was created with the software Flux$^{DSS/DESIGNER}$ which uses the code Floris^{2000} (Reichel et al. 2000) as computational core. Important factors in this choice were the ability to represent the operation of weirs and run-of-river power plants in the model and a history of successful applications in other forecast systems.

2.5.2 Data

To set up a 1D-hydrodynamic model, cross section data from the course of the river is needed to define the geometry. In the case of a natural or semi-natural river, the cross section data should not be too old, as the river bed is constantly changing due to erosion and deposition of bedload material.

To keep the forecast system up to date, the cross section data should be updated regularly.

For the Tyrolean section of the river Inn, more than 800 cross sections are available. Cross section surveys are carried out every seven to eight years, the last one after the flood event in 2005.

As input data the inflow hydrographs at the upstream boundary and from tributaries are used. Inflow from tributaries can be incorporated into the model between two cross sections. As the model represents a long reach of the river Inn, not all tributaries are considered individually. While large tributaries are represented separately, smaller ones are combined into groups on the basis of their size and the location of their catchment area. The definition of the represented tributaries is of course coordinated with the hydrological model (see section 2.3.3).

As boundary conditions at the downstream end of the model, a stage-discharge-relationship or a water level hydrograph is needed. Along the river Inn several gauging stations and weirs are located which can be used as a downstream boundary as well as for model control.

The upstream boundary is the gauging station at Martinsbruck at the Swiss-Austrian border, while the run-of-river power station Ebbs-Oberaudorf is used as a downstream boundary.

Data Preparation

Before incorporating the cross section data into the hydrodynamic model, they had to be prepared to ensure numerical stability and model accuracy.

The surveyed cross section points should be situated along a line perpendicular to the flow direction. If this was not the case, they were projected onto the perpendicular. The survey data of a cross section might include parts located much higher than the river which will never be below the water level. Those parts were eliminated as they would have decreased the spatial resolution of cross-section related data in the vertical direction. In cross sections extending into the floodplains the boundary between the river channel and the floodplain was defined. To avoid very low flow depths or completely dry cross sections at low flow conditions or in river branches with only residual flow, so called Preissmann-slots (Cunge et al. 1980) were inserted, which enable the use of a more efficient numerical scheme for the simulation (Fig. 2.10).

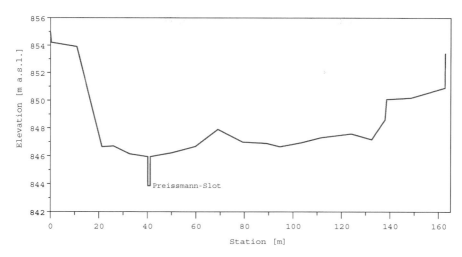

Fig. 2.10 Cross section geometry at river-km 382.009 with a Preissmann-slot inserted at the lowest point

2.5.3 Calibration

After setup the model has to be calibrated – certain model parameters are modified so that observed events can be reproduced as well as possible. In a 1D-hydrodynamic model the main calibration parameters are the roughness coefficients of the cross sections representing the flow resistance of the river bed. In the applied code, roughness coefficients given by the Gauckler-Manning-Strickler-Formula (Chanson 2004) (k_{st}) are used. If observed water level-data for a known discharge is available, the deviations between the calculated and the measured water level can be minimized by adjusting the roughness coefficients.

Method

Calibration can be done manually, usually through trial and error, which can be quite time consuming if a lot of observed water level data is available.

The software used to set up the model offers a so-called inverse modelling function which allows the automatic estimation of several model parameters, including the roughness coefficients, and therefore an automated calibration. Inverse modelling is an indirect method – the problem is not stated inversely but the equations are repeatedly solved with varying parameters. An objective function expressing the weighted residuals between measured and calculated data is minimized with the Marquardt-Levenberg-Algorithm (Nash 1990). The weights can be used to take measuring errors (Reichel and Baumhackl, 2000) into account. Furthermore, initial values for the roughness coefficients have to be defined. Usually, several inverse simulations are carried out consecutively, each using the results of the previous one as initial values.

Given the size of the model and the amount of available data, inverse modelling was chosen as the calibration method, rather than manual calibration. Leonhardt et al. (2006) compared the two methods using data from a reach of the river Inn and discussed the application of inverse modelling in cases of different availability of water level data.

For better handling, the model was divided into 10 sections bounded by gauging stations and weirs respectively. The inverse modelling function was then used to calibrate the model sections under steady state conditions. The calibrated roughness coefficients were transferred into the overall model.

The objective of the calibration was to achieve water level deviations lower than 0.2m.

Data

The first data used for calibration were water levels for design discharges (HQ_{30}, HQ_{100}) at all cross-sections obtained from another investigation with a numerical model (TIWAG 1999-2003). The values for the HQ_{30} discharge served as calibration data while the HQ_{100} discharges were used for model verification. Figure 2.11 shows the computed water level before and after calibration for the river section between Rotholz and Brixlegg.

The flood event in August 2005 provided a lot of important data to evaluate the hydrodynamic model. Not only discharge and water level data from the gauging stations was available, but also a survey of the marks of the maximum water levels at many cross sections was carried out a few days after the event.

The evaluation of the model with the data from the flood event proved that the flood wave propagation can be reproduced satisfactory. The deviations between the calculated and the observed maximum discharge at the gauging stations were small, considering that data from small tributaries was not available. Among the gauging stations which are not heavily influenced by run-of-river power plants, the maximum deviation was -5.3%. Looking at the hydrographs in Fig. 2.12 it is evident that the rise could be reproduced better than the decline.

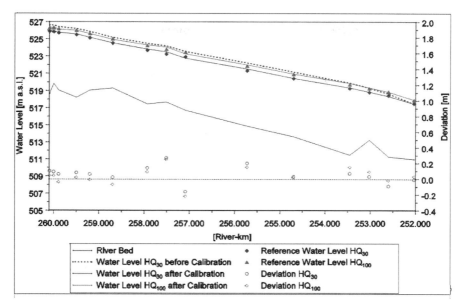

Fig. 2.11 Calibration of the section Rotholz – Brixlegg using inverse modelling. Calibration data: HQ_{30} design discharge; validation data: HQ_{100} discharge

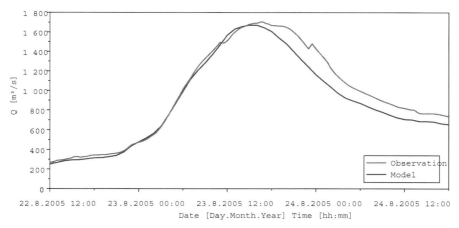

Fig. 2.12 Flood hydrograph at the gauging station at Rotholz on August 23rd, 2005 – observation and model output

Compared with the observed data the calculated maximum water level was too high at many cross sections. Simulation results showed deviations up to 1,5m. One reason for this could be bed degradation and the intense bedload transport occurring in the river Inn during flood events, which are not considered in the model. The occurrence of bed degradation can be proven by comparing different surveys of the same cross section (Fig. 2.13).

As a new cross section survey was conducted after the flood event in August 2005, it was decided that this would be a good time to develop the model further and update it accordingly. Therefore, a new calibration was necessary, which was carried out with the help of the data gained during and after the flood event.

The inverse modelling function was used again for calibration. Hydrographs from the flood event and the survey of the maximum water levels were used as calibration data. As inverse modelling only works satisfactorily in steady state simulations, corresponding inflows for the latter had to be determined. A simulation with the flood hydrographs as input was conducted to determine the maximum discharge for every cross section and to estimate the error caused by neglecting small tributaries without a gauging station. As the latter was quite small, no additional inflow was added for correction. Then the steady state inflow from all tributaries was determined, so that the steady state discharge at all cross sections was equal to the maximum flood discharge.

The calibration results were evaluated with steady state discharge as well as the flood hydrograph.

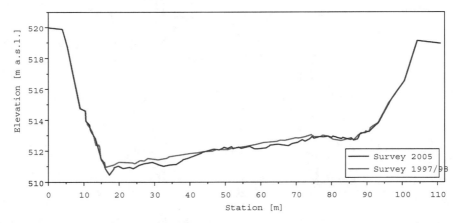

Fig. 2.13 Cross section at km 252.035; data from the cross section surveys conducted in 1997/98 and 2005 (after the flood event in August 2005)

Results

As already presented in the example by Leonhardt et al. (2006), satisfying results were achieved for most cross sections within a few simulations, while for a few cross sections water level deviations remained too large. For the latter, a manual refinement of the calibration was necessary, including adjustment of the roughness coefficients as well as modifications of the cross section geometry. Cross section modifications were carried out in the riverbed and sometimes in the floodplain. In most cases the riverbed had to be lowered, which is justifiable when modelling flood flow without considering bedload transport.

The roughness coefficients determined by inverse modelling are often very high (corresponding to a very smooth bed). Some of them are as high as $70 m^{1/3}/s$, values which certainly do not correspond to the natural condition of the riverbed. This might of course in part be a compensation for the neglect of bedload transport and erosion, but not completely. As a low flow resistance might cause supercritical flow (which does not occur in a natural river) and therefore numerical problems, an upper limit for the roughness coefficients was determined for different reaches and higher values were reduced to that limit. For the river Inn, those limits were set to $k_{St} = 40 m^{1/3}/s$ for the reach upstream of Innsbruck and $k_{St} = 50 m^{1/3}/s$ for the reach downstream, respectively. Although this was a major change to some values, the water level results were not significantly influenced and remained satisfactory (Fig. 2.14).

In the vicinity of run-of-river power stations, calibration had to be carried out taking weir operation into consideration. The special problems of those areas are discussed in Section 2.5.4.

Conclusions

Calibration of the hydrodynamic model is of major importance for a correct prediction of the water level, which still remains a challenging task.

Results of a calibration with inverse modelling are of equal quality with those of a manual calibration, but for large models and large amounts of calibration data they can be acquired with less time effort. A manual refinement is necessary and unrealistically smooth friction factors can be corrected without a major influence on the water level results (Leonhardt at al. 2006).

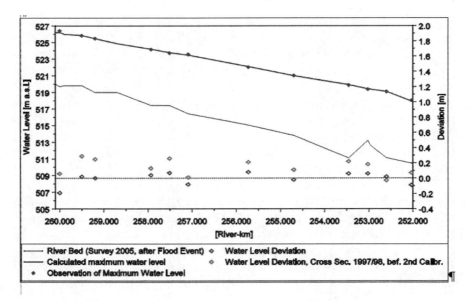

Fig. 2.14 Second calibration of the section Rotholz – Brixlegg using inverse modelling with manual refinements/upper limit of the roughness coefficients (Strickler). Data: Cross section survey 2005, August 2005 flood hydrograph, marks of maximum water level (for comparison the deviation of the maximum water level in the first model evaluation is included)

2.5.4 Rule-Based Operation of Power Plants

In the hydrodynamic model the management of the run-of-river and diversion power stations on the Inn can be taken into consideration. There are three power stations on the Tyrolean Inn that use weir plants and therefore reservoir storage in the river channel.

Power plants Kirchbichl and Langkampfen

The power plant at Kirchbichl, about 60km downstream of Innsbruck, is situated on a distinctive meander (Fig. 2.1). The headwater is conducted into the power plant via a canal that cuts off the river bend. The power plant does not make use of hydropeaking. The constant water level of 497m a.s.l. is regulated via turbine discharge of up to 300m^3/s (design discharge). If the discharge exceeds 300m^3/s the water level is regulated by the weir and water is released into the residual flow reach. If the discharge exceeds 700m³/s then the plant shuts down and free flow is established at the weir (Reindl 2004).

The low-pressure barrage plant at Langkampfen is situated 6.2km downstream of Kirchbichl (Fig. 2.1). The backwater reaches up to the bend at Kirchbichl and consequently is connected directly to the power plant there. The weir level is held at a constant 487,3m a.s.l. At a discharge of over 1,287m³/s the power plant shuts down and allows free flow.

Power plant Imst and weir plant Runserau

The weir plant Runserau lies 94km upstream from Innsbruck close to Landeck. The headwater is diverted at the weir and is conducted via a 12,5km long gallery to the power plant near Imst (Fig. 2.1). After its use in the power plant, at a design discharge of 85m³/s, the water is then released back into the Inn via a canal.

The maximum diverted amount at the weir Runserau lies between 75 and 80m³/s. A minimum instream flow of 1 to 3m³/s is maintained in the bypass reach of the river. If the discharge of the Inn exceeds 80m³/s at the gauge at Prutz, excess water is passed into the bypass reach. In the case of a further increase of discharge the water level at the weir is reduced from 858.5m a.s.l. to 855m a.s.l. At a discharge of 300m³/s the weir gates are opened and free flow is established. When the discharge is low, a portion of the reservoir storage is used for hydropeaking on a daily basis (Reindl 2004).

Table 2.3 Simplified weir operation Runserau

Discharge at Prutz Gauge	Water level weir Runserau	Discharge in bypass reach
Up to 80m³/s	858.5m a.s.l.	1 to 3m³/s
80 to 300m³/s	Lowering down to 855m a.s.l.	Amount of flow exceeding the design discharge of the intake
Over 300m³/s	Free flow at the weir	Total discharge

The following example shows the modelled power plant operation for the flood event on August 23, 2005 (Fig. 2.15). At the beginning of the event (up to 7.30 a.m., August 22) only the minimum acceptable flow was conducted into the bypass reach (green hydrograph in Fig. 2.15). The power plant flow rate (blue) was being regulated at this time so the waterlevel at the weir was being held at 858.5m a.s.l. When the design discharge in the tributary exceeded 80m³/s, the excess amount of water was released into the bypass reach (up to 2.30 a.m. on the 23rd). As soon as the discharge upstream of the weir exceeded 300m³/s (red hydrograph), operation at the power plant was shut down (from 2.30 a.m. on the 23rd onwards) and the weir gates were opened (see water level hydrograph, black dashes). As soon as the inflow was lower than 300m³/s, the weir level was raised and operation in the power plant started again (from 4 p.m. on the 23rd onwards).

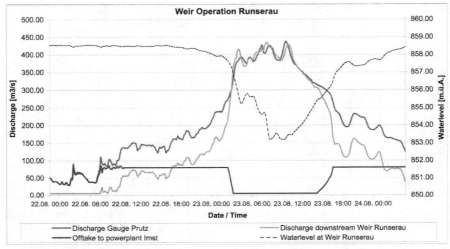

Fig. 2.15 Modelled operation at the Runserau weir for the flood in August 2005

Aim of modelling power plant operation

The aim of integrating the operation of power plants into the hydraulic model is to be able to incorporate their influence in a flood situation into the forecast. The actual chosen operational plan can vary from that of the model. Normally, when the final forecast is generated, there are no exact operational plans available, because these plans often have to be adjusted very quickly according to the approaching flood wave and changes in the discharge. For this reason, the standard operational plan that was chosen in the model offers a good assessment of the expected situation for the residents downstream of the power plant.

2.5.5 Hydrodynamic Model for the Bavarian Inn

The hydrodynamic model was generated for the Bavarian section of the Inn from Kufstein to the confluence with the Danube at Passau. This was done in cooperation with the Bavarian State Office for Environmental Protection (Bayerisches Landesamt für Umwelt) and SCIETEC River Management Corporation (SCIETEC Flussmanagement GmbH). This model differs from the Tyrolean one due to the large influence of run-of-river power plants. About 85-90% of the course of the river lies in the direct backwater areas of one of the 16 power plants (Fig. 2.16).

Additionally, the calibration of the model for the Bavarian Inn had to take other prerequisites and data into consideration. In contrast to the Tyrolean section of the Inn, the water levels and corresponding discharges were not available for every cross section. The calibration was based on only a few gauge measurements for every section, which is the reason why several cross sections in the model were calibrated with the same roughness coefficient. The maximum values (flood peak) of the measured discharge and stage hydrographs were used for the calibration. The available measurements for some sections of the highest water level during the passage of a flood wave correlated in several areas very badly with the corresponding gauge measurements. Consequently, these measurements were mostly not used for the calibration.

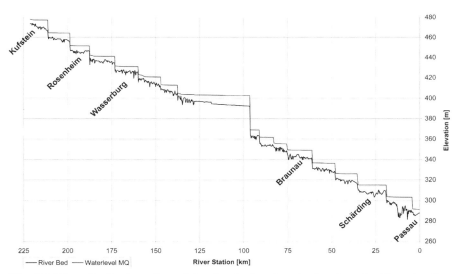

Fig. 2.16 Longitudinal section of the riverbed and water level on the Bavarian Inn at average discharge (MQ). The stepped water level course is created by the backwater from the power plants' weirs

One of the problems that arose were the high Strickler coefficients in the backwater of the power plants, which resulted from the calibration using the available water level data. The reason for these unrealistically smooth conditions was found in the change of the cross section geometry during a flood event due to the mobilisation of large amounts of bed material. Subsequently, erosion leads to a fall in the water level, which can be represented in a model using stable cross sections only by a rise of Strickler coefficients (reduction in roughness). In agreement with the model operators and after variation calculations for the estimation of sensitivity, a combination of an upper limit for the Strickler coefficients and the modification of the cross section geometry was chosen as the best approach.

2.5.6 Potential for the Optimization of the Operational Management of Alpine Reservoirs – Example Based on the Flood in August 2005

It was possible to reduce the flood peak in the lower Inn for the August 2005 flood by optimizing the operational management of the power plant reservoirs in the upper reaches of the river Inn. The reservoirs of the TIWAG Tiroler Wasserkraft AG (Tyrol electricity provider) (Kaunertal and Sellrain-Silz – Fig. 2.17) and the intakes of the Vorarlberger Illwerke AG (Vorarlberg electricity provider) from streams in the Paznaun-Region in the southwest of Tyrol caused a reduction in the maximum discharge of ca. $60 m^3/s$ for the Innsbruck gauge (Hofer 2005). The operational management of the TIWAG power plants Kaunertal and Sellrain-Silz was examined in more detail with the hydraulic model performing simulations of different scenarios.

It is important to mention that these ex-post analyses were made using observed gauge data which was not available to the power plant operators during the event. The forecast system was not yet ready in 2005. This survey was meant to show and enable a discussion of the potential of the optimized operational management.

In the future decisions could be based solely on flood forecasts. The quality of the forecast will influence the results of the optimization quite essentially.

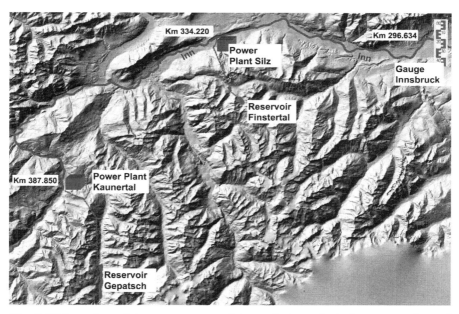

Fig. 2.17 Location of the components of the variation calculation: reservoirs and power plants Sellrain Silz and Kaunertal, river kilometre (Km) of the tailwater discharges and the gauge at Innsbruck

The analysis was performed with the hydrodynamic model developed for the forecast system. Observed hydrographs of the Inn and its tributaries were used as input data and the calculated output hydrograph for the Innsbruck gauge was used as the basis for the analysis. The operational management of the power plants Kaunertal and Sellrain-Silz was varied in numerous scenarios to see how the maximum discharge could be reduced. The figure below (Fig. 2.18) compares the observed hydrograph with the worst-case scenario (complete design discharge during the whole event) and the optimized operational management variant. The actual management strategy implemented a reduction of about 52m^3/s in the maximum discharge at Innsbruck in comparison with the worst case scenario. With the optimized variant the peak discharge at the gauge Innsbruck could have been reduced by another 43m^3/s (which equals about 9cm in water level). These marginal values could be quite significant when considering the small remaining freeboard in the centre of Innsbruck.

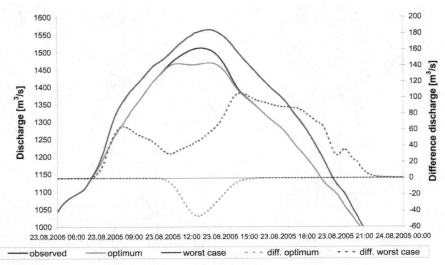

Fig. 2.18 Discharge hydrographs for the gauge in Innsbruck: observed, worst-case scenario, optimized variant

The example shows that, in the case of a relatively accurate forecast, it is possible, without needing a great deal of extra storage capacity, to make a decisive difference through the potential optimization of reservoir management. It is also possible to see how the quality of the forecast decisively influences the success of any measures taken. Optimum timing can only be found in very good forecasts. Through the operation of the model in the coming years it will be observable, whether the quality of the forecast can meet the requirements in the actual event.

2.6 Summary

A flood forecast system is an important contribution to effective flood risk-management. Particularly in the case of events potentially exceeding the design discharge of protection measures, it is essential to provide reliable information as early as possible.

While the first flood forecast systems were mostly based on empirical models, the hybrid approach using hydrological as well as hydrodynamic models is becoming more popular (Godina 2006). By using different models to describe the different processes contributing to the formation of a flood, it is possible to consider several regional distinctions. In the case of the river Inn, the alpine topography (snow line), glacier runoff and the influence of hydroelectric power plants (alpine reservoirs as well as run-of-river power stations) are important factors considered in the forecast system. In contrast to a detailed local analysis of a particular problem, different issues have to be emphasized when setting up models of the forecast system. For instance, the stability and the computing time of the hydrodynamic model is of greater relevance in the forecast system. Integrated and reliable data management is essential for large amounts of input data and to provide unproblematic data flow between the different models as well as for the visualization of the results.

The forecast system provides a lot of detailed information. Therefore, it must be operated by experts who have a good knowledge of the river Inn and its catchment area and are therefore able to interpret and review the results. Additionally, the system should be constantly enhanced and updated as every flood event provides new experiences and information.

As the presented examples show, not only the public authorities responsible for flood warning and protection, but also hydropower operators can benefit from the flood forecast system. The forthcoming test period will be the first challenge for the forecast system and its operators and will possibly show some potential for optimization.

References

Asztalos J (2004) Ein Schnee- und Eisschmelzmodell für vergletscherte Einzugsgebiete. Diplomarbeit Technische Universität Wien

BAFU - Schweizerische Eidgenossenschaft - Bundesamt für Umwelt-Abteilung Hydrolgie (2006) Hochwasserwahrscheinlichkeiten (Jahreshochwasser). http://www.bwg.admin.ch/lhg/hq/2067hq.pdf

BMLFUW (Bundesministerium für Land- und Forstwirtschaft, Umwelt und Wasserwirtschaft) (2005) Digitaler Hydrologischer Atlas Österreich (digHAO). Version 2.0.1. Österreichischer Kunst- und Kulturverlag, Wien

Baumgartner A, Liebscher H-J (1996) Lehrbuch der Hydrologie, Allgemeine Hydrologie, Quantitative Hydrologie, 2. Auflage. Gebrüder Borntraeger, Berlin Stuttgart

Blöschl G, Merz R, Humer G, Hofer M, Hochhold A, Wührer W (2006) HORA - Hochwasserrisikoflächen Österreich, Hydrologische Arbeiten, Endbericht an das Bundesministerium für Land- und Forstwirtschaft, Umwelt und Wasserwirtschaft, Sektion VII, 1030 Wien. Institut für Wasserbau und Ingenieurhydrologie, Technischen Universität Wien, Wien and Ingenieurbüro DI Günter Humer, Geboltskirchen

Buermann W, Dong J, Zeng X, Myneni R, Dickinson R (2001) Evaluation of the Utility of Satellite-Based Vegetation Leaf Area Index Data for Climate Simulations. Journal of Climate 14, pp 3536-3550

Chanson H (2004) The Hydraulics of Open Channel Flow: An Introduction, 2nd edn. Elsevier Butterworth-Heinemann, Amsterdam et al
Csekits C, Jann A, Wirth A, Zwatz-Meise V (2001) Nowcast Modules at the Austrian Meteorological Service - powerful tools for forecasters. Proceedings to the EGOWS (European Working Group for Operational Workstation Systems) Zürich, June 2001
Cunge JA, Holly F M, Verwey A (1980) Practical Aspects of Computational River Hydraulics. Pitman Publishing Ltd, London
Drabek U (2004) d.4.4 Teilmodell Brixentaler Ache. Internal Report, alpS – Zentrum für Naturgefahren Management Gmbh, Innsbruck
Federer CA and Lash D (1978) BROOK: A Hydrologic Simulation Model for Eastern Forests. Water Resource Research Centre, University of New Hampshire, Research Report No.19
Fritsche C (2001) Vergleichende Ereignisbezogene Modellierung der Abflussbildung mit PRMS/MMS und TOPMODEL in zwei Kleinen Quelleinzugsgebieten im Thüringer Wald. Diplomarbeit Friedrich-Schiller-Universität, Jena
Gattermayr W (2005) Hydrologische Übersicht August 2005, Landesbaudirektion - Abteilung Wasserwirtschaft - Sachgebiet Hydrographie, Innsbruck
Godina R (2006) Prognose und Hochwasserwarnung – Erfahrungen aus Sicht des HZB. In: Gutknecht D (ed) Wiener Mitteilungen: Hochwasservorhersage – Erfahrungen, Entwicklungen & Realität, vol 199. Institut für Wasserbau und Ingenieurhydrologie, Technische Universität Wien, pp 133-141
Hock R (2003) Temperature Index Melt Modelling in Mountain Areas. Journal of Hydrology 282, pp 104-115
Hofer B (2005) Dämpfende Auswirkungen der Speicherkraftwerksanlagen auf den Hochwasserabfluss am Inn, Untersuchungen der Ereignisse vom August 2005 und Juli 1987, Geoforum Umhausen 2005
HZB - Hydrographisches Zentralbüro (2006) Hydrographisches Jahrbuch von Österreich 2003, Abteilung Wasserhaushalt (Hydrographisches Zentralbüro) im Bundesministerium für Land- und Forstwirtschaft, Umwelt und Wasserwirtschaft, Wien
Kirnbauer R, Schönlaub H (2006) Vorhersage für den Inn. In: Gutknecht D (ed) Wiener Mitteilungen: Hochwasservorhersage – Erfahrungen, Entwicklungen & Realität, vol 199. Institut für Wasserbau und Ingenieurhydrologie, Technische Universität Wien, pp 69-84
Kleindienst H (1996) Erweiterung und Erprobung eines anwendungsorientierten hydrologischen Modells zur Gangliniensimulation in kleinen Wildbacheinzugsgebieten. Diplomarbeit, Institut für Geographie, Ludwig Maximilians Universität München
Leij F, Alves W, Van Genuchten M., Williams R (1999) The UNSODA Unsaturated Soil Hydraulic Database. Version 2.0
Leonhardt G, Senfter S, Schöberl S, Schönlaub H (2006) Ein hybrider Ansatz zur adäquaten Berücksichtigung des Betriebs von Wehr- und Kraftwerksanlagen im Rahmen des Hochwasservorhersagemodells Inn, In: Horlacher H-B, Graw K-U (eds) Wasserbauliche Mitteilungen - Strömungssimulation im Wasserbau

(Flow Simulation in Hydraulic Engineering), Vol 32, Institut für Wasserbau und Technische Hydromechanik, Technische Universität Dresden, pp 23-30

LFU - Bayerisches Landesamt für Umwelt (2006) Hochwassernachrichtendienst, Mittel- und Höchstwerte für den Pegel Oberaudorf/Inn, www.hnd.bayern.de

Lundberg A, Beringer J (2005) Albedo und Snowmelt Rates Across a Tundra-to-Forest Transition. 15th International Northern Research Basins Symposium and Workshop, Luleå to Kvikkjokk, Sweden, 29 Aug. – 2 Sept. 2005

Nash JC (1990) Compact Numerical Methods for Computers: liear algebra and function minimisation, 2nd edn. Adam Hilger, Bristol and New York

Reichel G, Baumhackl G (2000) A New Simulation Tool for Flood Routing in Mancontrolled River Systems Focusing on to the Needs of Operational Hydrology. International Symposium on Flood Defense, Kassel

Reichel G, Fäh R, Baumhackl G. (2000) FLORIS-2000: Ansätze zur 1.5D-Simulation des Sedimenttransportes im Rahmen der mathematischen Modellierung von Fließvorgängen. In: Heigerth G (ed) Symposium: Betrieb und Überwachung wasserbaulicher Anlagen, Graz, 19. - 20.10.2000. Eigenverlag Inst. f. Wasserbau u. Wasserwirtschaft, Techn. Univ. Graz, pp 485-494

Reindl R (2004) Protokoll der Besprechung zum Betrieb der Kraftwerke am Inn vom 21.04.2004, TIWAG Tiroler Wasserkraft AG, Innsbruck

Richard F, Lüscher P (1983) Physikalische Eigenschaften von Böden der Schweiz, Band 1-3. Eidg. Anstalt für das forstliche Versuchswesen, Birmensdorf

Richard F, Lüscher P (1987) Physikalische Eigenschaften von Böden der Schweiz, Band 4. Eidg. Anstalt für das forstliche Versuchswesen, Birmensdorf

Rickenmann D (1996) Fließgeschwindigkeit in Wildbächen und Gebirgsflüssen. Wasser-Energie-Luft 88 11/12, pp 298-303

Scurlock JMO, Asner GP, Gower ST (2001) Worldwide Historical Estimates of Leaf Area Index, 1932-2000. Tech Report ORNL/TM-2001/268, U.S. Department of Energy, Environmental Sciences Division

TIWAG - Tiroler Wasserkraftwerke Aktiengesellschaft (1999-2003) Schutzwasserwirtschaft Inn, Ermittlung der Ausuferungsbereiche bei einem HQ100 bzw. HQ30 Hochwasserereignis. Tech Report, TIWAG - Tiroler Wasserkraftwerke Aktiengesellschaft, Bautechnik - Abteilung Wasserbau, Innsbruck

Wang Y, Haiden T, Kann A (2006) The operational limited area modelling system at ZAMG: ALADIN-AUSTRIA. Österreichische Beiträge zur Meteorologie und Geophysik, Vol 37

Widmann, R (1989) Einfluß alpiner Speicher auf Hochwässer. In: Die Talsperren Österreichs, Vol 31

Wilson D, Western A, Grayson R (2005) A Terrain and Data-Based Method for Generating the Spatial Distribution of Soil Moisture. Advances in Water Resources 28, pp 43-54

3 Runoff and bedload transport modelling for flood hazard assessment in small alpine catchments - the PROMAB^{GIS} model

M. Rinderer, S. Jenewein, S. Senfter, D. Rickenmann, F. Schöberl, J. Stötter, C. Hegg

Abstract

Successful risk management of flood related hazards like extreme runoff and intense bedload transport in small alpine catchments is strongly dependent on accurate hazard assessment on a local scale. Some currently available methods used by practitioners are simple and therefore result in rather general evidence. Other methods – mostly scientific modelling approaches – are often too sophisticated to be applied for practical use with an economically justifiable effort. Thus, there is a need for a simple but reliable approach for flood related hazard assessment in small alpine catchments. As a consequence of this the model PROMAB^{GIS} (**PRO**cess orientated **MA**ss **B**alances) originally developed by Jenewein (2002) and Rinderer (2002) in cooperation with practitioners (i.n.n. ingenieurgesellschaft für naturraum-management GmbH & Co KG) has been further developed at alpS - Centre of Natural Hazard Management for estimating runoff and bedload transport in ungauged catchments of small size and alpine character. The model concept of both, the runoff and bedload transport module, are presented in this paper. Moreover first results of model validation using measurements of flood events in the Erlenbach catchment (Switzerland) give proof of the models potential for estimating runoff and bedload transport in small alpine catchments.

3.1 Introduction

Major flood events in small alpine catchments are normally associated with bedload transport and/or debris flows which pose a high threat to human life and property. This is different compared to flood hazards in major river basins where most damage is caused by inundation and suspended sediment accumulation. Therefore, methods for assessing flood hazards in small alpine catchments have to differ from those commonly used in large river basins. In particular, methods have to account for the available sediment potential and the specific transport processes that take place in torrents. Currently available methods are either simple and therefore result in rather general evidence or they are rather complex and therefore time-consuming in their application.

3.1.1 Simple approaches for total sediment load assessment

Most basic approaches for estimating total sediment load in torrential catchments are based on regression techniques or envelope-analyses using data of past events. They are missing a physical background and therefore sometimes are not consistent as far as dimensions are involved. In most cases only one parameter – the catchment size – can be chosen. Thus they are easy to use but result in rather general evidence. More enhanced approaches have basic options for regional adaptation but normally cannot take site-specific characteristics into account. The recurrence interval of calculated discharges and loads is not well defined in most cases. Formulas result in a total sediment load but can not gain hydro-sedigraphs showing the temporal distribution of runoff and bedload transport.

For examples of simple sediment load formulas see: Hampel (1977); Kronfellner-Kraus (1984, 1987); Zeller (1985); D'Agostino et al. (1996). For simple formulae of debris flows volume-estimation see Hampel (1980) or Rickenmann (1995). Zimmermann and Lehmann (1999) give a well structured overview.

Flood hazard assessment can be improved by applying simple formulae to individual reaches of the torrential channel. Focus is on quantitatively balancing the sediment availability and its accumulation within each channel reach. The actual process of bedload transport however is not modelled. For some examples of this type of approach see: Hungr et al. (1984); Scheuringer (1988); Spreafico et al. (1999) and D'Agostino and Marchi (2003).

Simple sediment delivery approaches focus on the source areas of the sediment to estimate total sediment load. Sediment routing within the channel is not modelled. Instead a complex set of rules determines how much of the total sediment potential of each channel section is able to reach the alluvial fan. For examples see: Brauner (2001) or Frick et al. (2006).

Massbalancing based on event-specific transport capacity marks a further improvement. The total sediment load is not the sum of all available sediment potential but is determined reach by reach using an empirical transport formula (e.g.: Rickenmann, 1990 or Smart and Jäggi, 1983). An event-specific input hydrograph necessary for the transport capacity calculation has to be determined for each channel reach with a rainfall-runoff model.

3.1.2 Simulation models for hydro-sedigraph estimation

Compared to the group of simple approaches –whether they are site- and event specific or not – more sophisticated approaches exist to deal with flood related hazard assessment in torrents: In most cases they are provided in a modelling framework based on semi-empirical findings and/or physical laws like conservation of mass and momentum and hydraulic principles. Solving complex equations for 1-, 2- or even 3- dimensions is involved with sophisticated numerical methods and long simulation duration. Moreover models often require a large number of input parameters for which data have to be gathered during extensive field-work. To gain good results many sophisticated models need to be calibrated using existing data measured during past events. However measurements are rather seldom for small alpine catchments and therefore some of theses model approaches can not be applied for practical hazard assessment.

In the following a short overview of models for sediment transfer simulation in mountain streams is presented:

The **PRO**cess based **MA**ss **B**alance model **PROMAB**^{GIS} Version 1.0 (Jenewein, 2002; Rinderer, 2002, Schöberl et al., 2004) was one of the first attempts to provide a modelling framework for practical flood hazard assessment in small alpine catchments. It is especially designed for spatially distributed modelling of overland flow and bedload transport trigged by short but intense rainfall events. As its name indicates it is implemented in the Geographical Information System software ArcView 3.x. PROMAB^{GIS} is capable of estimating sediment-laden peak discharge and total runoff and sediment load of steep alpine rivers up to slope gradients of 20 %. The watershed is represented by hydrological response units (not sub-catchments) and the channel network consists of linked, homogeneous channel sections. The estimation of direct runoff generated on each hydrological response unit relies upon design rainfall intensity and spatially distributed runoff coefficients. Runoff concentration is optionally modelled either by an isochrone approach or by a hydrological surface runoff routing. Within the channel network a hydrological routing method by Rickenmann (1996) (eq. 3.2) is used which is empirically derived from measurements in steep torrents and mountain-rivers. Bedload transport estimation is based on the transport formula by Rickenmann (1990) (eq. 3.6). Protective measures like flood retention basins can be taken into account. The PROMAB^{GIS} model was further developed at the alpS – Centre of Natural Hazard Management. The current state of this work is presented in this publication. Thereby the reader's attention is especially drawn on implemented approaches and selected results of model validation.

ETC (**E**rosion des **T**orrents en **C**rue) (Mathys et al., 2003) is a hydrological rainfall-runoff-erosion- and bedload transport model developed for alpine catchments up to 50 or 100 km². The runoff generation is based on the SCS-approach (Soil Conservation Service, 1972) used for calculation of the surface runoff of individual sub-catchments. Soil erosion can be taken into account by an amount of global abrasion or by incorporating results of an external, more sophisticated soil erosion model. Bedload transport capacity is calculated using semi-empirical equations (Rickenmann, 1990; Smart and Jäggi, 1983; Lefort, 1991) which are applicable to slope gradients between 3 and 20 % (Smart and Jäggi, 1983 and Rickenmann, 1990). Runoff and bedload material in the channel is routed using a hydrological routing methode.

SETRAC (**SE**diment **TRA**nsport model in alpine **C**atchments) (Friedl and Fuchs, 2004; Rickenmann et al., 2006; Chiari and Rickenmann, 2007) is a 1-dimensional sediment routing model especially developed for steep torrent channels up to a slope gradient of 20 %. It uses kinematic flood routing of the hydrograph, neglecting diffusive and convective terms of the full Saint Venant equations. Channel cross-sections can be of irregular shape and therefore represent natural inhomogenity of torrential channels. Erodable sediment potential can be assigned to each individual channel section. The model has to be fed with an input hydro- and/or sedigraph and bedload transport capacity is calculated with similar empirical equations for steep torrents as used in ETC but it can additionally account for form resistance losses.

MORMO (Zarn et al., 1995) is designed for major alpine rivers with rather moderate slope gradients like the upper reaches of the river Rhine. It is capable of modelling fluvial sediment transport, taking changes in grain size and slope gradient of the river bed into account. Implemented transport equations are the Meyer-Peter and Müller equation (1948) suitable for slope gradients between 0.04 to 2 % and the Smart and Jäggi equation (1983) which is valid for channel slopes between 3 and 20 %.

FLUMEN (**FLU**vial **M**odelling **EN**gine) (Beffa, 2004) is an extension to the Hydro2de software capable for complex hydraulic simulations including mobile bed sediment transport. The implemented sediment transport formulae are the Meyer-Peter and Müller equation (1948) suitable for slope gradients between 0.04 to 2 % and the Smart and Jäggi equation (Jäggi, 1992) valid for channel slopes between 3 and 20 %. Armouring of the channel bed is considered using a "two-grain-model" published by Günter (1971). As FLUMEN is no rainfall-runoff model its inflow has to be determined in advance using a different model.

FLO-2D (O'Brien, 2003) is a 2-dimensional flood routing model with a sediment transport module for small to moderate slope gradients (6 different equations for alluvial sediment transport of sand- and gravel bed rivers) and a mudflow module applicable for moderate to steep channel conditions. A number of agencies and engineers worldwide use FLO-2D for detailed flood hazard assessment. Surface runoff can be modelled using the Green and Ampt infiltration equation (Green and Ampt, 1911) and a hydraulic surface runoff routing. However, preprocessing and model setup is rather time-consuming as well as detailed simulations using small raster cells can cause long simulation duration.

HEC-RAS (**H**ydrologic **E**ngineering **C**enter – **R**iver **A**nalysis **S**ystem) (US Army Corps of Engineers, 2002) is a one-dimensional hydraulic model for steady and unsteady flow simulation, movable boundary sediment transport computation and water quality analysis. It is no rainfall-runoff model and therefore is dependent on an input-hydrograph. HEC-RAS is primarily designed for long-term simulations but modelling of single events is possible. As sediment transport is computed by grain size fraction, detailed hydraulic processes like sediment sorting or armouring can be simulated. Available sediment transport formulae are similar to those implemented in FLO-2D. Likewise they are all developed to simulate fine grained sediment transport under small to moderate slope gradients.

BASEMENT (**BA**sic **S**imution **E**nviron**MENT**) (Fäh et al., 2006) is a flexible, powerful environment for numerical modelling of fluvial processes in alpine rivers. The model is capable for computing 1-, 2- and 3 dimensional flow patterns as well as bedload and suspended sediment transport. However, the currently implemented formula for sediment transport (Meyer-Peter and Müller equation, 1948) is restricted to slope gradients between 0.04 and 2 %.

It appears that physically based, hydraulic runoff and bedload transport models which are applicable to steep torrents are rare and time-consuming for parameterisation, simulation and result evaluation. Most of them concentrate on hydro-sedigraph routing and therefore depend on other models for rainfall-runoff simulation. However, for detailed simulations, especially of more than one dimension, these models are the right choice.

Semi-empirical runoff and bedload transport models seem to meet most needs of practical flood hazard assessment in small alpine catchments. During the development of PROMAB^{GIS} it has been an important goal to optimise both quality of results and work effort for input data collection, model parameterisation and simulation for practical flood hazard assessment. The original modelling concept of the first version of the model developed by Jenewein (2002) und Rinderer (2002) has been further extended in the course of a research project undertaken by the alps- Center of Natural Hazard Management. The present status of the model and first validation results using data of a Swiss torrent are presented in this paper.

3.2 PROMAB^{GIS} a process-based approach for massbalances in small alpine catchments

PROMAB^{GIS} Version 2.0 comprises a standardised procedure of input data collection in the field, parameter derivation based on commercially available Digital Elevation Models and a modular modelling framework for scenario-based runoff and bedload transport simulation. The main focus is on fluvial processes in ungauged catchments of small size and alpine character mainly triggered by short but intense rainfall events. To optimise the workflow of data-preprocessing, simulation and result-evaluation an extension to a commonly used GIS-Software (Geographical Information System) has been developed providing user-friendly tools and routines embedded into a Graphical User Interface (GUI).
Modelling is based on an integrated inventory of spatially distributed information on geomorphology, vegetation, landuse as well as channel geometry, bedload characteristic and sediment potential.

3.3 Modell concept PROMAB^{GIS}

Since the application of computers in hydrological modelling in the 1960s the number of developed rainfall-runoff-models has increased rapidly and has reached a nearly unmanageable number (Nemec, 1993). In contrast there are only a comparatively small number of models for sediment transport quantification that are being used (see above). The classification of a model can be diverse depending on the criteria used. The definite allocation is further hindered by the fact that many models are of a hybrid-kind of type. An attempt to classify the PROMAB^{GIS} model Version 2.0 could be as follows (see table 3.1):

Table 3.1: Categorization of PROMAB^{GIS} according to established model classification (the applicable categories for PROMAB^{GIS} are highlighted in bold).

Criteria	Model design
Spatial representation of the catchment	Lumped model
	Semidistributed model
	Distributed model
Detailedness of the process description	Black-box model
	Grey-box model
	White-box model
Causality criteria	**Deterministic model**
	Stochastic model
	Hybrid model
Length of the observed time segment	**Event model**
	Continuous model

3.3.1 Runoff

A key-component of flood hazard assessment in torrential catchments is the determination of clear water runoff. Ultimately it is the surface runoff on the hill slopes and in the channels of the catchment that is causing sediment transport.

A flood event can be described by several attributes. Features such as peak discharge, time of rise, duration of the event, runoff volume and other values can be considered. For design purpose the peak discharge is of special importance. Thus, many procedures are focused on the estimation of peak discharge. For design engineering of flood retention basins the evaluation of the total runoff is significant. Therefore duration and temporal distribution of the flood discharge (flood hydrograph) are necessary.

In very simple model approaches (black-box models, lumped models) the components are reduced to only a few parameters such as for example the size of the catchment. More sophisticated hydrological approaches in regard to spatial discretisation or process description mostly consist of three components (a) runoff generation, (b) runoff concentration and (c) channel routing.

Runoff formation

Runoff formation describes the process of direct and delayed runoff due to a rainfall event. The part of the event precipitation that leads to direct surface runoff is called effective precipitation. The rest of the rainfall is delayed due to either initial abstraction (see details below) and/or infiltration into the soil or groundwater body and therefore does not directly contribute to the flood hydrograph (Ihringer, 1992).

Runoff is generated due to a combination of the processes interception, evapotranspiration, infiltration, snow melt, surface- and subsurface storage, as well as water percolation within the soil (Baumgartner and Liebscher, 1990). How much precipitation drains off or how long it is delayed depends on the retention capacity of the soil, the geomorphologic processes and the substratum. On barely permeable soil surface runoff is rapidly formed because of infiltration obstruction. On the other hand very permeable soil or extremely aerate substratum has to be saturated first before surface runoff or interflow is formed. So it depends primarily on the geo-inventory whether runoff forms instantly or delayed. According to Scherrer and Naef (2003) the individual runoff processes and their reactions to heavy rainfall can be depicted as follows (table 3.2):

The soil/vegetation complex is an essential factor for runoff formation. This is particularly the case when the runoff formation of a single spot or an individual slope segment is described. For this case numerous publications exist, in which the interrelation between the soil/vegetation complex and runoff (primarily surface runoff) is investigated in the course of sprinkling experiments. For alpine regions contributions by the Bundesamt und Forschungszentrum für Wald (BFW) and the Bayerischen Landesamt für Wasserwirtschaft (LfW) (Markart et al., 2004), have to be mentioned. In Switzerland Rickli and Forster (1997), Scherrer (1997) or Scherrer and Naef (2003) have been working on this issue. In these studies the test spots investigated during the sprinkling experiments are usually 100 m^2 or less.

For the modelling of the runoff generation for a whole catchment, larger spatial units need to be examined instead of small isolated slope sections. The sole consideration of the soil/vegetation complex is not enough in this circumstance. The influence of the prevailing geomorphologic processes or those possibly taking place during a heavy storm event as well as the geological substratum have to be considered.

Table 3.2: Runoff processes and their reaction to heavy rainfall.

Process	Type	Abbreviation	Reaction of the runoff process
Surface flow	Hortonian overland flow	HOF1	Rapid Hortonian overland flow as a consequence of infiltration obstruction
		HOF2	Delayed Hortonian overland flow as a consequence of infiltration obstruction
	Saturation overland flow	SOF1	Rapid overland flow as a consequence of soil saturation
		SOF2	Delayed overland flow ...
		SOF3	Highly delayed overland flow...
Subsurface flow	Lateral flow	SSF1	Rapid flow in the soil as a consequence of impermeable layers under highly permeable soils or pronounced macrospores
		SSF2	Delayed runoff ...
		SSF3	Highly delayed runoff ...
	Vertical flow	DP	Deep-seepage as a consequence of easily permeable soil and permeable geological substratum

Particularly in small alpine catchments, flooding is normally caused by heavy rainfall of short duration but high intensities. Therefore slowly responding storage components of the soil/vegetation complex are of minor importance. The direct runoff – in most cases generated by overland flow or return flow near the surface – is decisive for the flood discharge. Thus, modelling of runoff in PROMAB^{GIS} is concentrated on the simulation of surface runoff.

Consequently, in a first step runoff coefficient maps are created, that are based on the soil/vegetation complex and the geomorphologic processes of the catchment. For this purpose homogeneous soil/vegetation-units and geomorphologic process-units are mapped in the field (for the soil/vegetation-mapping guidelines see Markart et al., 2004). Then runoff- and roughness-coefficients are assigned to the individual soil/vegetation-units according to Markart et al (2004) and both datasets are intersected in a Geographic Information System (GIS) to generate a combined surface-runoff coefficient map. Thus this map is not only based on single-spot soil/vegetation surveys but also on an area-wide geomopholgical- and soil-substratum-mapping.

The expected surface runoff for each time-step can be derived from the spatially and temporally distributed design rainfall event and the surface-runoff coefficient of each hydrological response unit in which the whole catchment is spatially divided in the course of the GIS-intersection. Initial abstraction can be considered optionally (Markart et al., 2004). Initial abstraction is understood as the precipitation volume infiltrated until soil saturation is reached and overland flow initiates. This infiltration process at the beginning of a flood event leads to a delay in the rise of the flood peak.

Runoff concentration

Runoff concentration is the accumulation of runoff effective precipitation from the catchment slopes to an observed channel reach (Ihringer, 1992). The time needed for this process depends primarily on the distance, the slope gradient and the surface roughness. Besides the temporal shift of the effective precipitation (translation) the shape of the hydrograph is altered due to retention effects.

Thus modelling runoff concentration is concerned with the transformation of the spatially distributed effective precipitation on the hill slopes to a runoff hydrograph of a single point of interest.

In many approaches for describing runoff concentration on the hill slopes roughness parameters have to be defined. Besides the vegetation the soil substratum has an influence on the flow resistance. Mapping geomorphologic processes allows conclusions to be drawn on the surface morphology and therefore surface roughness of individual process units of a catchment.

Modelling of runoff concentration in the PROMAB^{GIS} model can optionally be undertaken by either an isochrones-approach or a cell-based hydrologic routing method: For the determination of the isochrones well-known empirical formulae for estimating runoff-concentration-duration as published by Izzard (1946) and Morgali and Linsley (1965) are provided. On the other hand the isochrones can be determined by assigning plausible flow velocities cited in the literature. These runoff velocities are further varied according to the topography, the runoff coefficient as well as the assigned surface roughness.

In the case of the cell-based hydrologic surface-routing of the PROMAB^{GIS} model the catchment is divided into surface raster pixels each considered as a storage element. Successive surface raster pixels form a storage cascade according to their flow-direction. The relationship between reservoir storage and outflow is described by the Strickler-formula (Strickler, 1923 cit. in Chanson, 2004). This procedure is more demanding in calculation compared to the isochrones-approach. However time-variable flow velocities are more realistic than mean static flow velocity values applied for the duration of the entire event. Calculation time has been tried to optimise by some simplifications: The succession of surface runoff is only considered to be in the direction of the steepest slope gradient (single flow-direction) and no hydraulic procedures for the description of the propagation of the flood wave is used.

Channel routing

For routing the runoff in the predefined channel reaches various channel routing procedures are implemented in the PROMAB^{GIS} model. An appropriate routing method can be chosen depending on effort/expense, data availability and precision demands.

In principle one distinguishes between hydrologic and hydraulic methods. All methods are similar in that they take the continuity equation into consideration. The differences lie in the assumptions of the simplification of impulse balance and in their handling.

For hydraulic approaches the propagation of the flood wave further downstream is calculated using the St. Venant-Equations (Chanson, 2004). For the solution of these equations various approximations are used simplifying some of the basic assumptions: Basically, one distinguishes between steady and unsteady flow conditions. In the case of steady flow conditions discharge and corresponding water depth are temporally invariant. Unsteady flow conditions can be depicted as Kinematic Wave, Diffuse Wave or Hydrodynamic Wave. The simplifications manifest themselves in the disregard of individual terms of the impulse equation (local and advection acceleration term, pressure term), whilst the continuity equation is equally valid for all the wave equations.

$$\underbrace{\underbrace{\underbrace{\frac{\partial Q}{\partial t}+\frac{\partial Q^2/A}{\partial x}}_{\text{local / advektive acceleration–term}}+\underbrace{g\cdot A\cdot\frac{\partial z}{\partial x}}_{\text{pressure–term}}+g\cdot A\cdot(J_E-J_0)}_{\text{kinematic–wave}}=0}_{\text{hydrodynamic–wave}}$$
$$\text{(diffusive–wave)}$$

(eq. 3.1)

Hydrologic approaches bridge physical laws with conceptual approaches to simplify the description of processes during the wave propagation in channels. The approaches reach from very simple estimations of peak discharge and time of the maximum peak, to detailed routing methods, which approximate the Diffusive Wave equation (e.g. Muskingum-Cunge-Routing; Cunge, 1969 cit. in Chanson, 2004).

For the very specific flow conditions in mountain torrent channels, which are characterised by their roughness and steepness, there are relatively few useful applicable approaches for the consideration of flow resistance. Most approaches are developed for rivers with low slope gradients in which the flow resistance is significantly dependant on the grain size distribution of the bed material. In steeper channels additionally bed morphology hast to be considered as further source of flow-energy losses.

The empirical approach by Rickenmann (1996) has been developed on the basis of natural measurements in mountain torrents:

$$v_m = \frac{0.37\cdot g^{0,33}\cdot Q^{0,34}\cdot J^{0,2}}{d_{90}^{0,35}}$$

(eq. 3.2)

Hereby v_m [m/s] is the mean velocity in the cross-section, g [m/s^2] the acceleration due to gravity, Q [m^3/s] the discharge, J [-] the mean slope gradient and d_{90} [m] the characteristic grain size, which 50 % of the material by weight is finer. Equation (3.2) is valid for a slope gradients of 0.006 < J < 0.63.

Since the flow depth does not have to be directly entered as an input parameter and the indication of the channel width suffices, the survey of the cross-section geometry is not compulsory.

The Manning-Strickler's equation (Strickler, 1923 cit. in Chanson 2004) is another very prevalent approach being used very frequently for the registration of flow resistance in practical hydraulic engineering:

$$v_m = k_{St}\cdot R_h^{2/3}\cdot J_E^{1/2}$$

(eq. 3.3)

In this formula k_{St} $[m^{1/3}/s^2]$ is the roughness coefficient according to Strickler, R_h [m] the hydraulic radius for the cross-section and J_E [-] the energy head line gradient.

The formula was in actual fact developed for lowland rivers, but its application is commonly extended to steeper channels by adequately adapting the roughness coefficient k_{St}. For the application of the Manning-Strickler formula knowledge of the cross-section geometry is necessary to be able to calculate the hydraulic radius. In equation (3.3) the Strickler-coefficient is a constant value. In fact it varies depending on the flow depth. Therefore flow depth is explicitly considerd in various approaches (e.g. Palt, 2001).

The approaches used in the current version of PROMAB^{GIS} structurally correspond with hydrological routing models (storage cascades). The link between storage volume and out flow from the storage is non-linear and for each channel section (represented by one storage element) there are several parameters available for calibration.

The first implemented approach describes the relationship of stage-discharge in a reservoir with the formula by Rickenmann (see eq. 3.2). The cross-section geometry is not needed explicitly. Besides the initial and boundary conditions only information about the grain roughness (d_{90}), the length of a channel and the mean slope gradient (J) are required.

The second implemented approach is the Manning-Strickler formula (see eq. 3.3). It needs additional data on the cross section geometry for the determination of the hydraulic radius. For this purpose the width of the channel bed and the gradient of both bank slopes need to be additionally surveyed in the field.

In general it must be noted that for the required task – under consideration of the large variation of input parameters and the resulting inaccuracies – the easier applicable hydrological routing procedures can be reasonably applied. Particularly in the channel reaches with large slope gradients the increase of precision through more complex procedures plays a subordinate role. For the PROMAB^{GIS} model it is much more important that for both implemented hydrological procedures no measured discharge hydrographs have to be used for calibration to achieve good results.

3.3.2 Bedload Transport

In small alpine rivers and steep torrents flooding is normally associated with bedload transport or even debris flows. Shear stress and drag forces generated by surface runoff can mobilise material within the channel or its adjacent banks resulting in bedload transport. Under low or mean water only particles with small grain size (silt, sand) can be mobilised and as time proceeds coarser grains tend to dominate the surface of the channel bed. These coarser grains can form an armour layer which prevents further sediment erosion under low and mean flow conditions. Bedload transport during a flood event is not initiated until flow conditions reach a certain threshold of shear stress which mobilises the coarse grains of the armour layer.

The PROMAB^{GIS} model calculates critical discharge which defines initiation of bedload motion based on results of the preceding runoff simulation. For rather smooth channel bed conditions without a significant armour layer an approach by Bathurst et al. (1987), later modified by Rickenmann (1990) (eq. 3.4) is used. For a rough streambed and channel slopes steeper than 0.05 the use of an approach by Whittaker and Jäggi (1986) (eq. 3.5) may be used.

$$q_{cr} = 0.065(s-1)^{1.67} g^{0.5} d_{50}^{1.5} S^{-1.12} \qquad \text{(eq. 3.4)}$$

$$q_{cr} = 0.143(s-1)^{1.67} g^{0.5} d_{65}^{1.5} S^{-1.67} \qquad \text{(eq. 3.5)}$$

q_{cr} is the critical discharge for initiation of bedload motion per unit width of the channel in [m^3 s^{-1} m^{-1}], s is the ratio between grain and fluid density, g is the acceleration due to gravity in [m s^{-2}], d_{50} in [m] is the characteristic grain size, which 50 % of the material by weight is finer and S (= *tan β*) is the slope gradient in [%, -].

If the critical discharge is exceeded, bedload transport initiates and bedload transport capacity is calculated in PROMAB^{GIS} with an equation by Rickenmann (1990):

$$q_s = \frac{12.6}{(s-1)^{1.6}} \left(\frac{d_{90}}{d_{30}}\right)^{0.2} (q - q_{cr}) S^{2.0} \qquad \text{(eq. 3.6)}$$

q_s is volumetric bedload transport rate per unit width of the channel in [m^3 s^{-1} m^{-1}], d_{30} respectively d_{90} in [m] is the characteristic grain size, which 30 % respectively 90 % of the material by weight is finer and q is the discharge per unit width of the channel in [m^3 s^{-1} m^{-1}].

Equation (eq. 3.6) was developed using results of laboratory experiments of bedload transport in a flume with slope gradients from 0.03 to 0.20. For steeper channel slopes the equation predicts debris-flow like sediment concentrations, which may lead to a flow-behaviour different from ordinary fluvial sediment transport; therefore the applicability of (eq. 3.6) is limited to the slope range mentioned before.

Unsaturated transport conditions

The transport formula by Rickenmann (1990) (eq. 3.6) characterises fully developed transport conditions. Field measurements in mountain streams however often indicate lower sediment transport rates than predicted by (eq. 3.6) (Rickenmann, 2001). Hegg and Rickenmann (2000) show that for a large number of flood events in the Erlenbach torrent (Switzerland) the transport conditions can rarely be characterised as saturated or fully developed.

Form resistance due to the rough channel bed is assumed to be one possible reason for unsaturated transport conditions. A part of the energy which causes bedload transport is dissipated by the turbulent currents of the water caused by the rough surface of the channel.

To account for this reduction of transport energy, approaches developed by Rickenmann (2005) and Rickenmann et al. (2006) are implemented in the model. Using flow velocity measurements in rivers and torrents up to slope gradients of 0.63 (Rickenmann, 1996), Rickenmann (2005) estimated the relative proportion of grain roughness to total roughness as:

$$\frac{n_r}{n_{tot}} = \frac{0.083 \cdot \left(\frac{h}{d_{90}}\right)^{0,33}}{S^{0,35}} \qquad \text{(eq. 3.7)}$$

n_r is the Manning-coefficient related to skin friction in [s m$^{-1/3}$], n_{tot} is the total friction in [s m$^{-1/3}$] and h is the flow depth in [m].

For uniform flow, the slope of the energy head line (S) can be approximated by the slope gradient of the channel. To account for form resistance losses, the energy slope can be partitioned into a fractional part S_{red} associated with skin friction only:

$$S_{red} = S \cdot \left(\frac{n_r}{n_{tot}}\right)^a \qquad \text{(eq. 3.8)}$$

For bedload transport capacity estimation considering form roughness losses, S_{red} is used instead of S in (eq. 3.6).Using the Manning-Strickler-equation to partition bed shear stress results in an exponent a = 2. Meyer-Peter and Müller (1948) applied a similar approach to their laboratory experiments and proposed a = 1.5. Based on reconstructed sediment transport loads during an extreme flood event in an Austrian torrent, Rickenmann et al. (2006); Rickenmann and Chiari (2007) and Chiari et al. (2008) back-calculated a best-fit exponent a = 1.

A second probable reason for unsaturated transport conditions may be a general or temporary shortage in erodable bedload material. The amount of available material depends on the sediment potential in the channel, its adjacent banks and hillslopes potentially providing material due to mass movements during the event. The sediment potential has to be estimated in the course of a geomorphologic field assessment. A possible sediment input from upstream affects transport conditions as well. Depending on whether the available amount of material exceeds the transport capacity or not the transport conditions will be saturated or unsaturated.

The possible input of material from a channel reach or tributary upstream, and the output of the channel section are balanced. If the input of sediment exceeds the output, material is accumulated in the channel section. If more material exits the channel section than enters it, material is eroded. If sediment input and output equals each other, there is no change in channel material storage for a particular time step of the simulation. Channel properties like the sediment potential have to be updated after each single time step to obtain new boundary conditions for the next simulation time step.

3.4 Case study Erlenbach

Discharge measurements in small mountain torrent catchments are rather seldom and if there are series of measurements available then they are often short and do not include major events. The PROMAB^{GIS} model was predominantly developed for the assessment of rare design events. Therefore events with a low probability of occurrence were chosen for validation. A further criterion was the availability of bedload measurements, which are scarce in steep alpine catchments. Finally, the Swiss Erlenbach was chosen in which the Swiss Federal Institute for Forest, Snow and Landscape Research (WSL) has been undertaking runoff- and sediment transport measurements for 20 years. Several runoff events with a high magnitude have been recorded and records of two of them were available for this assessment.

3.4.1 Overview

The Erlenbach catchment is 60 km south of Zurich in the Alptal/canton Schwyz (CH). The catchment size is 0.7 km². From the highest point, the Furggelenstock (1656 m a.s.l.) to the gauging station (1110 m a.s.l.) near the mouth of the catchment into the river Alp (local receiving stream) there is a difference in height of about 550m. The mean channel slope gradient is about 25%. The majority of the channels however is flatter (about 20%). The channel network itself is extremely dense and besides the natural channel network a ramified drainage system exists. The drainage density of the catchment lies at least at 16 km/km². The majority of the anthropogenic rifts were taken into consideration. In total for 11.4 km of channel length runoff and bedload transport were modelled.

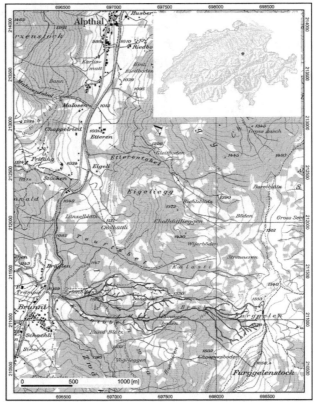

Fig.3.1 Map of the Erlenbach in the canton Schwyz (CH). Reproduced with the permission of swisstopo (BA068292)

3.4.2 Geology and Processes

For the geomorphologic map existing information was collected and examined. In addition, aerial photographs were stereoscopically assessed and results evaluated in the field.

The process dynamics in the Erlenbach catchment is strongly characterised by the geological setting. Indications of mass movements (such as sabre-like growth of trees, taut roots) predominately effecting the shallow weathering layers of the flysch can be found in the whole catchment. In the middle and upper parts of the channel bedrock is partly affected as well, whereby the ridge area of the Erlenbach does not show indications of movement and therefore was not classified as creeping. Apart from these ridge areas, most parts of the catchment show several indications of water logging that are especially pronounced in the area of the creeping mass movements.

The Erlenbach springs from the extensive water logging areas in the upper part of the catchment. Here numerous small channels have formed that subsequently get united. Large sections of these channels exhibit current erosion. Likewise, the subsequent main creek and its adjacent banks are effected by ongoing erosion and sliding processes.

3.4.3 Hydrological setting

As illustrated above, the catchment exhibits a large number of water logging areas. These areas in the middle and upper part of the catchment are tapped by an anthropogenic drainage system and so their runoff reaches the closest torrential channel with only a small delay.

The current anthropogenic usage of the catchment is restricted to seasonal alpine farming in the upper region "Furggelen" as well as farming in the lower parts of the catchment and at the valley bottom of the river Alp. In total, about a fifth of the catchment area is still being agriculturally used. In the past usage seemed to be even more intensive, which can be deduced from the numerous drainage ditches for dewatering the wet areas of the catchment.

Due to the numerous water logging areas and the limited infiltration capacity of the soil the general runoff disposition in the catchment is high. In areas used for seasonal alpine farming the infiltration capacity is further reduced by cattle compacting the soil. Therefore runoff coefficients are particularly high in these regions. Forest is represented by about 50 %. The composition varies, however the coniferous species dominate. Generally scrubs are well pronounced. Due to the unfavourable substratum water logging can be found in the forested area of the catchment as well. Thus in the Erlenbach catchment even forested habitats have to be considered as unfavourable in regard to surface runoff generation.

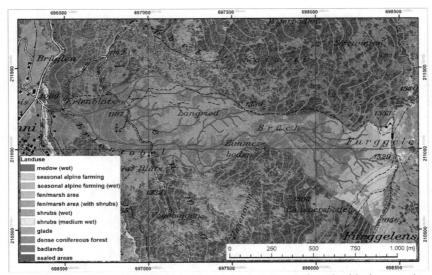

Fig. 3.2 Landuse map of the Erlenbach catchment. Reproduced with the permission of swisstopo (BA068292)

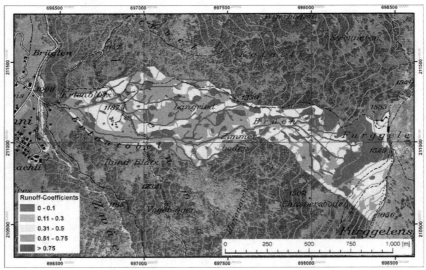

Fig. 3.3 Runoff coefficient map considering the soil/vegetation complex and geomorphologic processes. Reproduced with the permission of swisstopo (BA068292)

Surface roughness can also be considered as high in the Erlenbach catchment. The herb layer of the forest free area is composed of tightly rooted grasses and mosses. In the forest areas it is particularly dwarf-shrubs, such as the blueberry, and mosses that act as flow resistance and therefore retard surface runoff.

3.4.4 Modell validation

As already illustrated above the event based PROMAB^{GIS} model is primarily designed for rainfall-runoff-bedload transport modelling of short but intense rainfall events. Therefore two of the largest flood events in the Erlenbach were selected for model evaluation. These are the events of July the 14th 1995 and September the 12th 1997. For model testing precipitation measurements (rain gauge, situated in the lower part of the catchment), discharge measurements (gauge), as well as hydrophone measurements with a temporal resolution of one minute were available. To be able to draw conclusions on the antecedent system conditions (e.g.: soil moisture conditions before the events), the precipitation records of the days before the flood event were analysed.

The bedload measuring system in the Erlenbach, which has been in operation since 1986 registers the vibrations which in the case of bedload transport are caused by stones that are transported over a measurement cross-section. Hydrophones are attached to the underside of a metal plate that reaches over the measurement cross-section registering the impacts of transported bedload grains and transfering them into electromagnetic signals (piezoelectric bedload impact sensor impulses; for the test set-up see Rickenmann and McArdell, 2007). The measuring system is sensitive to grains with a minimum size of 1 to 3cm (Rickenmann and McArdell, 2007).

Analyses of Rickenmann (1997) have shown that the number of hydrophone-impulses is proportional to the bedload transport rate.

The hydrophone-impulses for the two available validation events were converted to a bedload transport rate using an empirical relation published by Rickenmann and McArdell (2007) (eq. 3.9).

$$F_p = 0.934 \cdot SP \qquad \text{(eq. 3.9)}$$

Fp in [m³] is the total bedload volume including fines (sand, silt, clay) of one or several flood events which were deposited in the sediment retention basin immediately downstream of the bedload measurement station. SP is the number of hydrophone-impulses of the associated event divided by 1000.

The correlation was developed based on 25 measurement periods between 1986 and 1992 with in total 287 individual events (Rickenmann and McArdell, 2007). For the comparison of the measured bedload transport intensity with the simulated ones the portion of suspended sediment is subtracted from the total sum as fines are not registered by the hydrophones. This percentage of the total sediment load is estimated to be 30 percent on average.

Results Clear Water Runoff

For both selected evaluation events the cell-based hydrological routing method was chosen for modelling the runoff concentration on the hill slopes. Results are shown in Figure 3.4.

When comparing the results of the two events the difference in precipitation intensity strikes out. The maximum values are 2.7 mm/min (1995) and 1.5 mm/min (1997) respectively. So both values lie well above the lower threshold of 0.5 mm/min conceived by Markart et al. (2004) as the limit for the validity of their mapping guidelines. The sum of precipitation till the peak discharge is reached is about 40 mm for the event in 1995 and about 45 mm for the event in 1997 respectively, whereby former rainfall event was particularly short and intense. Total sum of rainfall was about 50 mm for both storm events.

As discussed before not only the rainfall event actually triggering the flood event but also the antecedent hydrological system conditions play an important role in generation a high magnitude event. In this perspective the two evaluated storm events differ from each other: Whereas a sum of 35 mm of rainfall could be registered during 48 hours prior to the event in 1995 there were two days without any rainfall before the 1997-event.

For each of the selected flood events two different scenarios were considered: The first one neglects initial abstraction whereas the second one takes these preliminary losses into account. It could be shown, that modelling initial abstraction has a considerable influence on the quality of modelling results in cases where no rainfall directly prior to the actual flood event occurred.

3 Runoff and bedload transport modeling for flood hazard assessment 91

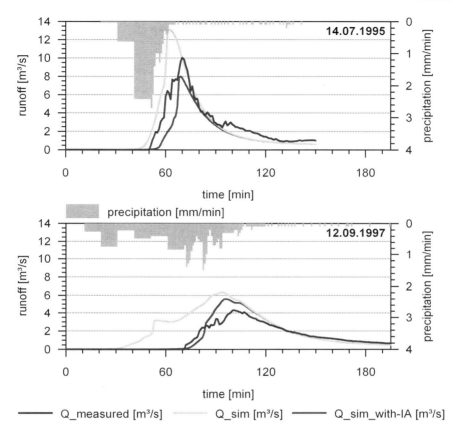

Fig. 3.4 Measured and simulated hydrographs for the two selected evaluation events of 14th of July 1995 and the 12th of September 1997 in the Erlenbach catchment

Table 3.3: Comparison of modelled and measured cumulative runoff volume [m³].

[m³]	1995	1997
measured	18500	15000
simulated without initial abstraction	20900	25300
difference to measured	+2400	+10300
simulated with initial abstraction	12600	16400
difference to measured	-5900	+1400

The simulation of the 1997-event considering initial abstraction results in much better representation of the actual hydrograph. The initial system-conditions were rather favourable due to dry weather conditions during the preceding days. From the records it can be derived that a considerable amount of precipitation could infiltrate at the beginning of the event till soils got saturated and began to drain. Taking this initial abstraction into account the rise and peak of the hydrograph can be simulated considerably well. In Comparison neglecting initial abstraction leads to a much too early rise of the flood hydrograph and an overestimation of the flood peak. Total Runoff volume is overestimated as well (see table 3.3).

For the event in 1995 the soil conditions can be expected to be almost saturated due to rainfall in the preceding days. Simulation neglecting initial abstraction leads to a rather good representation of the rising limb of the flood hydrograph jet overestimates the peak discharge. Better agreement between measured and simulated peak discharge can be gained when considering the initial losses however the initiation of surface runoff shows a delay. A possible reason for the fact that the two hydrographs of the both scenarios are much more equal as far as the delay in rise is considered may be explained by the wet system conditions prior to the event. Only a comparatively small amount of precipitation is needed to reach saturated soil moisture conditions which results in a short delay even considering initial abstraction. Total runoff volume is slightly overestimated without consideration of initial abstraction whereas it is underestimated them neglecting it (see table 3.3).

Results bedload transport

In a first step the implemented approaches for bedload transport calculation in mountain torrents during flood events were evaluated. Thereby, the focus was on the comparison of measured and simulated bedload hydrographs. To limit possible uncertainties due to variations of the measured and simulated clear water runoff the assessment of the implemented bedload transport modelling approaches were based on the measured clear water hydrographs.

The validation tests based on the measured hydrographs lead to the conclusion that the implemented bedload transport approach in the PROMAB^{GIS} model tends to overestimate the observed bedload transport (peak and load) when form resistance is neglected. On the other hand, when taking form resistance into account (eq. 3.7, eq. 3.8 with a=1) simulated results tends to underestimate bedload transport intensities and total load.

3 Runoff and bedload transport modeling for flood hazard assessment 93

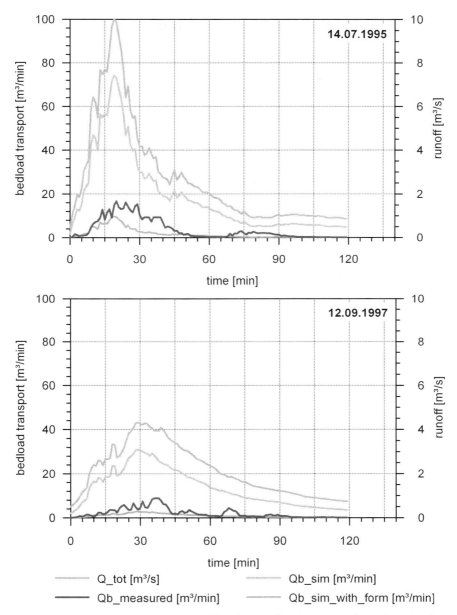

Fig. 3.5 Measured and simulated beldload rates for the two selected evaluation events of 14th of July 1995 and the 12th of September 1997 in the Erlenbach catchment. (Form resistance was taken into consideration with a=1 in eq. 3.8)

However, it is noted that the calculated transport loads including form resistance with a = 1 are much closer to observed loads. This is in agreement with results from simulations with the SETRAC model applied to several Austrian torrents (Rickenmann et al., 2006; Chiari and Rickenmann, 2007; Chiari et al., 2008).

Even if the difference between measured and simulated bedload transport rates is considerable, yet it appears that the implemented approaches are at least appropriate to mark an upper and lower bound of realistic bedload transport intensities. This restriction can therefore be meaningful as fluctuation is inherent to natural bedload transport in general and therefore no single correct result can be expected: Hegg and Rickenmann (2000) for example report a variation of more than two orders of magnitude for observed bedload transport intensities in the Erlenbach stream during periods of lower discharge and still a variation of one order of magnitude during higher discharge events. Therefore differences between simulated and measured bedload transport intensities lie within the same order of magnitude as natural observed variations.

Possible reasons for the variability of bedload transport can be the occasionally limited availability of material, the turbulent movement of the water and the changes of channel characteristics during the event (e.g.: cross section geometry, bed roughness, grain size).

During all events a large portion of the time unsaturated transport conditions seem to have prevailed. Inherently the form of the measured sedigraphs, showing alternating phases with high and low transport intensities, lead to this conclusion. Even for the event in 1995, that shows a (clear water) runoff peak of almost 10 m^3/s and therefore distinguishes itself in magnitude from the other tested event with a peak discharge of about 4 m^3/s, this is to be expected.

When evaluating the difference between simulated and measured bedlaod transport intensities uncertainties in both, the calculation approaches as well as reference data have to be kept in mind:

The transport formula by Rickenmann (1990) (eq. 3.6) was developed in the laboratory under idealised conditions. However even when defined basic parameters for the transport equation determination were used measured data was dispersed. When applying this laboratory-based formula to natural conditions one must assume that the uncertainties will even grow.

But also the reference data – the "measured" sedigraphs – are afflicted with errors: The hydrophone measurement method is well appropriate to register natural bedload transport intensities as influences due to the measurement device are limited (see Rickenmann and McArdell, 2007). Nevertheless, the measurement and conversion of the hydrophone-impulses into total bedload and transport intensities is linked with uncertainties: So for example signal intensity is depending on grain size and type of impact of the bedload grains on the metal plate with the hydrophone sensors underneath (see laboratory tests by Etter, 1996 cit. in Rickenmann and McArdell, 2007). Furthermore, the threshold value of minimum impact intensity necessary to register a transported grain moving over the metal plate holds certain fuzziness in recording actual bedload transport.

For converting the hydrophone impulses to bedload transport intensities equation 3.9 was used that was derived empirically from 25 measurement periods between 1986 and 1999 each including several events. The sum of hydrophone impulses of individual measuring periods and associated measured total sediment loads however do not coincide (see Rickenmann and McArdell, 2007 Fig. 3.5). As a consequence of this one can deduce that using equation 3.9 to determine total loads of individual flood events inevitably may lead to deviations from actual event loads. Rickenmann and McArdell (2007) reconstructed actual event loads of 287 individual events within the 25 measurement periods using different equations and could thereby determine variations of up to a factor of 5. They ascertained that the majority of the result-differences however lie about a factor of 2. This uncertainty involved with reconstructing individual event loads and herewith the uncertainty of the reference values for the model validation is within a similar dimension as the variation of errors between simulated and reconstructed sediment loads.

Furthermore, additional uncertainties arise due to simplified assumptions in regard to grain size distribution and suspended sediment load portion: The grain size distribution is, on the one hand, decisive for the volume estimation of the sediment load accumulated in the sediment retention basin and, on the other hand, for the estimation of the suspended sediment load portion of the total bedload. The later is insofar relevant as sediment with small grain size is not recorded by the hydrophones but contributes to total sediment load of an event. For the validation tests the reconstructions of all events are calculated with a mean grain size distribution which may not correspond with the actual distribution of the individual flood events examined.

Conclusions

To evaluate the quality of the results of the PROMAB^{GIS} model for practical hazard assessment the model needs to be tested on further extreme flood events. Nevertheless, the following facts can be derived from this first model validation:

Evaluating the runoff module of the PROMAB^{GIS} model three main influencing variables were indicated as crucial: (1) the runoff coefficients decisively influencing the quality of the correspondence between the measured and modelled cumulative runoff and peak discharge (2) the soil moisture conditions prior to the event – which can be taken into consideration in the model by initial abstraction – primarily influence the rise of the hydrograph. Depending on the system conditions, which are strongly influenced by antecedent rainfall events, better results could be achieved by considering or neglecting initial abstraction. For safety reasons simulations without taking initial abstraction into consideration seems to be the right choice. (3) Thirdly, the parameterisation of the surface routing influences the quality of the simulation result. In this paper the potential of hydrological routing methods for both runoff on hill slopes and in channels was evaluated.

When evaluating the bedload transport module modelling results may seem of be disappointing. However several great uncertainties, both in regard to the implemented approaches as well as the reference values have to be considered. Thus, the variations between the simulated and reconstructed bedload transport intensities relativise themselves. For practical hazard assessment the model has its relevance insofar as that it enables to mark an upper and lower bound of realistic bedload transport intensities. In case of doubt it is recommended to neglect form resistance as calculations than result in realistic maximum transport intensities. Additional mapping of the event-specific erosion potential can improve the quality of simulated erosion volumes.

From the previously described circumstances of model validation can be derived that the PROMAB^{GIS} model has a potential to make plausible estimations of runoff and bedload transport in small, steep torrential catchments. The results do not provide the quality of models that can be calibrated with data series. However in view of the great uncertainties both in the runoff simulation as well as in bedload transport simulation and - measuring these results are nevertheless helpful and suggestive for design questions. In addition there is great potential in further developing individual modules of the model.

Thanks to

Dr. Helmut Schönlaub (TIWAG Tiroler Wasserkraft AG, Innsbruck, Austria), DI. Alexander Ploner and Mag. Thomas Sönser (both i.n.n. ingenieurgesellschaft für naturraum-management GmbH & CoKG, Innsbruck, Austria) for finanzial support of the project „Massbalances in Alpine Catchments". Thanks to Dr. Hannes Kleindienst (GRID-IT Gesellschaft für angewandte Geoinformatik mbH, Innsbruck, Austria) scientific consultant of the project and to the Swiss Federal Research Institute WSL, Birmensdorf, Switzerland, providing data for model validation.

References

Baumgartner A, Liebscher H-J (1990) Lehrbuch der Hydrologie. Band 1: Allgemeine Hydrologie, Quantitative Hydrologie. Berlin

Bathurst JC, Graf WH, Cao HH (1987): Bed load discharge equations for steep mountain rivers. – In: Thorne CR, Bathurst JC, Hey RD (eds.): Sediment Transport in Gravel-Bed Rivers. New York pp 453-477

Beffa C (2004) FLUMEN User Manual v1.3. (http://www.fluvial.ch)

Brauner M (2001) Aufbau eines Expertensystems zur Erstellung einer ereignisbezogenen Feststoffbilanz in einem Wildbacheinzugsgebiet. Unveröffentlichte Dissertation am Institut für Forstliches Ingenieurwesen und Alpine Naturgefahren / Universität für Bodenkultur. Wien

Chanson H (2004) The Hydraulics of Open Channel Flow: An Introduction, 2nd edn. Amsterdam

Chiari M, Rickenmann D (2007) The influence of form roughness on modelling sediment transport at steep slopes. In: Proceedings of the International Conference Erosion and Torrent Control as a Factor in Sustainable River Basin Management, Belgrad, pp 25-38

Chiari M, Mair E, Rickenmann D (2008) Geschiebetransportmodellierung in Wildbächen und Vergleich der morphologischen Veränderung mit LiDAR Daten. In: Forschungsgesellschaft für vorbeugende Hochwasserbekämpfung (ed.): Proceedings International Symposium INTERPRAEVENT, Dornbirn

Cunge JA (1969) On the subject of a flood propagation computation method (Muskingum method). In: Journal of Hydraulic Research, IAHR 7(2), pp 205-230

D'Agostino V, Cerato M, Coali R (1996) Il trasporto solido di eventi nei torrenti del Trentino orientale. In: Forschungsgesellschaft für vorbeugende Hochwasserbekämpfung (ed.): Proceedings International Symposium INTERPRAEVENT, Garmisch-Partenkirchen. Vol. 1, pp 377-386

D'Agostion V, Marchi L (2003): Geomorphological estimation of debris-flow volumes in alpine basins. In: Rickenmann D, Chen, C (eds.): Proceedings of the International Symposium on Debris-Flow Hazards Mitigation: Mechanics, Prediction and Assessment, Davos. Vol. 2, pp 1097-1106

Etter M (1996) Zur Erfassung des Geschiebetransportes mit Hydrophonen. Unveröffentlichte Diplomarbeit am Geographischen Institut der Universität Bern and an der WSL Birmensdorf. Bern

Fäh R, Farshi D, Müller R, Rousselot P, Vetsch D (2006) System Manuals of BASEMENT Version 1.1, October 2006. Zürich

Frick E, Kienholz H, Romang H, Roth H (2006) SEDEX – a practical method to estimate sediment delivery in mountain torrents. In: European Geophysical

Union (ed.): Geophysical Research Abstracts Vol. 8. (Abstract-Number: EGU06-A-07435), EGU General Assembly 2006, Vienna

Friedl K, Fuchs H (2004): SETRAC – Berechnung und Simulation des Sedimenttransportes für alpine Einzugsgebiete. In: Wildbach- und Lawinenverbau 149, pp 71-78

GHO (1996) Empfehlung zur Abschätzung von Feststofffrachten in Wildbächen = Mitteilungen der Arbeitsgruppe für operationelle Hydrologie (GHO) 4. Bern

Green W H, Ampt G A (1911) Studies on soil physics, part I: The flow of air and water through soils. In: Journal of Agriculture Science

Günter A (1971) Die kritische mittlere Sohlenschubspannung bei Geschiebemischungen unter Berücksichtigung der Deckschichtbildung und der turbulenzbedingten Sohlenschubspannungsschwankungen. = Mitteilungen der Versuchsanstalt für Wasserbau, Hydrologie und Glaziologie der ETH-Zürich 3. Zürich

Hampel R (1977) Geschiebebewirtschaftung in Wildbächen. In: Wildbach- und Lawinenverbau 41(1), pp 3-34

Hampel R (1980) Die Murenfracht von Katastrophenhochwässern. In: Wildbach- und Lawinenverbau 44(2), pp 71-102

Hegg C, Rickenmann D (2000) Geschiebetransport in Wildbächen – Vergleich zwischen Feldmessungen und einer Laborformel. In: Forschungsgesellschaft für vorbeugende Hochwasserbekämpfung (ed.): Proceedings International Symposium INTERPRAEVENT, Villach. Vol. 3, pp 117-127

Hofbauer R (1916) Eine neue Formel für die Ermittlung der größten Hochwassermengen. In: Österreichische Wochenschrift für den öffentlichen Baudienst. pp 38-40

Hungr O, Morgan G C, Kellerhals R (1984) Quantitative analysis of debris torrent hazards for design of remedial measures. In: Canadian Geotechnical Journal 21(4), pp 663-677

Ihringer J (1992) Regionalisierung des Abflussbeiwertes. In: DFG (ed.): Mitteilung der Senatskommission für Wasserforschung der DFG 11. Weinheim, pp 304-316,

Izzard CF (1946) Hydraulics of runoff from developed surfaces. In: Proc. Highway Research Board, 26, pp 129-150

Jäggi M (1992): Sedimenthaushalt und Stabilität von Flussbauten. = Mitteilungen der Versuchsanstalt für Wasserbau, Hydrologie und Glaziologie der ETH-Zürich 119. Zürich

Jenewein S (2002) Entwicklung einer GIS-basierten Applikation (PROMABGIS) für die Berechnung von Abfluss und Geschiebefrachten in Wildbacheinzugsgebieten unter Verwendung des prozessorientierten Ansatzes PROMAB. Unveröffentlichte Diplomarbeit am Institut für Geographie der Universität Innsbruck. Innsbruck

Kronfellner-Kraus G (1984) Extreme Feststofffrachten und Grabenbildung von Wildbächen. In: Forschungsgesellschaft für vorbeugende Hochwasserbekämpfung (ed.): Proceedings International Symposium INTERPRAEVENT, Villach. Vol. 2, pp 109-118

Kronfellner-Kraus G (1987) Zur Anwendung der Schätzformel für extreme Wildbach-Feststofffrachten im Süden und Osten Österreichs. In: Wildbach- und Lawinenverbau 51 (106), pp 187-200

Lefort P (1991) Transport solide dans le lit des cours d'eau-Dynamique fluviale. – Sogreah-Enshmg-Inpg-Grenoble.

Markart G, Kohl B, Sotier B, Schauer T, Bunza G, Stern R (2004) Provisorische Geländeanleitung zur Abschätzung des Oberflächenabflussbeiwertes auf alpinen Boden-/Vegetationseinheiten bei konvektiven Starkregen (Version 1.0). – BFW-Dokumentation, (= Schriftenreihe des Bundesamtes und Forschungszentrums für Wald 3). Wien

Mathys N, Brochot S, Meunier M, Richard D (2003) Erosion quantification in the small marly experimental catchments of Draix (Alpes de Haute Provence, France). Calibration of the ETC rainfall-runoff-erosion model. In: Catena 50, pp 527-548

Meyer-Peter E, Müller R (1948) Formulas for bedload transport. In: o.A., (ed.): Third Conference of the International Association of Hydraulic Research, Stockholm. Appendix 2, pp 39-64

Morgali J, Linsley R (1965) Computer Analysis of Oberland Flow. In: Journ. Hydraul. Dev., Proc. ASCE 91, pp 81-100

O'Brien, J S (2003) FLO-2D User Manual Version 2003.06. Nutrioso.

Palt S M (2001) Sedimenttransportprozesse im Himalaya-Karakorum und ihre Bedeutung für Wasserkraftanlagen. – Dissertation an der Fakultät für Bauingenieur- und Vermessugswesen der Universität Fridericiana zu Karlsruhe (TH), Karlsruhe

Rickenmann D (1990) Bedload transport capacity of slurry flows at steep slopes. = Mitteilungen der Versuchsanstalt für Wasserbau, Hydrologie und Glaziologie der ETH-Zürich 103. Zürich

Rickenmann D (1995) Beurteilung von Murgängen. In: Scheizer Ingenieur und Architekt 48, pp 1104-1108

Rickenmann D (1996) Fliessgeschwindigkeit in Wildbächen und Gebirgsflüssen. In: Wasser, Energie, Luft 88 (11/12), pp 298-304

Rickenmann D (2001) Comparison of bed load transport in torrents and gravel bed streams. In: Water Resources Research 37 (12), pp 3295-3305

Rickenmann D (2005) Geschiebetransport bei steilen Gefällen. In: Minor H E (ed.): = Mitteilungen der Versuchsanstalt für Wasserbau, Hydrologie und Glaziologie der ETH-Zürich 190. Festschrift zum Festkolloquium 75 Jahre VAW., Zürich, pp 107-119

Rickenmann D, Chiari M, Friedl K (2006) SETRAC – A sediment routing model for steep torrent channels. – In: Ferreira R, Alves E, Leal J, Cardoso A (eds.): Proceedings of the RiverFlow - conference 2006, Lisbon. pp 843-852

Rickenmann D, McArdell B (2007) Continuous measurement of sediment transport in the Erlenbach stream using piezoelectric bedload impact sensors. – In: Earth Surface Processes and Landforms 32, pp 1362-1378

Rickli C, Forster F (1997) Einfluss verschiedener Standorteigenschaften auf die Schätzung von Hochwasserabflüssen in kleinen Einzugsgebieten. – In: Schweizerische Zeitschrift für Forstwesen 148 (5), pp 367-385

Rinderer M (2002) Entwicklung, Optimierung und Anwendung des numerischen Computermodells BEDLOAD^{GIS} zur Simulation des Geschiebetransportes in Wildbächen. Beitrag zur Verbesserung des GIS-gestützten Massenbilanzierungsmodells PROMAB^{GIS}. – Unveröffentlichte Diplomarbeit am Institut für Geographie der Universität Innsbruck. Innsbruck

Scherrer S (1997) Abflussbildung bei Starkniederschlägen. Identifikation von Abflussprozessen mittels künstlicher Niederschläge. = Mitteilungen der Versuchsanstalt für Wasserbau, Hydrologie und Glaziologie der Eidgenössischen Technischen Hochschule Zürich 147 Zürich

Scherrer S, Naef F (2003) A decision scheme to indicate dominant hydrological flow processes on temperate grassland. In: Hydrological Processes 17, pp 391-401

Scheuringer E (1988) Ermittlung der maßgeblichen Geschiebefracht aus Wildbach-Oberläufen. In: Wildbach- und Lawinenverbau 52 (107), pp 87-95

Schöberl F, Stötter J, Schönlaub H, Ploner A, Sönser T, Jenewein S, Rinderer M (2004) PROMABGIS: A GIS-basiertes Werkzeug für die Ermittlung von Massenbilanzen in Alpinen Einzugsgebieten. In: Forschungsgesellschaft für vorbeugende Hochwasserbekämpfung (ed.): Proceedings International Symposium INTERPRAEVENT, Riva del Garda. Vol. 3, pp 271-282

Smart G, Jäggi M (1983) Sedimenttransport in steilen Gerinnen. = Mitteilungen der Versuchsanstalt für Wasserbau, Hydrologie und Glaziologie der ETH-Zürich 64. Zürich

Soil Conservation Service (1972) National Engineering Handbook, Section 4 Hydrology – Washington DC

Spreafico M, Lehmann C, Neaf O (1999) Recommandations concernant l'estimation de la charge sédimentaire dans les torrents. Groupe de travail pour l'hydrologie opérationnelle ed.). Bern

Strickler A (1923) Beiträge zur Frage der Geschwindigkeitsformel und der Rauhigkeitszahl für Ströme, Kanäle und geschlossene Leitungen. = Mitteilungen des Eidgenössischen Amtes für Wasserwirtschaft Bern 16. Bern

Swartz M, Mc Ardell B, Bartelt P, Christen M (2004) Evaluation of a two-phase debris flow model using field data from the Swiss Alps. In: Forschungsgesellschaft für vorbeugende Hochwasserbekämpfung (ed.): Proceedings International Symposium INTERPRAEVENT, Riva del Garda. Vol. 3, pp 319-329

US Army Corps of Engineers (2002) HEC-RAS River Analysis System. Hydraulic Reference Manual Version 3.1. Davis

Whittaker J G, Jäggi M N R (1986) Blockschwellen. = Mitteilungen der Versuchsanstalt für Wasserbau, Hydrologie und Glaziologie der ETH-Zürich 91. Zürich

Wundt W (1953) Gewässerkunde Berlin

Zarn B (1995) Geschiebehaushalt Alpenrhein: neue Erkenntnisse und Prognosen über die Sohlenveränderungen und den Geschiebetransport = Mitteilungen der Versuchsanstalt für Wasserbau, Hydrologie und Glaziologie an der ETH-Zürich 139. Zürich

Zeller J (1985) Feststoffmessungen in kleinen Gebirgseinzugsgebieten. – In: Wasser, Energie, Luft 77 (7/8), pp 246-251

Zimmermann M, Lehmann C (1999) Geschiebefracht in Wildbächen: Grundlagen und Schätzverfahren. In: Wasser, Energie, Luft 91 (7/8), pp 189-194

4 Modelling peak runoff in small Alpine catchments based on area properties and system status

C. Geitner, M. Mergili, J. Lammel, A. Moran, C. Oberparleiter, G. Meißl, H. Stötter

4.1 Introduction

4.1.1 Background and motivation

Floods are an often occurring natural hazard in the Alps and other mountainous regions. Thereby, also small mountain torrents can pose a major threat. Torrent catchments normally incorporate several altitudinal belts and often exhibit a steep relief and react rapidly to precipitation, particularly to small scale events of high intensity in the form of convection cells. If a catchment is affected by such an event then increased surface runoff and rapid subsurface flow take place and consequentially induce a rapid and strong rise in discharge.

Extensive flooding or debris flow activity often have the greatest impact where these mountain torrents reach the nearest large valleys. This most often occurs on alluvial fans and at the bottom of the valleys of the receiving streams. Residential buildings, infrastructure as well as industrial buildings are often located in these areas, thus causing considerable local damage.

This study deals with small hydrological catchments in the Alps with an area of less than 25 km² each. In catchments of this dimension area properties play a particularly important role in runoff formation. However, in larger catchments the effects of the specific hydrological properties of subcatchments are often levelled out (Spreafico et al. 2003, Grayson and Blöschl 2005).

Even though many hydrological investigations have been conducted in the Alps, there are only a few small catchments with corresponding high-resolution data on precipitation and discharge (e.g. Löhnersbach, Schmittenbach, Spissibach, Erlenbach, Lainbach; see e.g. Kirnbauer et al. 2004, Hagen 2003, Kienholz et al. 1996, Hegg et al. 2006, Wetzel 2001,

2003). Thus, numerous questions regarding the runoff formation have still not been answered satisfactorily despite long-term research. However, runoff formation processes need to be understood in more detail for a better estimation of extreme runoff in areas without hydrological data. These estimation procedures enable the planning and implementation of appropriate protective measures, be it the dimension of protective structures or the designation of hazard zones. Hence, this project aims at contributing to natural hazard management.

4.1.2 Challenges and objectives

The runoff generation in a hydrological catchment is the result of complex processes that consist of numerous sub-processes. These in turn are decisively controlled by a series of natural and anthropogenic conditioned factors such as geological substratum, relief, soil, vegetation and land use. Due to the small-scale changing interaction of these factors in the Alps, the spatial variability of runoff processes is very high. This can be seen, for example, in the comparison of the specific runoffs of three catchments of the Val Poschiavo in Switzerland during a long lasting precipitation event of high intensity in June 1987 (Scherrer 1997); despite similar characteristics of the precipitation event in all three areas the maximum specific discharge ranged between about 0.3 und 1.2 $m^3 s^{-1} km^{-2}$. The differences can be explained by the respective properties of an area, which determine the ratio between precipitation and runoff.

Nevertheless, within an individual catchment the relationship between precipitation and runoff is subject to certain fluctuations as well. This is because besides the relatively permanent area properties the runoff generation is also controlled by the current system status of a catchment. Among the variable factors are the seasonal conditions of the vegetation, the soil moisture as well as frozen soil or snow cover. Each of these factors displays a typical range of possible values and a typical temporal progression over the course of a year. This knowledge is important for the estimation of possible runoff events.

Figure 4.1 based on the hydrological analysis of the Stampfangertal catchment (see Figure 4.2 and tables 4.1 and 4.2), clearly shows that similar precipitation events can cause different runoff reactions. Due to different previous precipitation, varied soil moistures prevailed preceding the event. The left graph depicts the discharge reaction in the case of a short and intense precipitation event following a seven-day dry period. The other event in the right graph took place only 24 hours after the first event. Consequently, the soil's storage capacity was already partially exhausted and accordingly in the second case the surface runoff and resulting peak discharge are higher. It was proven in the project at hand for another study area in the montane altitudinal belt (Ruggbach catchment, see Moran et al. 2005) that the accumulated precipitation of the previous days has a noticeable influence on the runoff development.

However, it should be noted that in very high precipitation events, especially of long duration, the soil moisture and other variable factors play a subordinate role. Furthermore, Naef (2007) points out that in extreme precipitation events additional processes supporting the increased runoff can occur, which can hardly be considered by conventional estimations based on catchment properties.

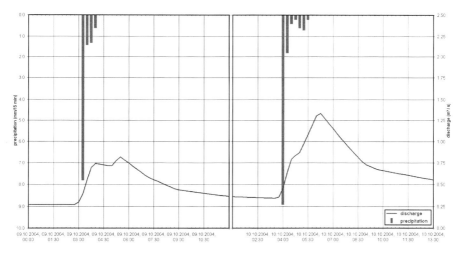

Fig. 4.1. Two similar precipitation events with their corresponding discharges in the Stampfangertal catchment (9-10 October 2004)

Based on these considerations, the objective of the project was the development and testing of a deterministic, distributed rainfall-runoff model for small Alpine catchments that calculates the discharge hydrograph dependent on precipitation, area properties as well as the current system status (Geitner et al. 2005). Herewith changes in the catchment area such as land use and climatic conditions can be taken into consideration and their effects on flood discharge can be modelled and analysed. The model should meet the following requirements in regard to data input and transferability in order to be used in practice by experts:
- Because data for the regional environment of Alpine catchments (geological substratum, relief, soil, vegetation, land use, precipitation

and discharge) are only limitedly available, the model should be able to run without a great deal of input data.
- Instead of conducting new data surveys the already existing information should be adapted to the model requirements and thus optimally utilized.
- Missing, but required data should be obtainable with manageable investigation expense.
- The model should be developed and tested in catchments with a sufficient amount of data.
- The model should preferably be transferable to other Alpine catchments. The possibilities and boundaries of transferability should be shown and requirements for future optimisation should be stated.

4.1.3 State of the art

Today a greater number of individual studies regarding runoff development in small Alpine catchments are available (e.g. Kienholz et al. 1996, Kirnbauer and Haas 1998, Kirnbauer et al. 2001, Kirnbauer et al. 2004, Markart et al. 1997, Kohl et al. 2004, Markart and Kohl 2004, Naef et al. 1998, Scherrer and Naef 2003, Wetzel 2001, 2003). Besides the complex analysis of precipitation and discharge values and often extensive surveys of the area properties these studies also partially make use of experimental techniques to increase the understanding of processes. In particular, rain simulations are an important experimental method in which artificial rainfall with varying intensity is created on test plots with different properties. The results of these experiments (e.g. Bunza and Schauer 1989, Markart et al. 2004, Scherrer 1997, Stepanek et al. 2004, Scherrer et al. 2007) have contributed considerably to the understanding of how the surface properties influence the runoff formation. However, since these tests are extremely time-consuming they are only available for a limited number of surface types and rarely repeated on the same plot under differing preconditions. In addition, the transfer of the results of a small test plot to catchment scale creates a number of methodological questions that should not be left unconsidered. On the basis of about 700 rain simulations Markart et al. (2004) developed a simple code of practice for the assessment of surface runoff coefficients for Alpine soil-vegetation units in torrential rain, which aims at being used in the field. These runoff coefficients were also used in the current project to aid the estimation of the infiltration capacity of hydrological response units (see chapter 4.3.3).

Based on these studies, approaches have been developed in the last years that aim at identifying dominant runoff processes on homogenous surfaces (Peschke et al. 1999, Tilch et al. 2002, Uhlenbrook 2003, Waldenmeyer 2003, Scherrer and Naef 2003, Schmocker-Fackel et al. 2007). However, for Alpine catchments only a few attempts have been made so far to determine dominant runoff processes (Kirnbauer et al. 2004, Tilch et al. 2006). The method developed by the authors mentioned above, provide a process orientated description of homogenous surfaces. By assigning soil hydrological parameters to them, they can be applied for the modelling of the runoff process (Schmocker-Fackel et al. 2007). The development of rainfall-runoff models has a history of more than 100 years and has become more dynamic in the past years due to the increase in modern computer technology. Beven (2001, 2005) provides a comprehensive overview of the status of hydrological modelling and the remaining unsolved questions.

The hitherto developed and implemented models for Alpine catchments differ in regard to data input and algorithms (e.g. PREVAH, WaSim-ETH, HQsim). An overview would go beyond the scope of this article. Nevertheless, it can be stated that hydrological models based on data obtained from certain areas led to good results. However, problems often occur when the models are transferred to other catchments. Essential findings for the optimisation of modelling could be won if different models were used in the same catchment area and the results were critically analysed (as e.g. in Seyhan and Van de Griend 1998). Unfortunately, hardly any such systematic surveys exist.

4.2 Study areas

The selection of the study areas for this project took place primarily according to size (<25 km^2) and the available meteorological and hydrological data. In addition, the test areas should cover a certain variety of Alpine catchments in regard to location, geology, relief and vegetation (Moran et al. 2005). The influence of karst formation was excluded. In the following section two catchments with very different area properties are introduced and their runoff dynamics discussed.

4.2.1 Location and catchment characteristics

Figure 4.2 shows the location of the test areas Stampfangertal and Längental in the Austrian Alps. Table 4.1 compiles important catchment characteristics. There is a distinct difference in altitude; the Stampfangertal catchment represents the montane and subalpine altitudinal belt and the

Längental catchment the subalpine, alpine and nival belt. Accordingly the vegetation and land use differ (see Figure 4.3). Three quarters of the surface in Längental consists of bedrock and almost unvegetated coarse debris, whilst nearly half of the Stampfangertal catchment is forested.

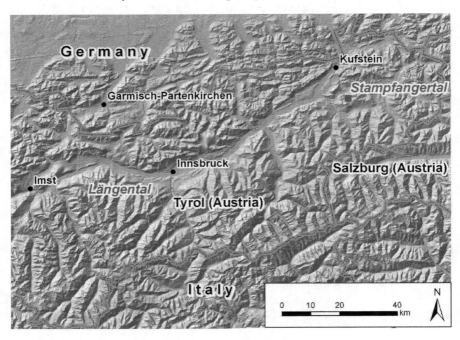

Fig. 4.2 Location of study areas in the Austrian Alps

Table 4.1 Characteristics of the study areas

	Längental	Stampfangertal
Area [km²]	9.2	22.3
Altitude (m a.s.l.)	1,900 – 3,010	640 -1,830
Altitudinal belt	subalpine, alpine, nival	montane, subalpine
Geology	Crystalline basement	Greywacke

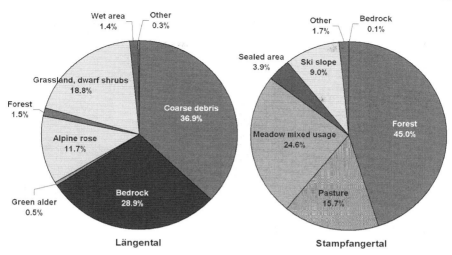

Fig. 4.3 Landcover in the study areas

4.2.2 Data basis and hydrological comparison

Table 4.2 provides information on the existing meteorological and hydrological stations as well as characteristic meteorological and hydrological values in the study areas. The meteorological station lies within the Stampfangertal whereas the station in Längental lies just outside the catchment boundary.

As confirmed by the hydrological and meteorological data the following main differences can be established:
- The annual sums of precipitation as well as the frequency of heavy rainfall decrease from the Alpine fringe (Stampfangertal) to the central Alps (Längental).
- Due to the altitude of the Längental, runoff is controlled much more by snowfall and snowmelt. Thus, the highest discharge is more likely to occur in early and midsummer.
- The maximum specific discharge ($m^3s^{-1}km^{-2}$) is higher in the Stampfangertal. The highest peak value has a much shorter return period than in Längental. The different reaction of both study areas can be ascribed to the great diversity in geological and geomorphological conditions (see chapters 4.4.1 and 4.4.2).

Table 4.2 Information about the observation stations and the meteorological and hydrologic characteristic values of the study areas

	Längental	Stampfangertal
Gauge	Längental, 1,933 m a.s.l.	Söll, 628 m a.s.l.
Series of measurements since	1981	1987
Max. discharge [$m^3 s^{-1}$]	7.0 (23.08.2005)	18.5 (22.07.1998)
Max. specific discharge [$m^3 s^{-1} km^{-2}$]	0.76	0.83
Return period, Gumbel distribution [Years]	~120 (7.0 $m^3 s^{-1}$ ± 2 $m^3 s^{-1}$, d = 90 %)	~19 Years (18.5 $m^3 s^{-1}$ ± 5 $m^3 s^{-1}$, d = 90 %)
Months with most discharge maxima	May, June, July	July, August, September
Rain Gauge	Kühtai Kraftwerk, 1,970 m a.s.l.	Bromberg, 1,180 m a.s.l.
Series of measurements since	1987	2004 (daily data since 1986)
Annual precipitation [mm]	1,200	1,450
Months with the highest precipitation	June, July, August	June, July, August
Max. precipitation rate [mm h^{-1}]	22.2 (27.06.1998)	33.6 (29.06.2006)

The analysis of the largest runoff events in both catchments shows, that these can be triggered through both short, intense, as well as significant, long-lasting precipitations. In August 2005 long-lasting and heavy precipitation led to maximum specific discharge in many areas of the Alps which caused unprecedented water levels (BWG 2005, BMLFUW 2006). In Längental the highest discharge value in the existing series of measurements was also reached during this event. The precipitation in Stampfangertal was not as heavy and the discharge was only the third highest in the series of measurements. This special event is of great interest not only for statistical evaluations but also for validating model results.

4 Modelling peak runoff in small Alpine catchments   111

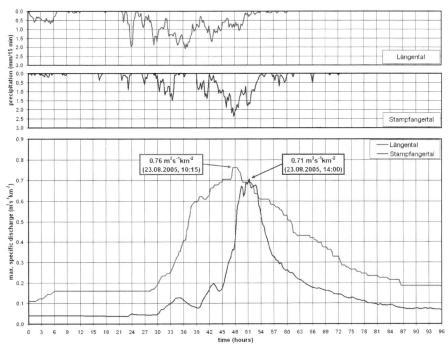

Fig. 4.4 Comparison of the development of the maximum specific discharges in 2005 (cumulative precipitation and discharge during the depicted period: Stampfangertal, 87 mm, $1.15*10^6$ m^3; Längental, 121 mm, $1.05*10^6$ m^3)

Figure 4.4 illustrates the specific discharges in each catchment for this special event. The hydrograph of Längental does not only exhibit the higher specific discharge, but remains longer on a higher level during the event, which is typical for this catchment. In addition, the maximum discharge follows the maximum precipitation with a greater delay than in the Stampfangertal. For this event the values of the maximum specific discharge in both catchments lie on about the same level. This is an exception because in most of the analysed events the specific discharge in the Längental is only half as high as in the Stampfangertal.

The unusual form of the hydrograph in the Längental can be explained to a certain extent by the long stretched geometrical form of the catchment area but also indicates a significant temporary storage capacity in the talus slopes of the catchment. Also summertime snowfields at higher altitudes can serve as a temporary storage and contribute to a prolonged runoff.

4.3 Model

4.3.1 General model layout

The model framework SYCOSIM (SYstem COnditions Spatial SIMulation Framework) was developed in order to allow for the data-extensive assessment of the hydrological response to rainstorms and rapid snowmelt, taking into account the status of the hydrological system at the start of the considered event (Mergili et al. 2006a). Topographical, landcover, pedological, geological and geomorphological information is required as input as well as meteorological datasets. The model operates with the following steps (see 4.5):

1. Parameter preparation: the input datasets are converted into raster maps representing hydrologically relevant information, including the water storing capacities (pool sizes) of vegetation, litter and soil. Some simple sets of rules are applied as well as a more advanced soil model (compare chapter 4.3.2 and 4.3.3).
2. System status model (raster-based): the status of the pools (degree of filling, remaining capacity) is computed in daily steps over a sufficient time span to account for long-term processes (e.g. the development of snow cover and the status of soil freezing). Rainfall and snowmelt serve as input to the pools, evapotranspiration, seepage and runoff as output (compare chapter 4.3.4).
3. Surface runoff model (raster-based): the status of the pools is computed in hourly steps over the period of a selected event. The surplus of available water being captured by neither of the pools (overflow) is considered as surface runoff of the corresponding pixel (compare chapter 4.3.5). This surface runoff includes the subsurface flow near the surface, as it is usually also measured in experimental rain simulations (Markart et al. 2004). The rest of the subsurface flow was not taken into consideration.
4. The surface runoff is routed through a stream network to the gauge. The model result is validated against the discharge measured at the gauge (compare chapter 4.3.6).

SYCOSIM is fully based on software products under the GNU licence (open source). GIS operations are implemented via GRASS (grass.itc.it), statistical operations were designed using RProject (www.r-project.org). Some operations were outsourced to the Python scripting language. The background code constitutes a system of shell scripts running under UNIX systems.

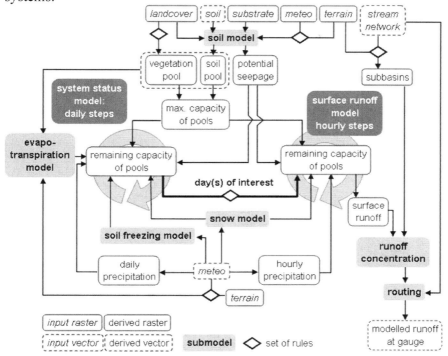

Fig. 4.5 The general layout of the SYCOSIM model (The system status model and the surface runoff model are illustrated in Fig. 4.6)

4.3.2 Input data and parameters

In this chapter all input data and parameters applied in the model are explained. A corresponding overview can be obtained in Table 4.3.

Table 4.3 The most important input data and parameters for the SYCOSIM model. Parameters written in bold indicate basic input datasets from which others are derived. All datasets have to be spatially distributed as raster, except those written in italic letters.

topographic parameters	**elevation z [m]**	input elevation map from official datasets	required
	local slope α [deg]	derived from elevation map (r.slope.aspect)	derived automatically
	solar irradiation R [Wh/m²]	derived from elevation map (r.sun)	
	topographic Index (rel)	derived from elevation map (r.topidx)	
meteorological data (daily resolution)	*precipitation P_d [mm]*	obtained from meteorological records	required
	temperature at defined elevation $T0$ [°C]	obtained from meteorological records	
	vertical temperature gradient $\Delta T/\Delta z$ [°C/m]	from records or standardized values	optional
	vapor pressure e [hPa]	from met. records or assumption that $e=0.5e_s$ on days with $P_d=0$, else $e=e_s$ (e_s = saturated vapor pressure)	
	temperature T [°C]	derived from T_0, $\Delta T/\Delta z$ and R	derived automatically
meteorological data (hourly resolution)	*precipitation P_h [mm]*	obtained from meteorological records	required
land cover parameters	**land cover class** (nominal)	existing map or mapping in field or from orthophotos	required
	carbon-nitrogen-ratio of the litter	derived from literature values for each land cover class	required
	interception capacity of stems and leaves IC_{veg} [mm]	derived from literature values for each land cover class	derived automatically
	interception capacity of litter IC_{lit} [mm]		
soil parameters	*total soil depth d [m]*	from field investigations	required
	depth of A horizon d_A [m]		
	average grain size [μm]		
	bulk density ρ [kg/m³]		
	organic content c_{org} (%)		

	skeleton content c_s (vol-%)		
	field capacity Θ_s (vol-%)	derived from soil parameters using predefined rules	derived automatically
substrate parameters	**chemical properties of bedrock or sediment cover** (ordinal)	existing map or mapping in field or from orthophotos	required
	physical substrate properties (ordinal)		
complex parameters	diffuse loss *dl* [mm]	estimate of drainage through soil, derived from parameters above	derived automatically
stream channel network	*vector geometry*	coordinates of the endpoint of each channel segment from elevation map and mapping in field or from orthophotos	required
	channel segment slope [deg]	from field measurement or slope map	
	channel segment width [m]	from field investigation	
	d90 grain parameter [µm]	from field investigation	

Terrain information

The DEMs with 20 m resolution were available for both study areas. Several derivates like slope, aspect, and watershed delineation were automatically computed from the DEMs during the process of calculation (see below).

Meteorological and hydrological information

Information about temperature, precipitation and runoff was obtained from the Hydrographic Service (HD) and the Austrian Meteorological Service (ZAMG) (see Table 4.2). For the Stampfangertal catchment daily precipitation data, namely sums from one permanent and two temporary gauges, were available. As daily precipitation values are insufficient for runoff modelling a temporary station (summer 2004 and 2005) was installed. Temperature data was computed from nearby stations. For the Längental daily temperature and hourly precipitation sums were available from a closely adjacent station (Kühtai). For both catchments daily snow height measurements, required for the validation of the snow model, were available from the vicinity of the rain gauges. Unfortunately no information on humidity, evapotranspiration and wind speed was obtainable.

Hydrological information was required for model validation. High resolution discharge values were available from the HD for the outlets of the study catchments.

Landcover and soil information

For selected parts of the research areas landcover (a combination of vegetation and land use) was mapped directly from aerial imagery and field studies. Additionally, the following datasets were available for use: (1) the Seger (2000) landcover dataset, representing a thematically highly detailed vector map for the entire territory of Austria at a scale of 1:100,000; and (2) the Schiechtl (1987) forest map for Tyrol, designed at a scale of 1:25,000.

Hydrologically relevant soil parameters (soil depth, depth of A horizon, skeleton content, grain size, bulk density and organic content; see Table 4.3 and Mergili et al. 2006b) were investigated at selected profiles using simple field methods (Schlichting et al. 1995).

Alpine soils offer some peculiarities (e.g. Geitner 2007) that ought to be considered. Their features are very variable and can change abruptly at small scales, making detailed maps highly laborious. Moreover, some typical characteristics of alpine soils e.g. high skeleton content and thick organic layers are difficult to evaluate in regard to their hydrological effects. Precise knowledge of the soil profile and the near surface substratum are especially necessary for the quantitative estimation of the subsurface runoff. As this project concentrates on surface runoff, the main focus of the soil examinations was placed on the upper part of the profiles in terms of infiltration and water storage capacities.

Geological and geomorphological information

A geological map at a reasonable scale (>1:100,000) was only available for the Längental. For the Stampfangertal catchment a geological survey was carried out leading to a map at a scale of 1:10,000. For the purpose of modelling, the geological maps were generalized to three classes on an ordinal scale, according to the chemical properties: (1) carbonate, (2) intermediate or mixed, and (3) silicate.

Geomorphological maps were not available for the study catchments. A survey resulting in a map at the 1:10,000 scale was carried out for the Längental. For the Stampfangertal catchment, the most important geomorphological features could be extracted from the geological map.

For the purpose of modelling the maps were simplified into three classes: (1) bare rock, (2) blocky debris, and (3) fine and intermediate debris (moraine, alluvium).

Stream network

Provisional stream network datasets for the study areas are derived from the DEMs. A field survey served for validating channel inclination as well as for the investigation of the d90-value, representing the bottom roughness (Rickenmann 1996) and the division of the networks into homogeneous segments according to all the relevant parameters.

4.3.3 Parameter estimation

Terrain and its derivates

Several terrain parameters were derived from the DEM:

- Slope and curvature (*r.slope.aspect* function of GRASS)
- Solar irradiation on a monthly basis (*r.sun* function of GRASS)
- A map of cumulative air temperature was prepared in order to account for the available energy for biological processes: The offset and the vertical gradient were calculated from the meteorological stations. The gradient according to the differences in solar irradiation (aspect) was computed following an approach developed by Welpmann (2003):

$$T = T_0 - \frac{0.005(R_0 - R)}{c}, \qquad \text{(eq. 4.1)}$$

where T [K] is the corrected air temperature, T_0 [K] represents the uncorrected air temperature, R [Wh m^{-2}] is the daily solar irradiation at the considered pixel, and R_0 [Wh m^{-2}] represents the daily irradiation in the same place if it would be flat. c [-] is the degree of cloudiness. In the present study R_0 was taken as the average irradiation over the study area under investigation. Due to a lack of data, a complete cloud cover was assumed, as otherwise the influence of the aspect seemed to be too high. The temperature correction was computed for the 15th day of each month, using monthly averages of the temperature and summed up over the year.

- A subbasin was computed for each segment of the stream network, in order to assign the runoff from each single pixel to one of the segments.

Landcover

The classes of all the landcover datasets were automatically referenced to predefined hydrological vegetation units according to Markart et al. (2004). The carbon-nitrogen-ratio of the litter, as a major factor for the biological activity of the soils was derived for each landcover unit, according to literature values (Blume et al. 2002).

Leaf area indices *LAI* and interception capacities *IC* of the foliage and the litter were estimated on a monthly basis for each landcover unit according to literature values (Frey and Lösch 1998, Larcher 2003, Mendel 2000) and expert opinions. Root depth was computed according to Hörmann (2003):

$$a = 1 - b^z, \qquad \text{(eq 4.2)}$$

where a [-] represents the share of the root network that is above the depth z [m]. b depends on the type of vegetation. The depth for which $a=0.9$ was considered as root depth.

Soil

As soil characteristics – in contrast to landcover characteristics – are not accessible for direct mapping a suitable model approach had to be chosen for the estimation of the spatial distribution of soil hydrological characteristics, based on the soil parameters investigated at selected profiles.

A three-step regression model was used for this purpose, refining the method presented by Mergili et al. (2006b). Raster maps representing the potential determinants for the spatial distribution of the soil variables were used as predictors. Thereby the following factors were taken into consideration: cumulative air temperature of the growing season, physical and chemical type of the substrate, slope inclination, topographic index, carbon-nitrogen-ratio of the litter and anthropogenic disturbance. The values of all predictor variables at the study sites were extracted from the corresponding maps (*r.what* function of GRASS) and joined to the soil variables table. The regression model framework operates as follows:

1. A regression that can be linear or nonlinear is performed separately for each predictor (e.g. substratum, slope, cumulative air temperature) and each soil variable (e.g. depth, organic content, skeleton content). This is

in order to (1) investigate the significance of the relationships between predictors and variables, analogous to an ANOVA (analysis of variance), and (2) to create linear relationships between the soil variables and derivates of the predictor variables, in order to facilitate the next step:
2. Multiple linear regression. For each soil variable the corresponding derivates of the significant (p>0.05) predictor variables are included in the regression equation.
3. Maps of field capacity FC [Vol. %], storage capacity SC [mm], pore volume V_p [Vol. %], hydraulic conductivity k_f [cm day^{-1}] and permanent wilting point PWP [Vol. %] are computed from the soil parameter maps obtained in step two using relationships derived from literature values (Schlichtung et al. 1995; Frey and Lösch 1998; see Mergili et al. 2006b). The reduction point of the soil PR [Vol. %], defining the water content below which soil evaporation is reduced, is estimated from PWP and FC.

Seepage through the soil: the concept of "diffuse loss"

Seepage is a complex issue that requires advanced measuring techniques and equipment as well as complex models in order to be fully accounted for in a deterministic way. Therefore a simplified approach had to be applied for the purpose of this study. Markart et al. (2004) compiled a large dataset relating seepage to an array of environmental variables, based on decades of measurements by the Federal Research and Training Centre for Forests, Natural Hazards and Landscape (BFW). This method requires the following information (Markart et al. 2004): vegetation type, water balance by means of indicator plants, grain size and soil density and the effects of land use (mainly pastures). For the investigation at hand the soil data was generated in the field, whereas the other information was derived from already existing datasets based on profound area knowledge. All this information was converted into maps of potential seepage for the study catchments.

The evaluation of the numerous rain simulations conducted by Markart et al. (2004, 2006) indicates that saturation of the soil does not completely impede infiltration, especially in steep terrain (Wetzel 2003). Instead, the infiltration rate, normalized to rainfall intensity, remains constant at sufficient high rainfall intensities, only decreasing at lower intensities. This situation implies that the same amount of water has to leave the system in order to maintain equilibrium conditions. As this concept represents a lumping of different processes (interflow, deep percolation) it was called "diffuse loss".

4.3.4 System status model

General aspects

This part of the model framework simulates the development of the status of each system component (vegetation, snow cover, and soil) in daily steps over a time period that can be chosen by the user. Generally, it is appropriate to use the start of the Alpine hydrological year (October 1) as starting date as the pools are usually least filled around this date and the program assumes empty pools at the start of the calculation.

A simple pool system and overflow model approach was used. The real capacities of the pools (vegetation, litter, soil) are dynamically modelled, based on the seasonally variable interception capacities (vegetation, litter) and the static field capacity (soil). Submodels were used for computing potential evapotranspiration, snowmelt and soil freezing/thawing. Figure 4.6 illustrates the major characteristics of the model.

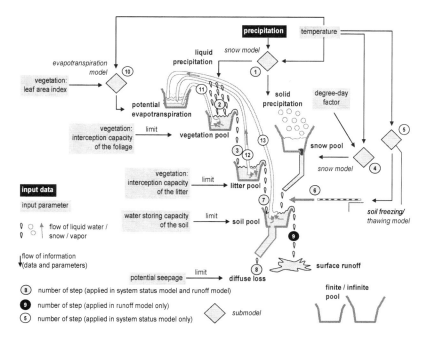

Fig. 4.6 Workflow of the system status and the runoff models. The diagram represents one daily step (system status model), or on hourly step (runoff model), respectively

Submodels

- Potential evapotranspiration *ETP* [mm] is computed using a common approach developed by Haude (1958). The simplified version used in this study,

$$ETP = \left[\frac{1 - 0.7^{LAI}}{2} + \frac{0.7^{LAI}(1 - 0.05n)}{4}\right](e_s - e), \quad \text{(eq 4.3)}$$

requires saturation vapour pressure deficit $e_s - e$ [hPa], leaf area index *LAI* [m^2 m^{-2}] and the number of previous days without precipitation n [-] as input. The relative humidity was generally set at 50 %. *ETP* was set to zero for days with precipitation. These simplifications had to be made due to a lack of appropriate data. Moreover it must be considered that the application of this method is questionable in windy, high altitudes.

- Melting snow constitutes an important contributor to runoff. For a study as presented here it is therefore required to apply a reliable model of snow cover build-up and melting. Data intensive models, as have already been applied in the Längental test area by Blöschl and Kirnbauer (1991), were not considered in this project. For due to the limited availability of data the snow model had to be kept as simple as possible, only considering water equivalent rather than actual snow height. Snow accumulation was computed by considering precipitation as snowfall if the average daily temperature remained under a critical value T_0 that was set to 273.15 K. Snowmelt was accounted for using a degree day factor DDF that was calculated based on temperature, precipitation and snow cover data from 88 stations in Tyrol (between 420 and 2245 m.a.sl.) by finding the best fit of the convergence of snow accumulation and melting. A vertical gradient of DDF was found and conceptualized by introducing four elevation classes for DDF (Table 4.4) (Geitner et al. 2006). No relationships were found between DDF and irradiation. Snowmelt SM was calculated using the relationship

$$SM = DDF(T - T_0), \qquad (eq. 4.4)$$

where T [K] is the daily average temperature and T_0 [K] is the temperature threshold for snowmelt, which was set to 273.15 K.

Table 4.4 Elevation classes of the degree day factor DDF

elevation [m]	<750	>750 – 1100	>1100-1800	>1800
DDF [mm K^{-1} d^{-1}]	0.75	1.25	2.00	3.00

- Soil freezing and thawing are complex issues. Schroeder et al. (1997), as part of the HELP model, have presented a relatively simple approach to cope with this problem: freezing occurs as soon as the average temperature of the previous 30 days remains below 273.15 K. Melting is treated in a more complex way, depending on the surplus of days above this threshold compared to days below. The required surplus DFS can be estimated using the relationship

$$DFS = 35.4 - 6.62 \cdot 10^{-3}(R_s - R_{so}), \qquad (eq. 4.5)$$

where R_s [Wh m^{-2}] is the solar irradiation on December 1 of the first year of the calculation, and R_{so} [Wh m^{-2}] represents the average daily sum of potential irradiation in December.

4.3.5 Pool model

Both filling and evaporating from the pools follow from top to bottom. Precipitation hits the foliage. The available interception pool is filled, the rest of the water continues to the litter pool, which is also filled and the rest of water infiltrates into the soil. Seepage to interflow and groundwater was computed using a concept related to that of diffuse loss.

After the precipitated water has been properly allocated to the pools and seepage has been estimated, evaporation comes into action. Water is removed from the foliage until potential evapotranspiration is reached, or until the foliage has dried up. In the latter case, water is evaporated from the litter, and then from the soil. At soil moisture levels below the reduction point soil evapotranspiration is reduced according to the following relationship:

$$ET = ETP\frac{SM - PWP}{RP - PWP} \qquad \text{(eq. 4.6)}$$

where ET [mm] is evapotranspiration, ETP [mm] is potential evapotranspiration, SM [Vol. %] is soil moisture, PWP [Vol. %] is the permanent wilting point, and RP [Vol. %] is the reduction point.

If precipitation occurs on the considered day, evapotranspiration is reduced. When all the pools are emptied, or the potential evapotranspiration has been reached, the pool capacities are updated and the whole procedure is repeated for the following day, or the system stati are fed into the runoff model.

The procedure described above is followed as long as none of the three following situations occurs:

1. Precipitation occurs as snowfall: the water equivalent of the snowfall is added to the water equivalent of the already existing snow cover while the other pools remain unchanged.
2. A snow cover exists, but the temperature is above the critical temperature for melting: the water equivalent of the melting snow is added to the soil pool, until the soil pool is filled. If rainfall occurs on the snow cover, it is directly traced to the soil pool. No sublimation from the snow cover is taken into account.
3. The soil is frozen: the soil pool is locked and its status remains unchanged – it can neither take up water, nor evaporate it.

4.3.6 Runoff model

Surface runoff generation

The runoff model works analogous to the system status model (see Figure 4.6), with the following exceptions:

- It operates in hourly steps over a range of days that can be chosen by the user. The status of each pool is calculated for each time step.
- The amount of water not fitting into the pools is considered as runoff. This way a runoff height is assigned to each pixel for each time step.
- Evapotranspitation is ignored in this module as (1) it only plays a major role over longer time spans; and (2) days with high rainfall, being of particular interest for this application, are frequently characterized by low evapotranspiration.
- Soil freezing and thawing are not computed as these processes run over longer time spans. The status at the beginning of the model run is applied.
- Water only infiltrates into the soil as long as the infiltration capacity (derived from data from the BFW, Markart et al. 2004) has not been exceeded. The surplus runs off immediately.

Runoff concentration

Each pixel has been assigned to a stream segment where it drains to. Time of concentration (runoff delay) depends on the flow length from the considered pixel to the stream network, slope and surface roughness. Time of concentration was estimated as average for each sub-basin, according to the relationship

$$T_c = \sqrt{\sum_{i=1}^{i=n} a \cdot e^{-b\varphi_i}} ,$$ (eq. 4.7)

where T_c [min] is the time of concentration, φ_i [°] is the local slope angle at the considered pixel i, and a [-] and b [-] are parameters that have to be adapted. n is the size of the sub-catchment [m²]. The idea behind this concept is that runoff delay increases with decreasing φ and with increasing n, the square root was introduced in order to account for the linearity of runoff concentration. Information on the surface roughness is not available at a sufficient accuracy for being implemented into the approach.

Routing through the stream network

Routing a flow through a stream network requires the application of an algorithm relating the runoff out of a stream segment to the water content of the same segment. Rickenmann (1996) developed such an algorithm, particularly suitable for streams with considerable slope angle as present in the study catchments. Its application does not require advanced parameters of stream geometry, but only segment length, segment slope and bottom roughness, expressed as d90-value. A routine that can be run using the Rickenmann algorithm, and that constitutes part of the hydrological model HQsim (Kleindienst 1996), was applied for the routing of the flow through the stream network to the gauge.

On the basis of the described input parameters and model components a greater number of runoff events were calculated for the catchments. Hereby, the peak discharges were of particular interest, because during their formation the rapid runoff components were dominant. In order for minor discharge to be correctly modelled the long-term runoff components should be taken into consideration, which was not the aim of the current study.

4.4 Results

In this chapter the model results are compared with measured peak discharges in both study areas. Since the model was developed for extreme precipitation and discharge events, the highest known discharges in the test catchments were chosen. Below the event on August 23, 2005 is presented, which was characterised by long-lasting and heavy precipitation. It led to a maximum specific discharge, as was also the case in many other areas of the Alps, causing unprecedented water levels (BWG 2005, BMLFUW 2006).

4.4.1 Comparison of modelled and gauged discharges in the Stampfangertal catchment

Figure 4.7 shows the modelled and measured values for the Stampfangertal catchment after calibration. In the calibration phase optimal values were ascertained for the diffuse loss (controls the height of the peaks) and the parameters a and b of the runoff concentration (controls the course of the hydrograph). Consequently, the peak discharge of the main event could be well represented. The comparison of the whole discharge hydrograph shows however that not all processes could be realistically depicted in the model. Thus, firstly the small events are overestimated in the model. Secondly, the main event ends too abruptly which means that the discharge amount is underrepresented.

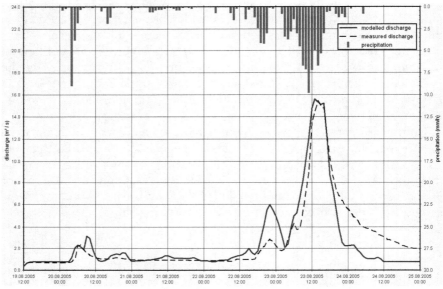

Fig. 4.7 Modelled and measured discharge hydrographs for the Stampfangertal catchment

4.4.2 Comparison of modelled and gauged discharges in the Längental catchment

In contrast to Stampfangertal, the modelled and measured discharge hydrograph in the Längental could not be brought into accord, despite various calibration attempts. Not only the amounts but also the peak discharge in Längental was underrepresented (see Figure 4.8). Only the first rise of the hydrograph was modelled well, however the modelled discharge hydrograph descends much too rapidly.

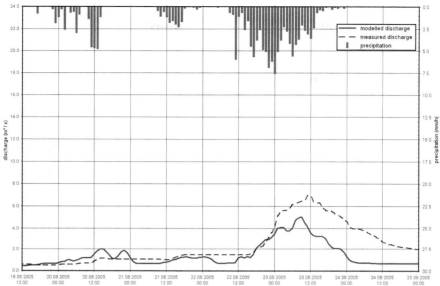

Fig. 4.8 Modelled and measured discharge hydrographs for the Längental catchment

Due to these results all data inputs and calculations were reassessed for the Längental catchment and compared to Stampangertal. Thereby the importance of certain phenomena for runoff generation became clearer, that had been inadequately considered by the model so far: namely the extensive talus slopes that cover about one third of the catchment (see Figure 4.9). In the case of a precipitation event or snowmelt, the water completely infiltrates the coarse debris and appears again at the foot of the slope with a certain delay. This process seems to be responsible for both, the time lag of the discharge peak, which took place a significant 12 hours after the maximum precipitation, and the slow decrease of the discharge hydrograph. This temporary water storage in the talus slopes after a precipitation event was repeatedly confirmed through field observations (see Figure 4.9). Since SYCOSIM only takes the rapid surface runoff components into consideration, the model is currently not suitable for application in areas such as Längental.

Fig. 4.9 The extensive talus slopes in the Längental dispense the infiltrated precipitation water with a delay at the foot of the slope

4.5 Conclusions and discussion

In this project a model prototype was developed for simulating peak discharge in small Alpine catchments. It calculates the rapid runoff components and takes both the area properties as well as differing system statuses of the catchments into consideration using only a relatively small amount of data. More comprehensive data acquisition in the catchments need only be made in regard to the soil and channel features.

The comparison of the first modelled results to the measured extreme discharges leads to the following conclusions, which ought to be subject to further investigation:
- In the Stampfangertal catchment the results of a calibrated model represented the peak discharge quite well. However, smaller discharge events are overestimated by the model. The total amount of the event's discharge was too low since the model only took the rapid runoff components into consideration.

- Despite calibration the model results in the Längental catchment did not represent the measured peak discharge. This was explained by the high cover of talus slopes that function as temporary storage and could not be modelled. Although their storages led to certain delays in runoff, they can still contribute to rapid runoff in the case of longer lasting events as was the case in 2005.

These conclusions lead to the following consequences for the application of the current model:

- At present it seems necessary to calibrate the model for each catchment. Without calibration a transferral to a different catchment ensues a great deal of uncertainty.
- The model SYCOSIM can only be applied in areas where peak discharge is mainly generated by surface runoff. Since the subsurface runoff can play a significant role in flood discharge (e.g. Wetzel 2003), it should be integrated in the model in the future.

The catchments of Stampfangertal and Längental represent part of the wide range of catchment types in the Alps. These differences are predominantly due to altitude. For instance, hardly any surface runoff occurs on the unvegetated talus slopes which can be wide-spread in the alpine and nival belt. To adequately represent these surfaces in the model, an additional module for different rapidly drained storages should be supplemented. These surfaces must also be dealt with differently in regard to system status. Due to the coarse substrates and the barely developed soils the influence of soil moisture can be neglected in this case.

This leads to the idea of classifying catchment types according to their dominant hydrologically relevant properties. Consequently, hydrological models can be adapted to and transferred within these catchment types. McDonnell and Woods (2004) also emphasise that catchment classification in the framework of hydrological research is generally an important methodological contribution. Primary criteria for this kind of classification in the Alps would be the altitudinal belts which are each normally characterised by a specific set of relief characteristics, substratum forms, soil types and vegetation. If the model components are aligned with these properties then the transferability of hydrological models in the Alps could be distinctly improved.

References

Beven KJ (2001) Rainfall – Runoff Modelling: The Primer. Wiley, Chichester
Beven KJ (2005) Rainfall – Runoff Modelling: Introduction. In: Anderson MG (ed) Encyclopedia of Hydrological Sciences. Wiley, Chichester, pp 1857-1868
Blöschl G, Kirnbauer R (1991) Distributed snowmelt simulations in an Alpine catchment. 1. Model evaluation on the basis of snow cover patterns. 2. Parameter study and model predictions. Water Resources Research 27(12) pp 3171-3179 and pp 3181-3188
Blume HP, Brümmer GW, Schwertmann U, Horn R, Kögel-Knabner I, Stahr K, Auerswald K, Beyer L, Hartmann A, Litz N, Scheinost A, Stanjek H, Welp G, Wilke BM (2002) Lehrbuch der Bodenkunde (Scheffer/Schachtschabel). Spektrum, Heidelberg Berlin
BMLFUW (Bundesministerium für Land- und Forstwirtschaft, Umwelt und Wasserwirtschaft) (2006) Hochwasser 2005. Ereignisdokumentation (Teilbericht der Wildbach- und Lawinenverbauung), Wien
Bunza G, Schauer T (1989) Der Einfluss von Vegetation, Geologie und Nutzung auf den Oberflächenabfluss bei künstlichen Starkregen in Wildbachgebieten der Bayerischen Alpen. Informationsbericht des Bayerischen Landesamts für Wasserwirtschaft 2/89, München
BWG (Bundesamt für Wasser und Geologie) (2005) Bericht über die Hochwasserereignisse 2005. (http://www.news-service.admin.ch/NSBSubscriber/message/attachments/1123.pdf, May 2007)
Frey W, Lösch R (1998) Lehrbuch der Geobotanik. Gustav Fischer, Stuttgart Jena Lübeck Ulm
Geitner C (2007) Böden in den Alpen – Ausgewählte Aspekte zur Vielfalt und Bedeutung einer wenig beachteten Ressource. In: Borsdorf A and Grabherr G (eds) Internationale Gebirgsforschung, IGF-Forschungsberichte, 1, pp 56-67, Innsbruck, Wien
Geitner C, Gerik A, Lammel J, Moran AP, Oberparleiter C (2005) Berücksichtigung von Systemzuständen und Unschärfen bei der Bemessung von Hochwasserereignissen in kleinen alpinen Einzugsgebieten. Konzeptionelle Überlegungen zum Aufbau eines Expertensystems. Geoforum Umhausen 4, pp 142-155
Geitner C, Lammel J, Moran AP, Oberparleiter C. (2006) alpS-Projekt A3.1: Ermittlung der abflusssteuernden Parameter und Prozesse in alpinen Einzugsgebieten auf der Basis von Systemzuständen und Wahrscheinlichkeiten. Unpublished Report, Innsbruck
Grayson RB, Blöschl G (eds) (2005) Spatial pattern in catchment hydrology: Observations and modelling. Cambridge University Press, Cambridge
Hagen K (2003) Wildbacheinzugsgebiet Schmittenbach (Salzburg), Analyse des Niederschlags- und Abflussgeschehens (Torrential Watershed of Schmittenbach (Salzburg): Analysis of Precipitation and Runoff) 1977-1998. BFW Berichte 129, Wien

Haude W (1958) Über die Verwendung verschiedener Klimafaktoren zur Berechnung der potentiellen Evaporation und Evapotranspiration. Meteorologische Rundschau 11, pp 96-99

Hegg C, McArdell BW, Badoux A. (2006) One hundred years of mountain hydrology in Switzerland by the WSL. Hydrological Processes 20, pp 371-376

Hörmann G, Scherzer J, Suckow F, Müller J, Wegehenkel M, Lukes M, Hammel K, Knieß A, Messenburg H (2003) Wasserhaushalt von Waldökosystemen: Methodenleitfaden zur Bestimmung der Wasserhaushaltskomponenten auf Level II-Flächen. Bundesministerium für Ernährung, Landwirtschaft und Forsten, Bonn

Kienholz H, Weingartner R, Hegg Ch (1996) Prozesse in Wildbächen – ein Beitrag zur Hochgebirgsforschung. Jahrbuch der Geographischen Gesellschaft Bern 59, pp 249–262

Kirnbauer R, Haas P (1998) Observations on runoff generation mechanisms in small Alpine catchments. IAHS Publication 148, pp 239-247

Kirnbauer R, Blöschl G, Haas P, Müller G, Merz B (2001) Space-time patterns of runoff generation in the Löhnersbach catchment. Runoff Generation and Implications for River Basin Modelling. Freiburger Schriften zur Hydrologie 13, pp 37-45

Kirnbauer R, Tilch N, Markart G, Zillgens B, Kohlbeck F, Leroch K, Seidler C, Haas P, Uhlenbrook S, Didszun J, Leibundgut C, Merz B, Chwatal W, Fürst J (2004) Hochwasserentstehung in der nördlichen Grauwackenzone der Alpen. Internationales Symposium Interpraevent 2004 1, pp 45-56

Kleindienst H (1996) Erweiterung und Erprobung eines anwendungsorientierten hydrologischen Modells zur Ganglinienssimulation in kleinen Wildbacheinzugsgebieten. Unpublished Diploma Thesis, University München

Kohl B, Sauermoser S, Frey D, Stepanek L, Markart, G (2004) Steuerung des Abflusses in Wildbacheinzugsgebieten über Flächenwirtschaftliche Maßnahmen. Proceedings of Interpraevent 2004 1, pp 159-169

Larcher W. (2003) Physiological Plant Ecology. Springer, Berlin Heidelberg New York

Markart G, Kohl B, Zanetti P. (1997) Oberflächenabfluß bei Starkregen. Abflußbildung auf Wald-, Weide- und Feuchtflächen (am Beispiel des oberen Einzugsgebietes der Schesa-Bürserberg, Vorarlberg). Centralblatt für das gesamte Forstwesen 114, pp 123-144

Markart G, Kohl B, Sotier B, Schauer T, Bunza G, Stern R (2004) Provisorische Geländeanleitung zur Abschätzung des Oberflächenabflussbeiwerts auf alpinen Boden-/Vegetationseinheiten bei konvektiven Starkregen (Version 1.0). BFW-Dokumentation 3/2004, Wien (http://bfw.ac.at/rz/bfwcms.web?dok=4343, November 2007)

Markart G, Kohl B (2004) Abflussverhalten in Wildbacheinzugsgebieten bei unterschiedlicher Landnutzung. Wildbach- und Lawinenverbau 149, pp 9-20

Markart G, Kohl B, Sotier B, Schauer Th, Bunza G, Stern R (2006) Geländeanleitung zur Abschätzung des Oberflächenabflussbeiwertes bei

Starkregen - Grundzüge und erste Erfahrungen. Wiener Mitteilungen Wasser, Abwasser, Gewässer 197, pp 159-178

McDonnell JJ, Woods R (2004) Editorial: On the need for catchment classification. Journal of Hydrology 299, pp 2-3

Mendel HG (2000) Elemente des Wasserkreislaufs. Eine kommentierte Bibliographie zur Abflussbildung. Bundesanstalt für Gewässerkunde (ed), Analytica, Berlin

Mergili M, Geitner C, Lammel J, Moran AP, Oberparleiter C, Gerik A, Meißl G, Stötter J (2006a) A GIS-based numerical model for predicting extreme runoff events in small alpine catchments. Geophysical Research Abstracts 8:05865 (CD)

Mergili M, Geitner C, Moran AP, Fecht M, Stötter J (2006b) SOILSIM – a GIS-based framework for data-extensive modelling of the spatial distribution of soil hydrological characteristics in small alpine catchments. In: Strobl J, Blaschke T (eds) Angewandte Geoinformatik 2006, Beiträge zum 18. AGIT-Symposium Salzburg. Wichmann, Heidelberg, pp 444-453

Moran AP, Lammel J, Geitner C, Gerik A, Oberparleiter C, Meißl G (2005) A conceptual approach for the development of an expert system designed to estimate runoff in small Alpine hydrological catchments. Landschaftsökologie und Umweltforschung 48, pp 199-210

Naef F, Scherrer S, Faeh A (1998) Die Auswirkungen des Rückhaltevermögens natürlicher Einzugsgebiete bei extremen Niederschlagsereignissen auf die Grösse extremer Hochwasser. Schlussbericht im Rahmen des Nationalen Forschungsprogrammes "Klimaänderungen und Naturkatastrophen" NFP 31, Zürich

Naef F (2007) Extreme Hochwasser verstehen – Beispiele aus der Schweiz. Wiener Mitteilungen Wasser, Abwasser, Gewässer 206, pp 59-68

Peschke G, Etzenberg C, Müller G, Töpfer J, Zimmermann S (1999) Das wissensbasierte System FLAB – ein Instrument zur rechnergestützten Bestimmung von Landschaftseinheiten mit gleicher Abflussbildung. IHI-Schriften 10, Zittau

Rickenmann D (1996) Fliessgeschwindigkeit in Wildbächen und Gebirgsflüssen. Wasser, Energie, Luft 88, pp 298-304

Scherrer S (1997) Abflussbildung bei Starkniederschlägen. Identifikation von Abflussprozessen mittels künstlicher Niederschläge. Mitteilungen der Versuchsanstalt für Wasserbau, Hydrologie und Glaziologie 147, Zürich

Scherrer S, Naef F (2003) A decision scheme to indicate dominant hydrological flow processes on temperate grassland. Hydrological processes 17, pp 391-401

Scherrer S, Naef F, Faeh A, Cordery I (2007) Formation of runoff at the hillslope scale during intense precipitation. Hydrology and Earth System Sciences 11, pp 907-922

Schiechtl H (1987) Die Vegetationskartierung in der Forstlichen Bundesversuchsanstalt. Informationsdienst der Forstlichen Bundesversuchsanstalt 239, pp 43-46

Schlichting E, Blume HP, Stahr K (1995) Bodenkundliches Praktikum. Blackwell, Berlin

Schmocker-Fackel P, Naef F, Scherrer S (2007) Identifying runoff processes on the plot and catchment scale. Hydrology and Earth System Sciences 11, pp 891-906

Schroeder PR, Dozier TS, Zappi PA, McEnroe BM, Sjostrom JW, Peyton RL (1997) The Hydrologic Evaluation of Landfill Performance (HELP) Model: Engineering Documentation for Version 3, EPA/600/R-94/168b, U.S. Environmental Protection Agency Office of Research and Development, Washington DC

Seger M (2000) Rauminformationssystem Österreich – digitaler thematischer Datensatz des Staatsgebietes fertiggestellt. In: Strobl J, Blaschke T, Griesebner G (eds) Angewandte Geographische Informationsverarbeitung XII., Beiträge zum AGIT-Symposium 2000, Salzburg. Wichmann, Heidelberg, pp 465-468

Seyhan E, Van de Griend AA (1998) Modelling Runoff from Alpine Headwater Catchments. IAHS Publication 248, pp 233-243

Spreafico M, Weingartner R, Barben M, Ryser A (2003) Hochwasserabschätzung in schweizerischen Einzugsgebieten. Berichte des Bundesamts für Wasser und Geologie, Serie Wasser 4, Bern

Stepanek L, Kohl B, Markart G (2004) Von der Starkregensimulation zum Spitzenabfluss. From Heavy Rain Simulation to High Water Discharge. Proceedings of Interpraevent 2004 1, pp 101-112

Tilch N, Uhlenbrook S, Leibundgut C (2002) Regionalisierungsverfahren zur Ausweisung von Hydrotopen in von periglazialem Hangschutt geprägten Gebieten. Grundwasser – Zeitschrift der Fachsektion Hydrogeologie 4/2002, pp 206-216

Tilch N, Zillgens B, Uhlenbrook S, Leibundgut Ch, Kirnbauer R, Merz B (2006) GIS-gestützte Ausweisung von hydrologischen Umsatzräumen und Prozessen im Löhnersbach-Einzugsgebiet (Nördliche Grauwackenzone, Salzburger Land). Österreichische Wasser- und Abfallwirtschaft 58, pp 41-51

Uhlenbrook S (2003) An empirical approach for delineating spatial units with the same dominating runoff generation processes. Physics and Chemistry of the Earth 28, pp 297-303

Waldenmeyer G (2003) Abflussbildung und Regionalisierung in einem forstlich genutzten Einzugsgebiet (Dürreychtal, Nordschwarzwald). Karlsruher Schriften zur Geographie und Geoökologie 20, Karlsruhe

Welpmann M (2003) Bodentemperaturmessungen und -simulationen im Lötschental (Schweizer Alpen). Doctor Thesis, University Bonn

Wetzel KF (2001) Die Prozesse der Abflussbildung in kleinen Hangeinzugsgebieten der nördlichen Kalkalpen bei unterschiedlichen Niederschlägen. Die Erde 132, pp 361-379

Wetzel KF (2003) Runoff production processes in small alpine catchments within the unconsolidated Pleistocene sediments of the Lainbach area (Upper Bavaria). Hydrological processes 17, pp 2463-2483

Cited models
HQsim: http://www.grid-it.at/home/produkt/modell/hqsim/
PREVAH: http://www.wsl.ch/hazards/prevah/
WaSim-ETH: http://homepage.hispeed.ch/wasim/index.html

5 Process-based investigations and monitoring of deep-seated landslides

C. Zangerl, C. Prager, W. Chwatal, S. Mertl, D. Renk, B. Schneider-Muntau, H. Kirschner, R. Brandner, E. Brückl, W. Fellin, E. Tentschert, S. Eder, G. Poscher, H. Schönlaub

5.1 Introduction

Through the consolidation of alpine settlement areas there have been an increasing number of incidents in recent years related to the activity of landslides in Northern Tyrol (Austria). This has led to humans, buildings, and communication and transportation routes being increasingly threatened. In 1999 a rockfall event in Huben (Ötztal, Austria) destroyed a wood mill and cut the main power supply for the inner Ötztal. In the same year increased deformation rates at the Eiblschrofen (Schwaz, Austria) induced reoccurring rockfall events. In early summer 2003, parts of the deep-seated Steinlehnen rockslide system (Gries i. Sellrain, Austria) were reactivated, causing an acceleration of a sliding slab (Henzinger 2005). Secondary events in the form of increased rockfall activity were the direct consequence of these slope movements and demanded temporary evacuations and roadblocks as immediate measure. In order to protect the road and settlement area permanently a safety dam was built. After the floods in Tyrol in August 2005, parts of the complex Zintlwald landslide system (Strengen, Austria) accelerated. This was triggered on the one hand by increased water infiltration of the slope and on the other hand by intense fluvial erosion of the slope foot. As a consequence important supra-regional infrastructure such as sections of the Arlberg national road were destroyed. In addition, the possibility was given that a rapid landslide could dam the river Rosanna. Considering that a collapse of this dam would entail a sudden flood event downstream, a monitoring and warning system has been installed.

This case study showed that slowly moving slopes can develop into rapid landslides with a high power of destruction. More often they can lead to differential block movements causing damage to the infrastructure on the surface and below the ground. For instance, long and wide cracks were discovered on buildings situated on the deep-seated Niedergallmigg-Matekopf landslide (Fließ, Austria, Kirschner and Gillarduzzi 2005).

Landslides that occur close to reservoirs possess a generally high risk potential and thus require detailed investigation and permanent monitoring. These landslides offer ideal conditions for the comprehensive study of the underlying slope mechanisms and processes, because they provide long-term deformation measurements, well documented in-situ investigations e.g. boreholes, investigation adits and geophysical surveys (e.g. Leobacher and Liegler 1998, Tentschert 1998, Brückl et al. 2004, Watson et al. 2004, Zangerl et al. 2007).
This paper focuses on basic mechanical processes, temporal activity distributions, applied geophysical investigations and monitoring methods of deep-seated landslides in fractured rock masses.

5.2 Landslide classifications

A simple and clear definition for the term "landslides" was proposed by Cruden (1991) and taken on by the International Geotechnical Society, UNESCO Working Party on World Landslide Inventory and represents the beginning of an international harmonising of the mass movement nomenclature: *"Landslide is a movement of a mass of rock, earth or debris down a slope"*. This definition includes debris flows but not ground subsidence or snow avalanches.
However literature offers a large number of different classification schemes for landslides. Due to the complexity and ambiguity of different terms, which is particularly the case in German-speaking countries, this paper refers to the practical and useful classification of Varnes (1978) and Cruden and Varnes (1996). In principal, these categorisations are based on the type of movement and type of material.
According to this, landslide movement types are termed as falls, topples, slides, spreads and flows, whereas complex landslide systems may be classified by a combination of different terms. In regard to the type of material it is possible to differentiate between rock and soil. For the further description of landslides Varnes' (1978) classification also includes a subdivision into different classes of acceleration, ranging from "extremely slow" for movement rates of under 16 mm/a to "extremely rapid" when there is acceleration of over 5 m/s.

5.3 Temporal distribution of dated landslides in the East Alpine region

Landslides are characterised by complex combinations of geological, hydrogeological, rock or soil mechanical and climatic processes. In order to understand potential causes, trigger mechanisms and the temporal distribution of deep-seated landslides geochronological data of fossil and active case studies were compiled. Age dating of landslides in the Alps was previously based on relative criteria such as geomorphological and lithostratigraphical field evidences. Consequently it was assumed that flucio-glacial erosion and glacier withdrawals triggered numerous prominent landslides in the Late-Glacial and early Post-Glacial (e.g. Abele 1974). But already early attempts of age dating, by means of pollen analysis of lake sediments that are genetically linked with landslides, indicated that some landslides took place in the Holocene (e.g. Sarnthein 1940). This was confirmed by first radiometric C-14 dating of Alpine landslides, e. g. the rocklslides at Molveno (Trentino, Italy), Köfels and Hochmais (both Tyrol, Austria), which yielded clearly Holocene ages for the slope failures (Marchesoni 1958, Heuberger 1966, Schmidegg 1966). Further dating showed that also several other deep-seated rockslides, e.g. in the Tschirgant (Tyrol; Patzelt and Poscher 1993) and Eibsee region (Bavaria; Jerz and Poschinger 1995) took place in the middle Holocene and not as previously believed in the Late-Glacial.

For the dating of Quaternary sediments and prehistoric landslide deposits there are a number of different radiometric analysis methods available (e.g. Geyh and Schleicher 1990, Lang et al. 1999). In the Eastern Alps, the majority of landslides were dated with the classic radiocarbon method, using organic remnants that are a) present in sediments buried by the rockslide (maximum age of the event), b) trapped within the rockslide debris (proxy for the event), and/or c) that were deposited in rockslide-dammed backwater deposits or lakes situated atop the rockslide mass (minimum age of the event). In recent years, new absolute age dating methods have been developed and increasingly applied. Among these are surfaces exposure dating using in-situ-produced cosmogenic radionuclides, applied e. g. at sliding planes and accumulated rockslide boulders of the Köfels landslide (Tyrol; Ivy-Ochs et al. 1998), optically stimulated luminescence (OSL) methods for dating landslide deposits and associated backwater sediments as well as U-/Th dating methods. At the prominent Fernpass rockslide, which is one of the largest landslide of the Alps, the field situation enabled the application of three individual dating methods on samples from geologically different localities (Prager et al. 2008a, Ostermann et al. 2007), which are a)

C-14 dating of organic material accumulated in rockslide dammed backwater deposits, b) Cl-36 exposure dating of sliding planes at the scarp (and accumulated rockslide boulders), and c) Th-230/U-234 disequilibrium dating of post-failure aragonite cements that precipitated in the pore-cavities of the landslide deposits. All dating data coincide well and indicate that the Fernpass rockslide took place in the middle Holocene at about 4200-4100 yrs.

Based on this, and in order to evaluate the spatial and temporal distribution of mass movements in Tyrol and its surrounding areas, a GIS linked geodatabase was set up. For the first time all available radiometric data of late-glacial and Holocene landslides in the East Alpine region were compiled (Fig. 5.1). Analyses of these data indicate that both rock slope failures (e. g. rockslides, rockfalls) and debris flows occurred rather continuously distributed in the Late-Glacial and Holocene. However, it is noticeable that there are two temporal clusters of increased landslide activities: a) in the early Holocene between 10500-9400 BP, comprising the large rockslides of Köfels, Kandertal and Flims, and b) in the middle Holocene between 4200-3000 BP, when several of the largest landslides in the Eastern Alps took place (e.g. Eibsee, Fernpass, Tschirgant, Haiming, Pletzachkogel).

Furthermore, the compiled age data show that several slopes were repeatedly reactivated in the Holocene. Multiple failure events in the Alpine region were, amongst others, documented in the following areas:

- Fernpass rockslide (Tyrol): a main large-scale failure event in the middle Holocene, a laterally adjacent secondary rockslide and the initial stage of a rockslide were observed (Prager et al. 2007);
- Tschirgant massif (Upper Inntal, Tyrol): two distinct scarp regions featuring at least five different failure events in the middle Holocene (Patzelt 2004);
- Köfels (Oetztal, Tyrol): one major rockslide in the early Holocene and at least one larger secondary event (Ivy-Ochs et al. 1998, Hermanns et al. 2006);
- Tumpen (Oetztal, Tyrol): several failure events, two of which dated roughly around the middle Holocene (Poscher & Patzelt 2000);
- Pletzachkogel (Lower Inntal, Tyrol): at least three temporally differing rockslide events in the Late-Glacial and middle Holocene, as well as smaller rockfall events in the 20th century (Patzelt 2004);
- Multiple reactivations or accelerations of pre-existing landslides in time intervals of a few thousand years were observed in Gepatsch-Hochmais (Kaunertal, Tyrol; Schmidegg 1966), Heinzenberg (Switzerland, Weidner 2000) and La Clapière (France, Cappa et al. 2004);

- Several cases of recent landslides, such as the in Val Pola (1987, Italy) and Randa (1991, Switzerland) have shown that failure was structurally controlled and that there were precursory events in (pre-historic times (Azzoni et al. 1992, Sartori et al. 2003).

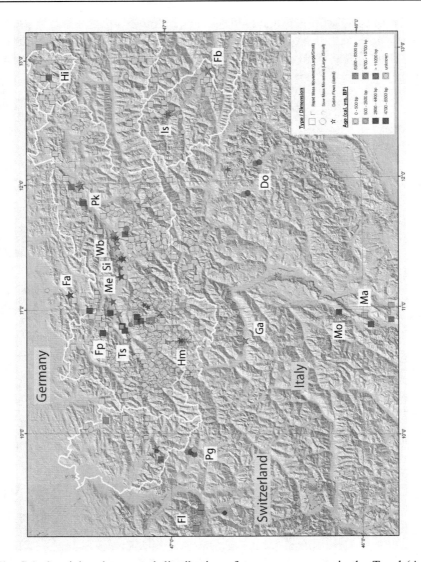

Fig. 5.1 Spatial and temporal distribution of mass movements in the Tyrol (Austria) and its surrounding areas (Prager et al. 2008b). Beyond Tyrol only dated fossil landslides are shown. *Do*: Dolomites rock slope failures, *Fa:* Farchant debris flow, *Fb*: Frauenbach debris flows, *Fl*: Flims rockslide, *Fp*: Fernpass rockslide, *Ga*: Gadria river, *Hi*: Hintersee rockslide, *Hm*: Hochmais rockslide, *Is*: Isel river, *Ma*: Marocche di Dro, *Mo*: Molveno rockslide, *Me*: Melach river, *Pg*: Prättigau rock slope failures, *Pk*: Pletzachkogel rockslides, *Si*: Sill river, *Ts*: Tschirgant rockslide, *Wb*: Weißenbach river

However, not only rock slope failures but also dated debris flows show fluctuating activities in the Holocene. Periods of increased debris accumulation rates were established for the Tyrolean Inn valley and its main tributaries, occurring at about 9400, between 7500-6000 and at around 3500 C-14 years (Patzelt 1987), and were probably climatically controlled by the amount of water in the catchment areas. These periods partially correlate temporally with glaciers advances in the Austrian Central Alps (Patzelt 1977) as well as with several other large landslides in the Eastern Alps (Fig. 5.2).

Hence, these data suggest that the stability conditions of numerous fossil landslides were affected by the climatically controlled water pressure distribution in the slopes. Similar phases of increased slope instabilities were previously detected in the Swiss and North Italian Alps (Fig. 5.2) and generally attributed to climate changes (Raetzo-Brülhart 1997, Dapples et al. 2003, Soldati et al. 2004).

The principal cause for Alpine landslides may be ascribed to glacier retreats, but this was not necessarily the direct trigger. In fact, the majority of slopes remained in a "stable" position for several thousands of years, after ice-withdrawal, before complete collapse set in. Detailed field studies and compiled geological data indicate that deep-seated slope deformations may generally be attributed to the propagation and coalescence of brittle discontinuities. Progressive failure is induced by complex interactions of different time-dependent processes such as a) stress redistributions due to glacial loading and unloading, b) subcritical crack growth, c) seismic activity and d) climatically controlled pore pressure changes (Prager et al. 2008b).

As a result, the analyses of the compiled data set shows that a) periods of significantly increased mass movement activity are distinguishable (Fig. 5.2), b) spatial accumulations of slope instabilities occur (Fig. 5.1) and c) predisposed areas were repeatedly (re)activated for mass movements and even posses high risk potential of future events. Consequently, well documented case studies indicate that the pre-failure mechanisms are of essential importance for the early detection of endangered settlement and economic areas.

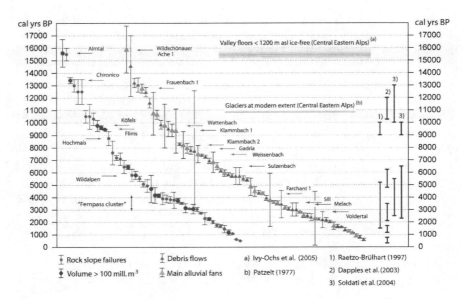

Fig. 5.2 Temporal distribution of fossil mass movements in Tyrol and the surrounding region (Prager et al. 2008b). Vertical axis = calibrated years before present (BP = before present), horizontal axis = dimensionless sequence of dated events

5.4 Basic principles of deformation and failure processes of landslides

For the estimation of the hazard potential of landslides it is decisive to understand the failure mechanisms, the deformation behaviours and the slope kinematics. Thus, for example, any sound forecast of slope instabilities is fundamentally affected by whether a translational or rotational slide prevails, or whether toppling or falling processes or complex combinations of both dominate the slope deformation. Only combined analyses of geological, hydrological, geomorphological, geodetical, geophysical and other exploratory data (e.g. boreholes and investigation adits) enables the development of a comprehensive model, which is able to represent the natural system characteristics as close as possible.

5.4.1 Fracture mechanical processes

The steeply inclined scarps on rockfall locations indicate that slope failures are induced by fracture mechanical processes such as fracture growth of new and coalescence of already existing fractures. Einstein et al. (1983) and Eberhardt et al. (2004) assume, that fully persistent discontinuities seldom occur in nature, and much rather that slope failures are induced by interactions of existing discontinuities and fracture propagation. In general a jointed rock mass consists of intact rock blocks (including microcracks and micropores), which are bound by discontinuities such as bedding planes, joints and fault zones. Intact rock bridges between these discontinuities increase the strength of a rock mass. This means that the stability of rock slopes is determined by the orientation, density and size of fractures in a rock mass (Einstein 1993).

Fracture propagation strongly depends on the existing stress field as well on the fracture geometry and network (Einstein and Stephansson 2000). In regard to the development of fractures there are three different basic fracture modes: mode I = opening, mode II = sliding, in-plane shear and mode III = scissoring, anti-plane shear. Naturally complex combinations of these three modes are likely to occur. Classical fracture mechanics postulates that in a linear, elastically solid body fracture propagation takes place with acoustic velocity of the medium when a critical stress intensity factor (K_{IC} for mode I, K_{IIC} for mode II) is reached in the crack tip. In the case of a stress intensity factor K_I (or K_{II}) below the critical stress intensity factor the crack remains stable.

However, there are physical-chemical processes in fractures that enable slow crack propagation even below this K_{IC}-threshold, referred to as subcritical crack growth (Fig. 5.3; Atkinson 1984). Due to the complex interactions of pore pressure, stress corrosion, dissolution, diffusion, ion exchange and microplasticity it is very difficult or nearly impossible to estimate the time factor for this mechanism. For example, the process of stress corrosion, which is characterised by weakening of crystal bonds through chemical fluid activities (e.g. water) in the crack tip, leads to slow crack propagation. The application of this fracture mechanical model on unstable slopes would, however, mean that over a longer time period the fracture density and persistence continuously increase. In the long term this leads to a continuous decrease in the slope stability and to a failure event when the strength threshold is exceeded.

Besides the sub-critical and critical crack growth, also temperature effects must be considered when dealing with rather shallow seated rockfall events. In-situ measurements at the rockfall Val d'Infern (Graubünden, Switzerland) show, that temperature induced stress changes affect the rock mass to depths of about 10 m. These cyclical loading conditions lead to progressive fracture displacements and thus reduce the long-term stability (Krähenbühel 2004). Furthermore the stability of Alpine rock flanks may also be influenced by permafrost conditions, depending on the exposition, inclination and elevation of the slope.

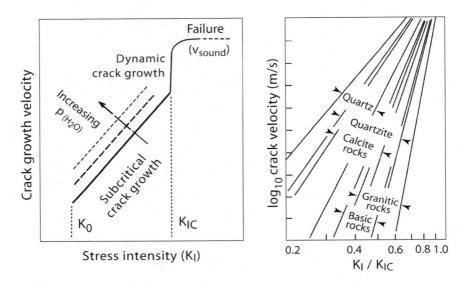

Fig. 5.3 Schematic illustrations showing (on the left) crack growth velocity versus stress intensity factor (K_I) for Mode I loading: sub-critical crack growth starts when driving forces exceed a threshold K_0. Approaching to a critical level K_{IC}, cracks propagate dynamically to near the velocity of sound in the rocks. Note that increasing pore pressures $p_{(H2O)}$ may significantly accelerate crack propagation velocities. On the right, variations of sub-critical tensile crack growth in different rock types are shown (log/log plot; arrows indicate range of experimentally obtained data; K_{IC}=fracture toughness, K_0=stress corrosion limit; modified after Atkinson 1984, 1987)

Given the extremely complex interactions of the physical and chemical processes discussed above, the time-dependent reduction of rock mass strength is difficult to forecast without any deformation monitoring data. So far only a few fracture mechanical based stability approaches are published. For example, a time-dependent cohesion loss model derived by Kemeny (2003) suggests that after a phase of progressive strength reduction slope failure occurs very rapidly. This and some case studies show that long lasting deformation phases preceded the failure event. For example, pre-failure deformation measurements in the region of a rockfall near the Jungfraujoch (Switzerland) proved that the actual event was preceded by rock deformation phases of at least one year (Keusen 1998). A high resolution monitoring device based on extensometers was installed 85 days before the failure event. Whereas during the first 2 months after installation a relatively linear displacement trend was measured (total displacement of 4 mm), the slope started to accelerate 2.5 weeks before failure and finally collapsed after another 10 mm of displacement. Also geodetic measurements at the rockfall Val d'Infern, which were performed between 1995 and 2006, showed increasing slope velocities before it collapsed (Krähenbühl 2006). Based on such acceleration patterns, the time of slope failure can be determined by analyses of inverse slope velocities versus time (Voight 1988).

5.4.2 Sliding processes

If a fully persistent sliding zone is formed through fracture mechanical processes, then further slope deformation is essentially determined by sliding processes resulting from material creep and/or shear slip in these zones. The sliding mass itself is mostly characterised by relative little internal deformation. Whether a fully persistent sliding zone can exist within a slowly creeping, deep-seated landslide has often been debated controversially, but is highly relevant for stability analysis. In some cases detailed field investigations show that surface deformation is obviously the result of flexural or block toppling mechanisms without any indications of a continuous sliding zone (Zischinsky 1969, Amann 2006). However, in many cases the slope kinematics is clearly characterised by sliding mechanisms. This is confirmed by several field observations and inclinometer measurements on landslides showing a) extensive sliding planes in the scarp area, b) a discrete offset between sliding masses and the stable bedrock unit, c) lateral exposures of sliding zones and d) kakirites (i.e. uncemented breccias and gouges) acting as active sliding zones. Also data obtained from boreholes and investigation adits in Switzerland (Noverraz 1996) show that the major deformation of deep-seated mass movements generally occurs along discrete sliding zones (Fig. 5.4).

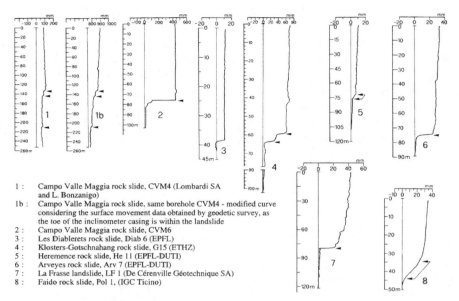

1 : Campo Valle Maggia rock slide, CVM4 (Lombardi SA and L. Bonzanigo)
1b : Campo Valle Maggia rock slide, same borehole CVM4 - modified curve considering the surface movement data obtained by geodetic survey, as the toe of the inclinometer casing is within the landslide
2 : Campo Valle Maggia rock slide, CVM6
3 : Les Diablerets rock slide, Diab 6 (EPFL)
4 : Klosters-Gotschnahang rock slide, G15 (ETHZ)
5 : Heremence rock slide, He 11 (EPFL-DUTI)
6 : Arveyes rock slide, Arv 7 (EPFL-DUTI)
7 : La Frasse landslide, LF 1 (De Cérenville Géotechnique SA)
8 : Faido rock slide, Pol 1, (IGC Ticino)

Fig. 5.4 Inclinometer data of seven deep-seated landslides in Switzerland showing major displacement along discrete sliding zones (Noverraz 1996)

Similar results were obtained from several rockslides in the Tyrolean Alps (e.g. Gepatsch/Kaunertal, Gries/Sellraintal), where one or more sliding zones control slope deformation and kinematics (Zangerl et al. 2007). Exploration adits, borehole data and field surveys show that these sliding zones can extend to thicknesses of several metres and consist of slope failure induced kakirites. These sliding zone kakirites are characterized by intensively fractured, fragmentised and triturated rocks which are generally difficult to distinguish from tectonically formed fault breccias and gouges. Several field studies show that already existing tectonic fault zones are often reactivated as sliding zones of landslides (Fig. 5.5). Conclusively it was found that many of the investigated landslides are in fact "slides" which are characterised by slipping along discrete shear planes and/or creeping within one or several sliding zones. In this case "creeping" is defined as a continuous material deformation under constant stress state (e.g. Hudson and Harrison 1997).

5 Process-based investigations and monitoring of deep-seated landslides

Fig. 5.5 a) Tectonically formed kakirite of a brittle fault zone acting as a mass movement sliding zone, b) slope failure induced slickenside striations (landslide system Steinlehnen, Tyrol; location see Fig. 5.13)

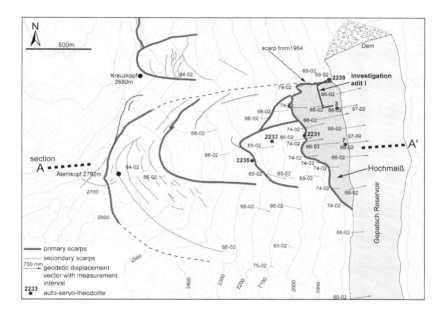

Fig. 5.6 Map of the Hochmais-Atemkopf rockslide (Kaunertal, Tyrol) and its monitoring system showing: a) primary and secondary scarps of different sliding slabs (red lines; grey coloured: Hochmais slab), b) episodic triangulation points (red arrows), c) automatic total station measurements (blue points), d) investigation adit that passes from the stable bedrock into the Hochmais slab (see Fig. 5.7)

In the Kaunertal, the active sliding slab Hochmais has slid over glacial till and talus material and continued to move towards the valley bottom (Fig. 5.6). Exploration adits and boreholes document that the sliding zone did not form directly between the sliding mass and its soil substrate beneath, but much rather a 4 to 5 m thick sliding zone developed within the glacial till itself (Fig. 5.7). In the exploration adit, which intersects the active movement zone, permanent measurements are taken with a wire extensometer in order to measure the horizontal displacement vector. In addition vertical displacements have been measured by means of episodic levelling. Based on a linear regression analysis of both data sets the E-W orientated displacement vector at the base of the sliding mass dips about 32° to the East (Fig. 5.8). Comparisons with levelling und triangulation measurements on the surface of the sliding mass show similar slope velocities and dip angles as observed in the sliding zone. Thus, the Hochmais rockslide represents a slide along a several meter thick distinct deformation zone, whereby the internal deformation of the sliding mass is comparatively small.

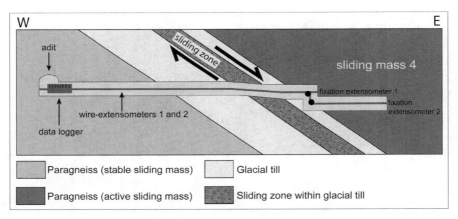

Fig. 5.7 Schematic illustration of the exploration adit which intersects both the stable bedrock and the active sliding slab Hochmais (Kaunertal, Tyrol). Along this adit interval, a wire extensometer device, a water level gauge and levelling points were installed

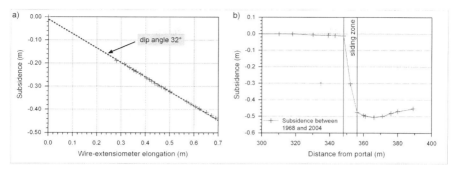

Fig. 5.8 a) Linear regression between levelling and wire extensometer measurements and the resulting dip angle of the displacement vector at the base of the active sliding slab Hochmais, b) calculated subsidence between both levelling measurements in 1968 and 2004 around the movement zone in the exploration adit. The levelling measurements in the region of the sliding zone showed subsidence whereby it could not definitely be determined in how far the total slip along the movement zone is due to pure material creep and/or accumulation of small shear slip movements on numerous discrete planes

5.4.3 Failure and temporal behaviour of sliding zone materials

Classical approaches to estimate the slope stabilities are based on the calculation of the safety factor, i.e. defined as the ratio between the driving and resisting forces. With this factor it is possible to estimate the present stability of a slope, even though time-dependant strength degradation processes are not considered. The shear strength of the sliding zone material, i.e. the resisting forces, is of decisive important for the determination of the stability conditions. A simple model for the estimation of the shear strength represents the Mohr-Coulomb model, which is based on two parameters only, the cohesion and the friction angle. These can be determined by means of triaxial compression or shear box tests in the laboratory.

Results of tests performed on different sliding zone materials, i.e. kakirite and glacial till samples, are given in Table 5.1 Irrespective of the applied laboratory testing method the friction angle of the material ranges between 31° and 39°, and the cohesion was found to be below 54 kN/m^2. Comparisons of these laboratory data with shear strength parameters obtained from back-calculations show that - for some investigated landslides - friction angles significantly below 31° and cohesions close to 0 are needed to induce slope failure. This discrepancy suggests that laboratory tests and in-situ back-calculation may differ due to scale effects, pore pressure influences and others.

Table 5.1 Mohr-Coulomb shear parameters of sliding zone samples

Sample number	Test type	Friction angle (°)	Cohesion (kN/m^2)
KG1	CD1	35,3	14,0
	RS-saturated	32,3	14,8
	RS-unsaturated	36,4	22,2
KG2	CD1	30,7	54,0
	RS-unsaturated	32,7	36,7
HM1-K	CD1	34,1	15,0
	RS-saturated	30,9	37,4
HM1-M1	CD1	36,0	18,0
	CD2	36,2	7,0
	RS-saturated	34,4	36,2
HM1-M2	CD1	39,1	24,0
	CD2	38,0	6,0
	RS-saturated	35,8	42,9

Samples KG1, KG2 and HM1-K: kakirite formed from paragneiss; samples HM1-M1 and HM1-M2: glacial till material from the sliding zone (see Figs. 5.7); CD: triaxial tests of cylinders with D x H = 10 x 20 cm (CD1) or 4 x 7.5 cm (CD2); RS: fully saturated and unsaturated shear box test.

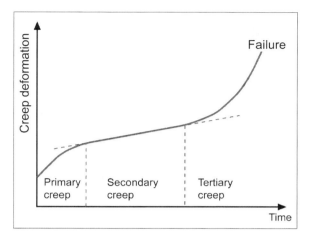

Fig. 5.9 Theoretical time-deformation behaviour (creeping) of material under constant stress

The deformation along an existing sliding zone is characterised by time-dependent processes such as material creep and/or friction along small-scale slip planes (Fig. 5.9). If creeping is the dominant mechanisms, the viscous deformation phase can be subdivided into three main regimes: primary, secondary and tertiary creep. So-called primary creep occurs simultaneously with the application of load and it is characterised by a monotone decrease in the creep rate. The secondary or steady-state creep is characterised by constant creep rates. If the loading is sufficiently high and enduring, an increase in the creep rate ensues until the material fails (tertiary creep).

The time-dependent (viscous) creep characteristics of sliding zone materials can be determined by laboratory tests. Therefore samples are loaded by constant stress over a longer period of time. Results from triaxial creep tests show that after consolidation under a defined stress condition all samples (approximately stationary) began to creep in a secondary manner (Fig. 5.10).

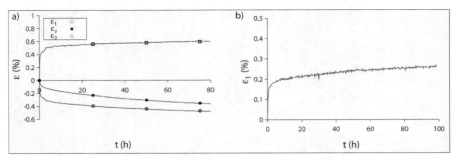

Fig. 5.10 Creep curves of a) a sample of glacial till material and b) a sample of kakirite rock

The evaluation of the laboratory tests was done with a linear-viscous and a non-linear-viscous model (Schneider-Muntau et al. 2006, Renk 2006), whereby the elastic deformations were analysed separately. In the case of the linear-viscous model according to Newton, for each test the viscosity of the material can be obtained from the applied stresses and the measured creep strain rates. The equivalent creep strain rate $\dot{\bar{\varepsilon}}$ can be determined from the strain rate tensor and the Mises equivalent stress from deviatoric stress tensor:

$$\dot{\bar{\varepsilon}} = \frac{1}{\eta}\bar{q} \qquad \text{(eq. 5.1)}$$

The assessment using a non-linear-viscous model was based on the following equation:

$$\dot{\bar{\varepsilon}} = \frac{1}{\eta}(\bar{q} - q_y)^n \qquad \text{(eq. 5.2)}$$

The crucial material parameters are the viscosity η, the yield stress q_y, which describes the stress state when the material begins to deform viscously, and the exponent n, that describes the non-linear behaviour. For the determination of these parameters several laboratory tests are needed. Based on the stress and creep strain rate of each single laboratory test a regression function can be calculated using the method of the least squares fit, which best describes the material behaviour according to the 3 parameters. For a non-linear-viscous model with n>1, the increase of the stress has a higher influence on the creep strain rate than in a purely linear-viscous model.

Seeing as triaxial tests only allow for relatively small displacements and strains, for the study of larger deformations ring shear experiments can be conducted. In addition, laboratory tests always include scale effects, which need to be taken into account in the transferral to real models. Nevertheless geotechnical laboratory tests can contribute to a better understanding of the deformation processes along sliding zones.

Because there are parallels discernable in the movement patterns of active brittle fault zones with the formation of kakirites and sliding zones from landslides, frictional approaches from earthquake mechanics can be transferred to landslides. Thus, through the application of empirical material laws, such as the state- and velocity dependent friction laws (Ruina 1983, Dieterich 1992), it is possible to describe acceleration phases with subsequent stabilising or complete failure of slopes (Helmstetter et al. 2004).

5.4.4 Temporal deformation behaviour of landslides

Alpine regions are characterised by rock falls, topples, flows and slides with velocities ranging from several mm/a to several m/s. Slope instabilities that end in a sudden collapse show a deformation behaviour that depicts an accelerated velocity curve (Fig. 5.11). However, some landslides are characterised by a base activity superimposed by episodic phases of a higher slope velocities, which can be observed particularly in spring due to snow melting and rainfalls (Fig. 5.11). These phases of acceleration and stabilisation can presumably be ascribed to pore pressure fluctuations, to stabilising effects in the slope foot area and/or to changes in the material properties of the sliding zone itself.

Time series of deformation measurements from two cases studies show significant acceleration phases and periods of minor activity (Figs. 5.12 and 5.14). Remarkably, the velocities of low and high activity phases can differ about a factor of up to 8000, as was observed at the Steinlehnen rockslide (Zangerl et al. 2007; Figs. 5.13 and 5.14). There at the end of June and beginning of July 2003, a maximum slope velocity of over 4 m per day was measured here with a terrestrial laser scanner. After a one-month acute phase, whereby the trigger of this acceleration phase is still unknown, a continuous decrease of the velocity ensued (Fig. 5.14a). In spring 2004, once again there was a phase of acceleration with a maximum slope velocity of up to 4 cm per day (about a factor of 100 lower), which was followed by a deceleration to about 0.5 mm per day in the autumn of 2004 (Fig. 5.14b,c). Even the strong rainfall in August 2005 could not lead to a remarkably reacceleration of the sliding slab (Fig. 5.14b, Zangerl et al. 2007).

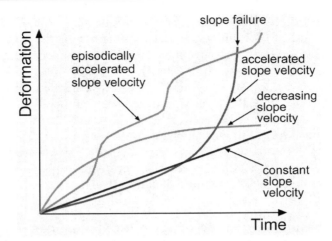

Fig. 5.11 Temporal deformation types: episodic accelerated velocity, constant slope velocity, decreasing slope velocity, accelerated slope velocity with failure (modified according to Keusen 1997)

Cyclical movements but in a much lower magnitude could also be observed at the Hochmais rockslide (Zangerl et al. 2007, Figs. 5.6 and 5.12). Exceptionally long time-series of about 40 years observation period show that the yearly slope velocity varies between 0.01 mm and 1 mm per day, i.e. a factor of 100. Every year an acceleration phase begins between January and March and lasts to the summer. This interval marks a period where precipitation in the form of rainfall does not occur. Instead, a regression analysis of the data together with that of the reservoir levels show a good temporal agreement between the slope velocity and the depletion of the Gepatsch reservoir (Figs. 5.7 and 5.12, Evers 2006, Zangerl et al. 2007). Hence for this case study the slope velocity is primarily influenced by the reservoir water level.

Similar velocity patterns, however not connected to reservoirs, could be observed at the landslides of La Clapière (France, Helmstetter et al. 2004), Triesenberg (Liechtenstein, Francois et al. 2006) or Gradenbach (Austria, Weidner 2000). On the basis of these case studies a relationship between hydrological or hydrogeological influence factors and slope activity could be shown.

5 Process-based investigations and monitoring of deep-seated landslides 155

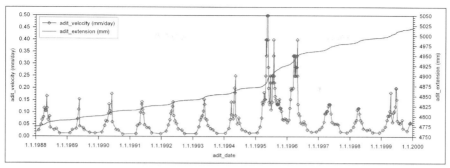

Fig. 5.12 Elongation of wire extensometer in the exploration adit I (see Fig. 5.6 and 5.7) and the derived velocity of the sliding slab Hochmais. The yearly acceleration phases occur in late winter and spring

Fig. 5.13 Overview of the Steinlehnen rockslide system (Gries i. Sellraintal, Tyrol) with a highly active sliding slab (Ortho-image: TIRIS Land Tirol). Red squares = laser scanner windows, yellow points = terrestrial survey points, measured from the opposite slope (cf. Fig. 5.14). The complete rockslide system shows a difference in height of about 800 m and a thickness of about 70 m. The highly active sliding mass reaches a thickness of about 20 m. The discontinuities which may induce these slope failures are not the foliation and compositional layering dipping flat into the slope, but one of the main fracture sets dipping moderately inclined down the slope

5 Process-based investigations and monitoring of deep-seated landslides 157

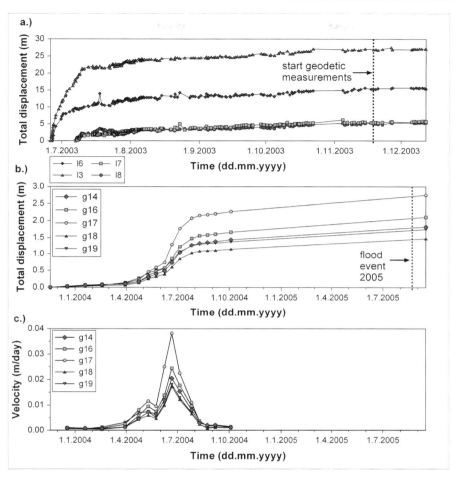

Fig. 5.14 Deformation monitoring of the highly active sliding slab Steinlehnen (cf. Fig. 5.13): a) total displacement of laser scanner windows measured from the opposite slope, b) total displacement of triangulation points measured from the opposite slope, and c) derived velocity pattern of triangulation points

5.5 Geophysical investigation methods

5.5.1 Active seismic methods

Active seismic methods can be applied to investigate the geometry, internal structure and material parameters, e.g., landslide thickness and boundaries, the basal sliding zone, degree of fracturing, sliding mass porosity (Brückl 2006). The objective of all active seismic methods focuses on the determination of the seismic velocity field in the subsurface through analysis of the wave propagation. The seismic velocity is related to the mechanical parameters of the rock mass, especially to the bulk modulus, shear modulus and the density (Ewing 1957).

Seismic waves can be generated with hammers, mechanically or pneumatically accelerated drop weights, vibrators or detonating explosives. The recording of the seismic waves can be done via geophones in a linear (2-D seismology) or an areal arrangements (3-D seismology). The signals of the geophones are stored in a portable registration unit. Given that the propagation of seismic waves represents an extremely complex process, several different seismic methods that focus on different types of waves are available (Kearey 2002). The three most important methods are described in the following paragraph:

a) Seismic reflection concentrates on the measuring and processing of seismic waves, which are reflected at boundaries characterised by a change of seismic impedance (i.e. the product of seismic wave velocity and density). Reflected waves never appear as first breaks; therefore an appropriate data acquisition and processing must be applied to extract these signals from the entire wave field. Most of the different reflection processing techniques (e.g. static correction, spike deconvolution, bandpass filter, NMO-correction, CDP-stacking, migration) were developed for the hydrocarbon exploration industry (Yilmaz 1987) and then transferred to engineering and environmental related tasks.

b) Seismic refraction is based on the analyses of critically refracted seismic waves, which appear as first breaks in the seismogram. Since the first breaks are clearly detectable, data acquisition and evaluation of signals is easier and less influenced by geological conditions. Data processing of seismic refraction measurements can be done based on different methods such as picking of first breaks, delay-time method and time to depth conversion. Hence a layered model characterised by a step-like velocity-depth function will be obtained. The fundamental condition that critically re-

5 Process-based investigations and monitoring of deep-seated landslides 159

fracted seismic waves can occur in a rock or soil is related to the seismic behaviour of the refractor. A seismic velocity of the refractor higher than the velocity of the overlaying layer should be given. In contrast to seismic reflection, velocity inversions (with an increase of depth) can not be resolved. Even though the penetration depth of seismic refraction is smaller than that of seismic reflection, the former provides better resolution of the velocity field.

c) Seismic tomography is based on the travel times of transmitted waves. These travel times contain integral information about the velocity along the seismic ray. Generally, the standard measurement geometry applied to seismic tomography is performed between two boreholes. Nevertheless the principles can also be transferred to surface investigations (seismic refraction tomography). This method is also based on observation and processing of the travel times of first breaks. Seismic tomography needs a continuous increase of the velocity with depth, because only then waves are able to return back to the surface (geophone). In spite of this basic condition, velocity inversions are possible under certain circumstances. The data processing (picking of first breaks, forward modelling with a initial velocity field, wave front inversion) results in a spatially distributed velocity field. The resolution of the tomographic method depends on the quantity of rays penetrating the subsurface and their direction. In the ideal case the rays cover directions from 0° to 180° in relation to the surface.

For the seismic methods, the penetration depths, the resolution of velocity and depth, and the accuracy mostly depends on the measuring geometry (geophone distance, profile length) and on the frequency domain of the source and the geophones. The penetration depth for seismic refraction is 1/4 to 1/3 of the total profile length and therefore the geophone distance is normally larger than for the seismic reflection, where the penetration depth is mainly influenced by the strength of the source and the subsurface velocity field. In order to get a good resolution a small geophone distance should be applied. The frequency domain of the seismic measurements also restricts the resolution, because the wavelength of the seismic wave is a product of the frequency and the velocity. So the size and thickness of a structure must be adequately large to be resolved. In general, for all methods it can be said that the velocity and depth can be determined with an accuracy of 15-20%. The classic seismic survey is performed along 2-D profiles, whereby subsurface mapping of complex 3-D structures remains difficult. For this reason an increasing number of 3-D seismic surveys were carried out, particularly in oil field exploration.

In the Tyrol a 3-D refraction seismic survey was carried out to explore the thickness and internal structure of the large-scale Niedergallmigg - Matekopf rockslide (Fig. 5.15, Chwatal et al. 2005, 2006). The upper part of the sliding mass is composed of paragneisses and schists, which were thrust on phyllitgneisses, phyllites and amphibolites that are encountered at the middle and lower part of the slope (Kirschner and Gillarduzzi 2005). Morphological features, in particular the size of the main scarp, indicate a total displacement of about 180 m. Results from geodetic measurements show annual surface displacements ranging between 5 to 10 centimetres for the active sliding mass.

Fig. 5.15 View of the Niedergallmigg-Matekopf landslide. Insert shows movement induced fracturing of a house on the landslide

In order to perform the 3-D refraction seismic survey 373 seismic stations were installed along 4 profiles comprising a total length of 7.5 km and a geophone distance of 15-25 m. 41 shots were recorded simultaneously by all receivers. This geophone set up enabled a 2-D analysis along the 4 profiles but also a 3-D inversion of the whole data set (inline and cross-line shots).

The seismic data show a vertical gradient of the velocity for the sliding mass and a nearly constant velocity for the underlying compact rock. Therefore, a combination of seismic refraction tomography for the landslide mass and standard seismic refraction method for the compact rock were applied. In order to analyse the measured data, 3-D processing techniques described by Brückl et al. (2003) and tomographic inversion algorithm published by Hole (1992) were used. As a result, the 3-D seismic velocity distribution of the landslide system and its stable surrounding is gained.

The P-wave velocities of the sliding mass are near the surface 1000-2000 m/s, at depths of 25-150 m 2000-3000 m/s and below 150 m 3000-4000 m/s. Further below velocities of 4800-5200 m/s were measured, which can be interpreted as the basis of the landslide (Fig. 5.16).

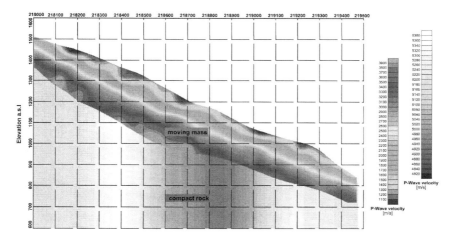

Fig. 5.16 S-N section of the 3-D velocity field of the Niedergallmigg-Matekopf landslide

Based on the seismic results a geometrical model showing the spatial thickness distribution of the Niedergallmigg-Matekopf landslide was constructed, featuring a maximum thickness of 320 m and a volume of 0.43 km^3 (Fig. 5.17). At the lateral boundaries, i.e. areas yielding little seismic information, morphological observations (e.g. scarp features) were added to establish the geometrical model. In addition, in areas without seismic data i.e. the upper region of the landslide, an interpolation technique was used.

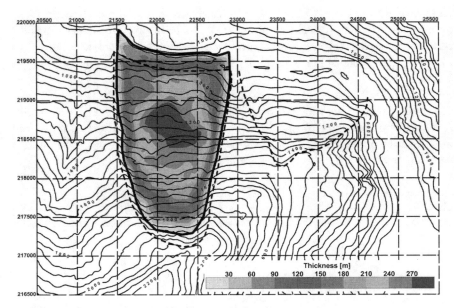

Fig. 5.17 Thickness map of the Niedergallmigg-Matekopf landslide

Plotting the P-wave velocities of the sliding mass versus depth an average velocity depth function that is based on the assumption of dry or drained slope conditions (Brückl and Parotidis 2005) can be fitted. The porosity of the fractured rock mass may be estimated from a relationship according to Gassmann (1951) and Watkins et al. (1972) and is based on P-wave velocities. For the Niedergallmigg-Matekopf landslide an average creeping rock mass porosity of 0.21 was estimated (Fig. 5.18)

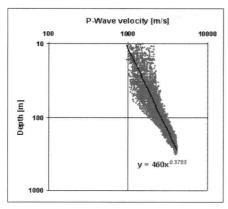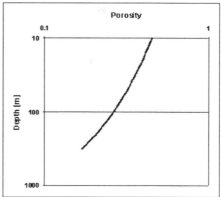

Fig. 5.18 Velocity-depth and derived porosity-depth function of the Niedergallmigg-Matekopf landslide

5.5.2 Ground Penetrating Radar

Ground Penetrating Radar (GPR) systems enable quick and nondestructive field explorations to depths of some tens of metres and were, in the Eastern Alps, already applied to rock-glaciers and water-unsaturated talus-deposits (e. g. Sass and Wollny 2001, Krainer et al. 2002). Now GPR measurements were successfully taken for near-subsurface investigations at different accumulation areas of fossil landslide deposits to analyse their thickness, internal structure and spatial distribution. Field surveys were carried out off-road at two lithologically different sites in Tyrol (Austria): firstly, at a rockslide in the Ötztal basement complex, and secondly, in the Northern Calcareous Alps at medial to distal accumulation areas of the prominent Fernpass rockslide. System parameters, measurement modes and data processing are documented in Prager et al. (2006).

Based on detailed field studies and drilling campaigns, at Fernpass down to depths of 14 m, the processed and topographically corrected GPR data can be well attributed to different depositional units. At both sites, the radargrams show several distinct reflectors with varying intensities and geometries, extending to depths of at least 20-30 m. Remarkably, the radar signals were not effectively shielded by the shallow-seated groundwater table, but penetrated down into deeper parts of the water-saturated rockslide deposits and their substrate. The combined field and subsurface data indicate that the distal Fernpass rockslide deposits spread upon groundwater-saturated, fluvio-lacustrine sediments and show strong variations in thickness. Measurements taken at a distance of about 11 km from the uppermost scarp area, clearly point to a maximum thickness of the distal slide debris of approx. 20-30 m. As a further result, the Toma, i.e. cone-shaped hills composed of rockslide debris, show deeper roots than the topographically less elevated rockslide deposits between them. These undulating basal reflectors indicate subsidence of the rockslide debris into the fine-grained substrate due to loading; a distinct sub-planar sliding plane is not detectable here. Further GPR measurements were taken at an alluvial plain, which is situated on top of the medial Fernpass rockslide deposits. Here the processed radargrams clearly show an on-lap of prograding debris flows onto the hummocky rockslide relief, a geometry which correlates well with the field situation and drilling data. Furthermore, the internal stratification of the fluvial debris is best recognisable in the radargrams and demonstrates the high resolution of the applied GPR system. Therefore, this is a useful tool to explore near-subsurface structures, even in groundwater-saturated environments and hardly accessible study-areas.

5.6 Monitoring of landslides

Even though falls, rapid topples and slides may occur seemingly unexpected and suddenly, such events are nearly always characterised by long-term preparation phases. The cause for pre-failure deformation can be found in the physical properties of soil and rock masses, which are able to accumulate elastic and plastic deformation before failure occurs. This characteristic material behaviour enables the application of monitoring and warning systems with the aim of an early detection of possible slope failure. In addition, surface and subsurface deformation monitoring of landslides are absolutely essential for the development of kinematic models.
Classic monitoring systems are based on the recordings of displacements and strains on the surface or in the depths of the landslides. In addition new monitoring systems based on micro seismic activity have been developed. These systems measure the seismic energy released through fracture and deformation processes within a landslide.

5.6.1 Deformation Monitoring

Deformation measurements aim to determine the kinematics and temporal velocity behaviour of unstable slopes. Although field observations can provide indications of the current movement status of a landslide, the question whether a slope is currently stabilised or moving can only be definitely answered with deformation measurements. Landslide deformation measurements provide data for the spatial distribution, orientation and magnitude of the displacement vectors on surface and subsurface and the temporal and spatial variation of slope velocities. Based on surface deformation data a spatial differentiation between stable bedrock units and active sliding masses as well as between variable active sub-slabs can be performed. These data are useful for the determination of the temporal slope activity, the landslide kinematics and the acceleration/trigger factors. In general, deformation measurements also form the basis for slope stability forecasts. The resulting difficulties in regard to the temporal development of a slope can clearly be seen in Fig. 5.11. This is especially the case for the prognosis of episodically accelerated slope velocities with a relatively linear base trend. In contrast, for accelerated slope velocities the trend and time of failure are easier to predict.

Slope deformation can be monitored a) pointwise, b) linewise and c) areal and can be measured in-situ and/or by remote sensing methods. Point data can be obtained through triangulation (x,y,z-coordinates), levelling (vertical z-coordinate), global positioning system (GPS, x,y,z-coordinates), wire extensometer (distance between 2 points), joint- or crackmeter (distance between 2 points), laser distance meter (distance between 2 points) and water level gauge measurements (vertical z-component between 2 points). Line data are the result of inclinometer measurements (Willenberg 2004), some types of extensometer- (Krähenbühl 2004) and Trivec measurements (Kovari 1988). Areal information about the deformation field on the surface of a landslide can be obtained by photogrammetry (Casson et al. 2003), terrestrial or satellite based radar interferometry (Rott et al. 1999, Fig. 5.19) and terrestrial or airborne laser scanning (Scheikl et al. 2000, Kemeny et al. 2006, Figs. 5.13 und 5.14).

Crucial parameters when planning a slope monitoring system are a) the number of observation points that are needed to monitor the whole landslide, b) the size and boundaries of a landslide, and c) the expected slope velocity and hence the required accuracy of the measurement method. Furthermore, it must be clarified in how far surface observations are representative for the understanding of landslide kinematics or whether borehole data (inclinometers) are needed to obtain the internal deformation behaviour. Moreover, the accessibility of the measuring area is decisive for the choice of an appropriate system. The frequency of the measurement campaigns should be defined when working with an episodical type of monitoring set-up. First evidences of the spatial distribution of slope deformation and its velocities may be given by structural and geomorphological features (open cracks, vegetation markers, etc.). Based on these field observations and the estimated slope velocity the time span between the first and second measuring campaign is defined. Subsequent monitoring data can be used to refine the measuring interval. At some landslides with high risk potential it is essential to install permanent monitoring systems that are able to record slope displacements continuously. Usually such monitoring systems can be upgraded with automatic warning devices, which set off an alarm when a predefined velocity threshold is reached.

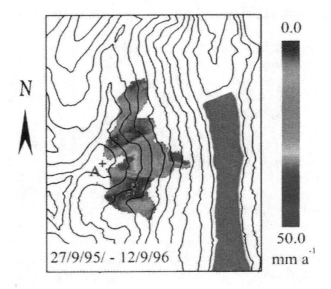

Fig. 5.19 Surface velocity pattern (mm/a) of the Hochmais-Atemkopf rockslide (Kaunertal, Tyrol), derived from radar interferogramms from 27.09.1995 and 12.09.1996. In the lower region of the slope no velocity values could be obtained due to forestation (from Rott et al. 1999)

The displacement vectors of points can be measured in form of absolute and relative coordinates. Absolute coordinates may be gained from Global Positioning Systems (GPS) and geodetic terrestrial methods. Latter may also be applied for relative displacement measurements between stable reference points outside and unstable points on the landslide (Fig. 5.20). This measurement configuration helps to avoid large measuring distances and height differences and therefore yields highly accurate measurement data. Other relative displacement measurements may result from wire extensometer or joint-meter installations. Given that these methods can only measure the distance between two points, a check by means of terrestrial geodetic methods to obtain the 3-D displacement vector of the stable and unstable point is necessary to avoid misinterpretations.

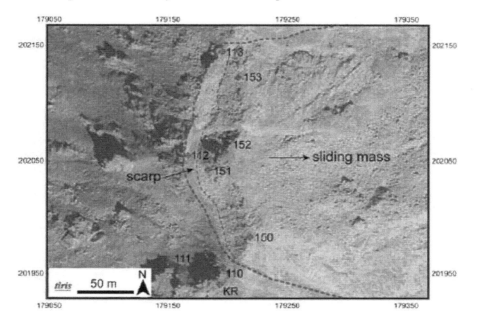

Fig. 5.20 Relative monitoring network around a mountain ridge to detect the rockslide activity (Kreuzkopf, Tyrol). Whilst the monitoring points 110, 111, 112, 113 are assumed to be stable, the points 150, 151, 152, 153 are on the sliding mass. Through the relatively short distance of less than 100 m and the minor height difference of less than 40 m possible displacements can be achieved with an accuracy of less than 2 mm (Orthofoto: Tiris Land Tirol)

For any deformation and risk analyses, data of the temporal variation of slope velocities are essential. In order to determine this, the measured displacement values must be numerically differentiated over time. The data thus obtained can be plotted in a velocity versus time diagram, whereby the velocity represents a derived value. Differential values such as the velocity of slope points are error-prone, especially when data points are temporally close together. Considering this and the fact that monitoring points contain a measurement error, the differentiation of such data sets can lead to great fluctuations in the calculated velocities. For example laser scanner raw data (Fig. 5.14a) from the Steinlehnen rockslide yielded widely scattering slope velocities, because of the relatively large measurement error and the high frequency of measurement campaigns. In order to avoid unrealistic velocity pattern fluctuations, a) the displacement curves may be smoothed out, b) the time intervall between measurement campaigns for differentiation may be increased and/or c) more accurate measurement methods such as terrestrial geodetic surveys may be applied. Based on precise triangulation measurements performed monthly at the Steinlehnen rockslide it was possible to calculate a velocity-time curve (Fig. 5.14b,c).

5.6.2 Seismic monitoring

Slope movements are characterised by the formation of new and coalescence of brittle fractures in the sliding mass and active sliding zones and presumably by "stick-slip" movements on existing shear planes. These fracturing and failure processes induce seismic energy that can be measured with a seismic monitoring network. The seismic events ("microearthquakes") can be analysed to determine the magnitude and source location of brittle deformation.

Since 2001 regular seismic monitoring campaigns were conducted on the deep-seated rockslides in Gradenbach (Carinthia, Austria) and Hochmais-Atemkopf (Tyrol, Austria). In 2005 the monitoring campaign was extended to the large-scale Niedergallmigg-Matekopf landslide (Fig. 5.15). Analyses of these monitoring data led to the design of a monitoring network which focused particularly on the observation of deep-seated mass movements (Fig. 5.21). The basic problems, when applying microseismic monitoring to landslides are: a) the signal strength of the events caused by landslides is usually very weak, b) settlements and infrastructure generate a high seismic noise level and c) there is still a lack of knowledge about the characteristic of the expected microseismic events.

Overall geophones register a variety of events (local earthquakes, events created by humans), that can not easily be classified due to their waveform and frequency characteristics. In order to allocate seismic events to landslides, the monitoring network should not only cover the landslide mass itself but also the surrounding areas (Fig. 5.21). Moreover, the seismological observation stations surrounding the landslide should be incorporated in the monitoring network.

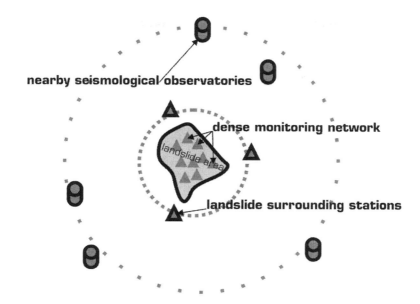

Fig. 5.21 Schematic set-up of a monitoring network for the observation of deep-seated landslides

Currently there is not enough information available about the characteristics of such landslide-induced seismic events (frequency pattern, wave form, duration of an event) to search specifically in the recorded data set. For the detection of events already known signals such as human generated „noise" as well as global, regional and local earthquakes are eliminated from the data. The remaining signals that can not be assigned to any of the known types of events are classified as "interesting". Earthquake catalogues and the surrounding seismological observatories aid in making a distinction between global, regional and local earthquakes. Spectro- and sonograms are used to detect weak signals and to visualise the frequency content (Joswig 1990). Different algorithms such as STA/LTA (Allen 1982, Allen 1978) and Principle Component Analysis (Magotra 1987, Wagner 1996) were tested for their automatic detection of events. The sensitivity of these detectors has to be adjusted to a very high level to be able to also detect weak events. This high sensitivity causes a great number of "error detections". For an improved and automatic detection of events "pattern recognition algorithms" can be applied (Joswig 1990). The disadvantage of these algorithms is that a profound knowledge about the frequency characteristics of the interesting events is presupposed, however, is currently unavailable.

The localisation of the events is an important indicator of whether a registered event is a micro-earthquake induced by landslides. For "interesting" events, showing a sufficient signal-noise-ratio, localisation can be conducted with the help of first arrival travel times and a 3-D P-wave velocity model. Using the NonLinLoc software the source coordinates can be determined (Podvin and Lecomte 1991, Lomax et al. 2005). This localisation method was tested with controlled seismic sources i.e. dynamite detonations at the study site Hochmais-Atemkopf and shows that through a 3-D velocity model reliable results can be achieved (Fig. 5.22). Generally it was found that localisation of the epicentre in regard to a transferred coordinate system parallel to the slope inclination are easier to determine than the focal depth.

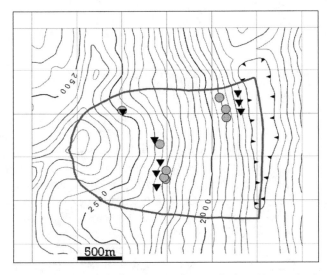

Fig. 5.22 Re-localisation of the detonations of the measurement campaign Hochmais-Atemkopf (Kaunertal, Tyrol). Black triangles = actual detonation points, grey circles = re-localised detonations

Currently there is not enough information available about the characteristics of such landslide induced events (frequency pattern, wave form, duration of an event) to search specifically in the recorded data set

However, events with low signal-noise ratio or emerging energy cannot be detected by standard detection and localization routines. Spectrogram and image processing routines are used to automatically detect and classify the recorded seismic events. Localization routines based on amplitude distribution are used to localise events with no clear first P-wave arrival (Mertl and Brückl 2007). Many events that could be classified as microearthquakes exhibit composite or multi-event characteristics. Such events were observed on all three examined slopes (Fig. 5.23). The total duration of micro-earthquakes lies between 5-20 seconds, whereby the main part of the frequency content is <30 Hz. The later events of a composite microearthquake mostly exhibit a lower frequency content than the primary event.

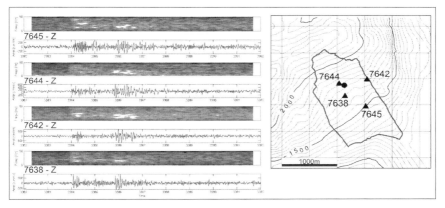

Fig. 5.23 Typical registration of a multi-event microearthquake, recorded on the landslide Gradenbach

Besides the composite and multi-events other types i.e. narrow frequency-band and low-frequency events were recorded on the three slopes. However, these could not distinctly be classified as landslide induced events. Figure 5.24 depicts an example of such a "narrow frequency" event.

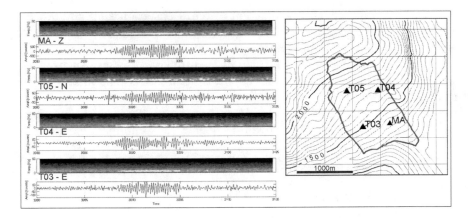

Fig. 5.24 Typical registration of a narrow-frequency event, recorded on the landslide Gradenbach

5.7 Summary

The complexity and variability of diverse kinematical landslide types requires multi-disciplinary approaches to analyse the underlying processes and mechanisms. Therefore disciplines from the field of geotechnics, geology, geomorphology, geodetics and geophysics should ideally interact to reach a high level of knowledge.

Age dating of fossil landslides helps to resolve the temporal and spatial interrelationship of slope instabilities in different geological settings and under different climatic conditions in the Holocene. Time periods of increased slope activity can be resolved and linked to feasible triggers. Analysis of age dating results shows that in several cases reactivation or multiple failures on sites occurred in the past. Thus an increasing compilation of landslide events from the past can provide data to establish probability-based estimations of the occurrence of slope failures for a given area. Such data are needed as input for further risk analyses approaches.

Fundamental principles from fracture and friction mechanics and material creep laws help to understand slope deformations in a kinematical, mechanical and temporal manner. Both the triggers that induce slope failures or accelerations and the factors which lead to slope stabilisation require detailed investigation and analyses to establish reliable forecasts. In addition the physical comprehension of slope failures and deformation processes provide the basis for the development and planning of monitoring systems, in-situ investigation methods and stability analyses based on limit equilibrium methods or numerical modelling approaches. Monitoring of landslide deformation and microseismic activity provide crucial information about the kinematics and the temporal deformation behaviour.

References

Abele G (1974) Bergstürze in den Alpen. Ihre Verbreitung, Morphologie und Folgeerscheinungen. Wiss. Alpenvereinshefte, München 25, pp 1-230

Allen R (1978) Automatic earthquake recognition and timing from single traces. Bulletin of the Seismological Society of America 68(5), pp 1521-1532

Allen R (1982) Automatic phase pickers: their present use and future prospects. Bulletin of the Seismological Society of America 72(6), pp 225-242

Amann F (2006) Großhangbewegung Cuolm da Vi (Graubünden, Schweiz). Geologisch-geotechnische Befunde und numerische Untersuchungen zur Klärung des Phänomens. Dissertation, Friedrich-Alexander Universität Erlangen-Nürnberg, p. 206

Atkinson BK (1984) Subcritical crack growth in geological materials. Journal of Geophysical Research 89(B6), pp 4077-4114

Atkinson BK (1987) Introduction to fracture mechanics and its geophysical applications. In: Atkinson BK (ed), Fracture mechanics of rock, Academic Press, pp 1-26

Azzoni A, Chiesa S, Frassoni A, Govi M (1992) The Val Pola Landslide. Engineering Geology 33, pp 59-70

Brückl E, Brückl J (2006) Geophysical models of the Lesachriegel and Gradenbach deep-seated mass movements (Schober range, Austria). Engineering Geology, 83(1-3), pp 254-272

Brückl E, Parotidis M (2005) Prediction of slope instabilities due to deep-seated gravitational creep. Natural Hazards and Earth System Sciences 5, pp 155-172

Brückl E, Zangerl C, Tentschert E (2004) Geometry and deformation mechanisms of a deep seated gravitational creep in crystalline rocks. In.: Schubert (ed), Proceedings EUROCK 2004 & 53rd Geomechanics Colloquium, pp 229-230

Brückl E, Behm M, Chwatal W (2003) The application of signal detection and stacking techniques to refraction seismic data. Oral Presentation at AGU, San Francisco, 08-12 December 2003

Cappa F, Guglielmi Y, Soukatchoff VM, Mudry J, Bertrand C, Charmoille A (2004) Hydromechanical modeling of a large moving rock slope inferred from slope levelling coupled to spring long-term hydrochemical monitoring: example of the La Clapière landslide (Southern Alps, France). Journal of Hydrology 291(1-2), pp 67-90

Casson B, Delacourt C, Baratoux D, Allemand P (2003) Seventeen years of the ''La Clapière'' landslide evolution analysed from ortho-rectified aerial photographs. Engineering Geology 68, pp 123-139

Chwatal W, Kirschner H, Brückl E, Zangerl C (2005) Geology and 3D seismic structure of the Niedergallmigg-Matekopf mass-movement, Tyrol, Austria. EGU 2005, Wien, Geophys. Res. Abst. 7, 02566

Chwatal W, Kirschner H, Brückl E, Zangerl C (2006) Kinematics and Hazard of the Niedergallmigg-Matekopf mass movement. EGU 2006, Wien, Geophys. Res. Abst. 8, 05998

Cruden DM (1991) A simple definition of a landslide. Bulletin International Association for Engineering Geology 43, pp 27-29

Cruden DM, Varnes DJ (1996) Landslide Types and Processes. In: Turner AK, Schuster RL (ed) Landslides: investigation and mitigation (Spec. Rep. 247), National Academy Press, Washington D.C., pp 36-75

Dapples F, Oswald D, Raetzo H, Lardelli T, Zwahlen P (2003) New records of Holocene landslide activity in the Western and Eastern Swiss Alps: Implication of climate and vegetation changes. Ecl. Geol. Helv. 96, pp 1-9

Dieterich J (1992) Earthquake nucleation on faults with rate- and state-dependent strength. Tectonophysics 211, pp 115-134

Eberhardt E, Stead D, Coggan JS (2004) Numerical analysis of initiation and progressive failure in natural rock slopes-the 1991 Randa rockslide. Int. J. Rock Mech. Min. Sci. 41, pp 69-87

Einstein HH (1993) Modern developments in discontinuity analysis - the persistence-connectivity problem. In: Hudson JA (ed) Comprehensive Rock Engineering, Volume 3, Pergamon Press, Oxford, pp 193-213

Einstein HH, Stephansson O (2000) Fracture systems, fracture propagation and coalescence, Issue Paper, Proc. GeoEng 2000, Melbourne

Einstein HH, Veneziano D, Baecher GB, O'Reilly KJ (1983) The effect of discontinuity persistence on rock slope stability. Int. J. Rock Mech. Min. Sci. Geomech. Abstr. 20(5), pp 227-236

Evers H (2006) Geodätisches Monitoring und einfache statistische Auswertungsmöglichkeiten für Massenbewegungen an Hängen. Master Thesis, HTWK Leipzig, p 118

Ewing WM, Jardetzky WS, Press F (1957) Elastic waves in layered media, Mc Graw-Hill Book Company, New York

François B, Tacher L, Bonnard Ch, Laloui L, Triguero V (2007) Numerical modelling of the hydrogeological and geomechanical behaviour of a large slope movement: the Triesenberg landslide (Liechtenstein). Canadian Geotechnical Journal 44, pp 840-857

Gassmann F (1951) Über die Elastizität poröser Medien. Vierteljahrsschrift der Naturforschenden Gesellschaft in Zürich 96(1), pp 1-23

Geyh MA, Schleicher H (1990) Absolute age determination: physical and chemical dating methods and their application. Springer, p 503

Helmstetter A, Sornette D, Grasso JR, Andersen JV, Gluzman S, Pisarenko V (2004) Slider block friction model for landslides: Application to Vaiont and La Clapière landslides. Journal of Geophysical Research 109(B02409), pp 1-15

Henzinger J (2005) Massenbewegung Steinlehne-Gries im Sellrain. In: Heissel G, Mostler H (ed) Geoforum Umhausen Band 3, pp 71-82

Hermanns RL, Blikra L H, Naumann M, Nilsen B, Panthi KK, Stromeyer D, Longva O (2006) Examples of multiple rock-slope collapses from Köfels (Ötz valley, Austria) and western Norway, Engineering Geology 83, pp 94-108

Heuberger H (1966) Gletschergeschichtliche Untersuchungen in den Zentralalpen zwischen Sellrain und Ötztal. Innsbruck, Wiss. Alpenvereinshefte 20, 1-126

Hole JA (1992) Nonlinear high-resolution three-dimensional seismic travel time tomography. Journal of Geophysical Research 97, pp 6553-6562

Hudson JA, Harrison JP (1997) Engineering rock mechanics. Elsevier Science Ltd., UK, p. 444

Ivy-Ochs S, Heuberger H, Kubik PW, Kerschner H, Bonani G, Frank, M, Schlüchter C (1998) The age of the Köfels event. Relative, 14C and cosmogenic isotope dating of an early Holocene landslide in the Central Alps (Tyrol, Austria). Zs. Gletscherkd. Glazialgeol. 34 (1), pp 57-68

Jerz H, v. Poschinger A (1995) Neueste Ergebnisse zum Bergsturz Eibsee-Grainau. Geol. Bavarica 99, pp 383-398

Joswig M (1990) Pattern Recognition for earthquake detection. Bulletin of the Seismological Society of America 80(1), pp 170-186

Kearey P (2002) An introduction to geophysical exploration, Brooks und I. Hill, p 262

Kemeny J (2003) The Time-Dependent Reduction of Sliding Cohesion due toRock Bridges Along Discontinuities: A Fracture Mechanics Approach. Rock Mechanics Rock Engineering 36 (1), pp 27-38

Kemeny J, Norton B, Turner K (2006) Rock Slope Stability Analysis Utilizing Ground-based LIDAR and Digital Image Processing. Felsbau 24(3), pp 8-15

Keusen HR (1998) Warn- und Überwachungssysteme (Frühwarndienste), Fan-Forum, Zollikofen, 1-40

Kirschner H, Gillarduzzi K (2005) Geodätisches Monitoring und Modellierung instabiler Hänge. In: Chesi G,Weinold T (ed) Internationale geodätische Woche Obergurgl 2005, Austria, pp 193-197

Kovári K (1988) General report: Methods of monitoring landslides. In: Bonnard C. (ed), Land-slides, Proceedings of the 5th International Symposium on Landslides, Lausanne, Switzerland, Balkema, Vol 3, pp 1421-1433

Krähenbühl R (2004) Temperatur und Kluftwasser als Ursachen von Felssturz. Bull. Angew. Geol. 9(1), pp 19-35

Krähenbühl R (2006) Der Felssturz, der sich auf die Stunde genau ankündigte. Bull. Angew. Geol. 11(1), pp 49-63

Krainer K, Mostler W, Span N (2002) A glacierderived, ice-cored rock glacier in the western Stubai Alps (Austria): evidence from ice exposures and Ground Penetrating Radar investigation. Zs. f. Gletscherkunde u. Glazialgeologie, Innsbruck, 38(1), pp 21-34

Lang A, Moya J, Corominas J, Schrott L, Dikau R (1999) Classic and new dating methods for assessing the temporal occurrence of mass movements. Geomorphology 30, pp 33-52

Leobacher A, Liegler K (1998) Langzeitkontrolle von Massenbewegungen der Stauraumhänge des Speichers Durlaßboden. Felsbau 16(3), pp 184-193

Lomax A, Virieux J, Volant P, Berge C (2000) Probabilistic earthquake location in 3D and layered models: Introduction of a Metropolis-Gibbs method and comparison with linear locations. In: Thurber CH, Rabinowitz N (ed) Advances in Seismic Event Location, Kluwer, Amsterdam, pp 101-134

Magotra N, Ahme N, Chael E (1987) Seismic event detection and source location using single-station (three-component) data. Bulletin of the Seismological Society of America, 77(3), pp 958-971

Marchesoni V (1958) La datazione col metodo del carbonio 14 del Lago di Molveno e dei resti vegetali riemersi in seguito allo svaso. Studi trentini d scienze naturali, Trento, (cit. in: Abele 1974), pp 95-98

Mertl S, Brückl E (2007) Detection and localization of micro-earthquakes on deep-seated mass movements. Geophysical Research Abstracts, Vol. 9, 07187

Noverraz FI (1996) Sagging or deep-seated creep: Fiction or reality? In. Senneset (ed) Proceedings of the 7th International Symposium on Landslides, Balkema, Rotterdam, pp 821-828

Ostermann M, Sanders D, Prager C, Kramers J (2007) Aragonite and calcite cement "boulder-controlled" meteoric environments on the Fern Pass rockslide (Austria): implications for radiometric age-dating of catastrophic mass movements, Facies 53, pp 189-208

Patzelt G (1977) Der zeitliche Ablauf und das Ausmass postglazialer Klimaschwankungen in den Alpen. In: Frenzel B (ed) Dendrochronologie und postglaziale Klimaschwankungen in Europa, Steiner, Wiesbaden, pp 248-259

Patzelt G (1987) Untersuchungen zur nacheiszeitlichen Schwemmkegel- und Talentwicklung in Tirol. Veröff. Mus. Ferdinandeum, Innsbruck 67, pp 93-123

Patzelt G, Poscher G (1993) Der Tschirgant-Bergsturz. Arbeitstagung 1993 Geol. B.-A., Geologie des Oberinntaler Raumes, Schwerpunkt Blatt 144 Landeck, Exkursion D: Bemerkenswerte Geologische und Quartärgeologische Punkte im Oberinntal und aus dem äußerem Ötztal, pp 206-213

Patzelt, G., 2004: Tschirgant-Haiming-Pletzachkogel. Datierte Bergsturzereignisse im Inntal und ihre talgeschichtlichen Folgen. Öffentl. Vortrag, 13.10.2004, alpS Symposium Naturgefahren Management 2004, Galtür

Podvin P, Lecomte I (1991) Finite difference computation of traveltimes in very contrasted velocity models: a massively parallel approach and its associated tools. Geophys. J. Int. 105, pp 271-284

Poscher G, Patzelt G (2000) Sink-hole Collapses in Soft Rocks. Felsbau, Rock and Soil Engineering 18(1), pp 36-40

Prager C, Krainer K, Seidl V, Chwatal W (2006) Spatial features of Holocene Sturzstrom-deposits inferred from subsurface investigations (Fernpass rockslide, Tyrol, Austria). Geo.Alp 3, pp 147-166

Prager C, Zangerl C, Brandner R & Patzelt G (2007) Increased rockslide activity in the Middle Holocene ? New evidences from the Tyrolean Alps (Austria). In: McInnes R, Jakeways J, Fairbank H & Mathie E (eds), Landslides and Climate Change, Challenges and Solutions, Taylor & Francis, pp 25-34

Prager C, Ivy-Ochs S, Ostermann M, Synal HA, Patzelt G (2008a) Geology and radiometric 14C-, 36Cl- and Th-/U-dating of the Fernpass rockslide (Tyrol, Austria), Geomorphology, in press

Prager C, Zangerl C, Patzelt G, Brandner R. (2008b) Age distribution of fossil landslides in the Tyrol (Austria) and surrounding areas. Nat. Hazards Earth Syst. Sci., in press

Raetzo-Brülhart H (1997) Massenbewegungen im Gurnigelflysch und Einfluss der Klimaänderung. Arb.-Ber. NFP 31, Hochsch.-Verlag. ETH Zürich

Renk D (2006) Geotechnische Untersuchungen von Gleitzonenmaterialien großer Hangbewegungen. Master thesis, Universität Karlsruhe (TH), Universität Innsbruck

Rott H, Scheuchl B, Siegel A, Grasemann B (1999) Monitoring very slow slope movements by means of SAR interferometry: A case study from a mass waste above a reservoir in the Ötztal Alps, Austria. Geophysical Res. Letters 26(11), pp 1629-1632

Ruina A (1983) Slip instability and state variable friction laws. Journal of Geophysical Research 88, pp 10359-10370

Sarnthein v.R (1940) Moor- und Seeablagerungen aus den Tiroler Alpen und ihre waldgeschichtliche Bedeutung. II. Teil: Seen der Nordtiroler Kalkalpen. Beih. Botan. Zentralblatt, LX, Abt. B (3), pp 437-492

Sartori M, Baillifard F, Jaboyedoff M, Rouille JD (2003) Kinematics of the 1991 Randa rockslides (Valais, Switzerland). Natural Hazards and Earth System Sciences 3, pp 423-433

Sass O, Wollny K (2001) Investigations regarding alpine talus slopes using ground-penetrating radar (GPR) in the Bavarian Alps, Germany. Earth Surf. Process. Landforms 26, pp 1071-1086

Scheikl M, Angerer H, Dölzlmüller J, Poisel R, Poscher G (2000) Multidisclinary Monitoring Demonstrated in the Case Study of the Eiblschrofen Rock fall. Felsbau 18(1), pp 24-29

Schmidegg O (1966) Bericht Staudamm Gepatsch, Geologie im Speicherbecken. Unveröffentlichter Bericht, K13-392, TIWAG Innsbruck

Schneider-Muntau B, Renk D, Marcher T, Fellin W (2006) The Importance of Laboratory Experiments in Landslide Investigation. In: Nadim F, Pöttler R, Einstein H, Klapperich H, Kramer S (ed) Geohazards, ECI Symposium Series, Volume P7 (2006). http://services.bepress.com/eci/geohazards/12

Soldati M, Corsini A, Pasuto A (2004) Landslides and climate change in the Italian Dolomites since the Late glacial. Catena 55, pp 141-161

Tentschert E (1998) Das Langzeitverhalten der Sackungshänge im Speicher Gepatsch (Tirol, Österreich). Felsbau 16(3), pp 194-200

Varnes DJ (1978) Slope Movement Types and Processes. In: Schuster RL, Krizek RJ (ed) Landslides-Analysis and Control, Special Report 176 (2), Washington D.C. (National Academy of Sciences), pp 11-33

Voight B (1988) Material science law applies to time forecasts of slope failure. In: Bonnard C. (ed), Landslides, Proceedings of the 5th International Symposium on Landslides, Lausanne, Switzerland, Balkema, Vol 3, pp 1471-1472

Wagner GS, Owens TJ (1996) Signal detection using multi-channel seismic data. Bulletin of the Seismological Society of America, 86(1A), pp 221-231

Watkins JS, Walters LA, Godso LA (1972) Dependence of in-situ compressional wave velocity on porosity in unsaturated rocks, Geophysics 37(1), pp 29-35

Watson AD, Martin CD, Moore DP, Stewart, TWG, Lorig LJ (2006) Integration of Geology, Monitoring and Modelling to Assess Rockslide Risk. Felsbau 24(3), pp 50-58

Weidner S (2000) Kinematik und Mechanismus tiefgreifender alpiner Hangdeformationen unter besonderer Berücksichtigung der hydrogeologischen Verhältnisse. Dissertation, Friedrich-Alexander Universität Erlangen-Nürnberg

Willenberg H (2004) Geologic and kinematic model of a complex landslide in crystalline rock (Randa, Switzerland). Dissertation Thesis, ETH Zurich, No. 15581, p 184

Yilmaz O (1987) Seismic Data Processing, Seismic data processing: Soc. of Expl. Geophys, p 526

Zangerl C, Eberhardt E, Schönlaub H, Anegg J (2007) Deformation behaviour of deep-seated rockslides in crystalline rock. In: Proceeding of the 1st Canada - U.S. Rock Mechanics Symposium, Vancouver, Canada, pp 901-908

Zischinsky U (1969) Über Sackungen. Rock Mechanics, 1, pp 30-52

6 Alpine tourist destinations – a safe haven in turbulent times? – Exploring travellers' perception of risks and events of damage

C. Eitzinger, P.M. Wiedemann

6.1 Introduction

Risk is a prominent issue in modern society and in our daily lives (Banse 1996; Beck 2007). There is talk of financial risks (e.g. insolvency) political risks (e.g. terrorism), recreational risks (e.g. extreme sports), lifestyle risks (e.g. smoking), health risks (e.g. SARS), traffic risks (e.g. driving a car), and technological risks (e.g. nuclear power). Of special interest at this point however, are travel or tourism risks. Tourism research demonstrates that travelling exposes people to varying degrees of risk (Bentley and Page 2006; Page, Bentley and Meyer 2003; Hunter-Jones 2000; Page and Meyer 1996; Phillip and Hodgkinson 1994) and shows that destination choice is not only based on price and destination image, but also on perceived personal safety and security (Pizam, Tarlow and Bloom 1997).

Risk can be and is conceptualized in a number of ways (Holzheu and Wiedemann 1993). Besides the differences in the risk definitions, another important discrepancy exists: the discrepancy between experts' risk judgements on the one hand and those of laypersons on the other. Expert-lay differences significantly contribute to controversies and conflicts regarding risks and consequently to public risk debates. In risk perception research the term risk perception is used to describe laypersons' risk judgements. This term is misleading insofar as, according to the social-constructivist perspective represented by the authors, risks are not directly perceivable or observable. They are rather the result of subjective attribution processes. Therefore, it has to be explicitly stated that the term risk perception involves intuitive judgements of and subjective attitudes towards risks (Slovic 1992) and not only perception in a narrower sense. In contrast to experts' methodological and analytical risk assessment, laypersons' risk judgements are based on media information, intuitions and vicarious as well as personal experiences.

6.2 Theoretical approaches in risk perception research

Risk perception research originates in the attempt to explain the expert-lay difference in the perception of and the reaction to natural hazards and risks of modern technology. In the following section the history of this relatively new field of research and its central theoretical approaches are discussed.

6.2.1 Bounded rationality, heuristics und biases

Starting from the observation that, at least from an expert's point of view, people often respond to risks in an irrational way, risk perception research initially concentrated on two fields of interest. On the one hand, studies in the area of natural hazards proved that persons affected by natural hazards misjudged the likelihood of occurrence of such events. Thus, a comparison of statistical data on the probability of occurrence of natural disasters with lay people's assessments of the probability of occurrence of such events showed that natural hazards are often underestimated or even denied (Slovic, Kunreuther and White 1974). On the other hand, differences between experts' and lay people's judgements of risks associated with modern technologies prompted the development of this field of research. Compared to experts, laypeople found modern technologies, particularly nuclear energy, to be more risky. Accordingly, they reacted with greater rejection than experts (Gardner and Gould 1989). Based on Simon's (1955) concept of bounded rationality as well as on findings on heuristics and biases (Kahneman, Slovic and Tversky 1982) laypeople's risk perceptions where explained by cognitive limitations of human beings (Slovic, Fischhoff and Lichtenstein, 1977).

This view was accepted with respect to natural hazards. In the context of technological risks, more precisely in the context of nuclear energy however, the argumentation of limited cognitive abilities even fostered opposition. As a consequence, a new and contrary view of lay risk perception arose: one that refuses the notion of irrational and faulty public perceptions and instead postulates that laypeople, compared to experts, have a more comprehensive understanding of risk. This new perspective on public risk perception also gained attention in the scientific debate on the expert- lay difference. In his frequently cited Science article, Starr (1969) argues that public risk acceptance is determined by the following three factors: (1) the voluntariness of risk exposure, (2) the social benefit of risk exposure as well as (3) the number of people who are exposed to a risk. Starr's (1969) paper paved the way for the development of the Psychometric Paradigm of risk perception by Slovic and his co-workers (Slovic, Fischhoff and Lichtenstein 1980; Fischhoff, Slovic, Liechtenstein, Read and Combs 1978).

6.2.2 The Psychometric Paradigm

The Psychometric Paradigm of risk perception can be seen as an expansion and empirical examination of Starr's (1969) assumptions. This approach is the most influential one in risk perception research to date, even though it has also been criticised in recent years (e.g. Sjöberg and Moen 2004). The success of the Psychometric Paradigm is essentially based on the works of Fischhoff, Slovic, Liechtenstein, Read and Combs (1978), even though a few studies on risk perception were already conducted in the 1960s (e.g. Bauer 1960; Slovic 1962). Fischhoff et al. (1978) did not deduce lay risk perception from the actual individual behaviour reflected in historical accident and fatality records (method of revealed preferences), but used questionnaires to directly ask people about their subjective risk perceptions (method of expressed preferences). Within these questionnaire studies, subjects had to rate a large number of risks on nine qualitative risk characteristics that were derived from the literature by Fischhoff et al. (1978). These risk characteristics or dimensions include the (1) voluntariness of risk exposure, (2) the immediacy of effect, (3) the extent to which a risk is known to the people exposed, (4) the chronic vs. catastrophic potential of a risk, (5) the extent to which it is a common or a dread risk, (6) the severity of consequences of a risk, (7) the extent to which a risk is known to science, (8) the controllability of a risk and (9) the newness of a risk. By means of factor analysis these qualitative risk characteristics can be reduced to two (sometimes also three) central factors determining perceived risk. These factors are "dread risk" and "unknown risk".

6.2.3 The Cultural Theory

Whilst the psychometric approach stresses the importance of subjective cognitive and emotional processes in risk perception, the Cultural Theory (Thompson, Ellis and Wildavsky 1990; Douglas and Wildavsky 1982) focuses on the socio-cultural context of risks. Cultural Theory postulates that within a society, different social systems, different ways of life exist. These are thought to influence people's worldviews or cultural biases. The four central worldviews are the fatalistic, the individualistic, the hierarchic and the egalitarian worldview. According to Cultural Theory these worldviews allow for the explanation of differences in risk perception. Dependent on their respective worldview, people are supposed to fear different threats. Hierarchists for example, are thought to especially fear threats to their central values authority, superiority/subordination and group cohesion. Egalitarians, who in contrast seek equity for all people, should fear risks like discrimination, limitation of democratic rights or the lack of participation possibilities. To summarize, each social system should fear those risks that are perceived to pose a threat to its central worldviews. Empirical evaluations of the Cultural Theory show that it explains only 5-10% of the total variance in risk perception (Sjöberg 2000).

6.2.4 Context dependency of risk perception

Laypeople's intuitive risk judgements are additionally dependent on the social context, within which risks are discussed. Of particular importance in this regard is the emotion-laden media coverage of risk issues. Using risk dramas, the media puts risks in a context of social action (Dunwoody 1992) Within these risk dramas, roles (culprit/victim) as well as (dishonest) motives are ascribed to the acting persons (Palmlund 1992). Aspects like these are particularly taken into account in the Risk Story Approach to risk perception (Wiedemann, Clauberg and Schütz 2003). Experimental studies based on this approach confirm that the perception of one and the same risk differs, depending on whether the risk is described in a neutral, objective manner or imbedded in a negative, emotion-laden context. Further evidence for the context dependency of risk perception is provided by findings on framing effects (Tversky and Kahneman 1981; Sandman, Miller, Johnson and Weinstein 1993; Wang 1996; Levin, Schneider and Gaeth 1998).

6.2.5 The role of affective processes

Empirical analyses applying the Risk Story Model as well as the emotion-laden "dread" factor identified in research on the Psychometric Paradigm indicate that emotions and emotional processes implicitly have always played a role in risk perception research. Considering that risk perception research originates in the attempt to explain the intuitive, irrational risk judgements of laypeople, this is not really surprising. In recent years however, the significance of emotions for risk perception has been explicitly addressed within concepts such as the Risk-as-Feelings hypothesis (Loewenstein, Weber, Hsee and Welch 2001) or the Affect Heuristic (Slovic, Finucane, Peters and MacGregor 2002). According to these concepts, risk judgements are influenced by emotions, by negative and positive feelings towards a risk source. Positive feelings are supposed to lead to lower, negative feelings to higher risk perception (Finucane, Alhakami, Slovic and Johnson 2000).

6.3 Risk and damage perception in alpine tourist destinations

The concept of "perceived risk" is also applied in tourism research. For the tourism industry risk perception is relevant because perceived destination safety influences the travel decision-making process, the scope of activities tourists engage in at the destination and the intention to return to or recommend a destination to others (George 2003).

Tourism risk perception studies predominantly sought to identify holiday risks, which are of general relevance in connection with travelling (e.g. Roehl and Fesenmaier 1992, Sönmez and Graefe 1998a, Lepp and Gibson 2003) and to provide information on the existence or non-existence of holiday risks at a certain destination. Moreover, they intended to analyze the influence of selected holiday risks (e.g. terrorism, crime) on the travel decision-making process (Sönmez and Graefe 1998b, Mitchell and Vassos 1997) and on the on-site behaviour of tourists (e.g. Mawby 2000). Risks of alpine tourist destinations however have been of little interest to tourism researchers (Pikkemaat and Weiermair 2003). Furthermore, tourism research did not take into account approaches of psychological risk perception research, such as the Psychometric Paradigm or the Risk Story Model

To fill this research gap, the following section summarizes own empirical studies of holiday risks in alpine destinations, which are primarily based on psychological approaches to risk perception. Using psychological concepts and approaches to risk perception, one gets additional information about the subjective characteristics that influence perceived destination risk as well as about social context factors that influence risk and damage perception. The overview of the author's research on the perception of holiday risks starts with interviews that aimed at identifying typical holiday risks in alpine destinations. Hereafter an online survey and a psychometric study on risk perception in alpine tourist destinations are presented. Finally, findings of two experiments based on the risk story approach will be discussed.

6.3.1 Interviews: What are typical risks of alpine destinations?

In order to identify typical holiday risks of the alpine tourist destination Tyrol interviews based on card sorting techniques (Coxon 1999) were conducted. A sample of residents (N=103) as well as a tourist sample (N=132) were interviewed. The interviewees had to judge each risk as to whether it is typical, rather untypical, or a risk that is of no relevance with respect to holiday-making in Tyrol. Risk judgments were made by assignment of each of 20 holiday risks to one of the following three categories: (1) typical, (2) rather untypical, (3) of no relevance for a holiday in Tyrol. The prompted risks were thought to represent the broad variety of risks from the beginning of a journey to its end. They are listed in table 6.1.

Table 6.1 Prompted Risks

01. plane crash	11. terrorist attack on the Europa-bridge
02. cable car accident	12. avalanche
03. bus accident	13. breaking of an embankment dam
04. derailing of a train	14. car accident on the highway
05. food poisoning	15. chemical accident
06. potable water poisoning	16. breakdown in a nuclear power plant
07. fire in a hotel	17. skiing accident
08. electrical power outage	18. to get lost on a ski tour
09. mass movement	19. theft
10. thunder storm	20. alcoholic intoxication

Interview results demonstrate a high agreement between locals and tourists regarding the question of typical tourism risks within alpine destinations. Risks judged to be typical for Tyrol are winter sports risks on the one hand and natural hazards on the other. Thus, in both samples "cable car accident", "skiing accident", "to get lost on a ski tour" and "avalanche" were perceived as typical holiday risks. In the local sample also "thunder storm" was judged to be a typical risk of a mountain resort. Moreover, both sample groups found "car crash on the highway" to be a typical risk of alpine destinations. The latter finding might have to do with the fact that a large number of interviewees were from Germany or Austrian federal states and consequently often bring their own car[1].

---

[1] Based on a guest questionnaire compiled by Tirol Werbung (visitor questionnaire T-Mona, Winter 2004/2005), 87% of the holiday makers travel by car.

6.3.2 Online questionnaire: Risk perception and trust building safety measures in alpine tourist destinations

Based on the interview results described before, an online survey was conducted to examine whether winter sport risks and natural hazards still prove to be central risks of alpine destinations, when classical tourism risks derived from tourism literature (Cheron and Ritchie 1982; Roehl and Fesenmaier 1992, Sönmez and Graefe 1998a, Fuchs and Peters 2005) are considered. Furthermore, the survey intended to analyze the impact of various tourism risks on the travel decision making-process.

An online questionnaire placed on the homepage of the Tiscover Tourism Platform was thought to answer the research questions. All in all 640 users participated. The according results are shown in figure 6.1.

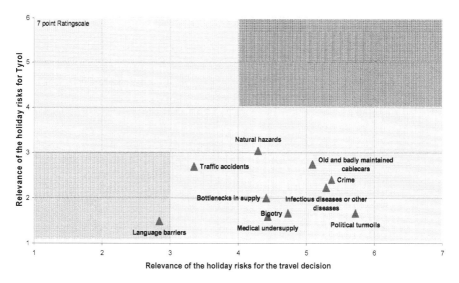

Fig. 6.1 Relevance of the holiday risks for the travel decision and for Tyrol

The findings depicted in figure 6.1 point out that there are no risks that are at the same time perceived to be of highest relevance for the travel decision-making process *and* for the destination Tyrol. In accordance with other empirical studies – the risks "political turmoil" (Ioannides and Apostolopoulos 1999; Seddighi, Nuttall and Theocharous 2001), "crime" (Mawby 2000; George 2003) and "infectious diseases or other diseases"

(Page and Meyer 1996; Wilks and Page 2003) seem to play the central role in the travel decision-making process. For the destination Tyrol – like in the interviews – "natural hazards", winter sports risks ("old and badly maintained cable cars") and "traffic accidents" were perceived to be of highest relevance.

With mean values ranging from 1.49 to 3.04 on a seven point rating scale, all risks were perceived to be of lower importance for the destination Tyrol. In spite of this, destination management should, besides natural hazards, consider the issue of felt safety in risk management. Measures with respect to felt safety should tackle both the external appearance of tourism infrastructure ("old and badly maintained cable cars") and protection against crime. Management of these risks is of decisive importance insofar as "old and badly maintained cable cars" and "crime" are both risks that are perceived to be of comparably high relevance for the travel decision-making process *and* at the same time for the destination Tyrol. In this regard the crucial question is not whether objective risk assessment and subjective risk perception do concur or not, because: Perception builds reality! In other words, what people subjectively perceive and suppose to be true has an impact on destination choice and on tourist behaviour on-site.

Besides analyzing perceived risk, the online questionnaire aimed at identifying measures and conditions that create trust in the safety of the tourist destination Tyrol in the case of their provision, and mistrust in the case of their absence.

Due to the study findings, the top 5 trust increasing measures or conditions in alpine destinations are "clearly visible signs on ski slopes", "detailed information on snow conditions, danger of avalanches and weather", "sufficient avalanche barriers", "ski slopes and ski lifts are in a good condition" and "strict closure of ski slopes in the face of high avalanche danger". Measures and conditions with the highest impact on mistrust are: "reluctant to closure of ski slopes even in the face of high avalanche danger", "insufficient avalanche barriers", "badly visible signs on ski slopes", "infrequent control runs on ski slopes", "ski slopes and ski lifts are in a bad condition", and "infrequent trainings of tourism employees regarding critical incidents". All of the above mentioned safety measures and conditions refer to safety and security on ski slopes on the one hand and protection against natural hazards on the other. The fact that the highest trust and mistrust impact is attributed to safety measures and conditions pertaining to safety on ski slopes and protection against natural hazards is in line with the finding that risk perception regarding the destination Tyrol is highest for risks associated with winter sports and natural hazards. Seemingly, safety measures and contrariwise a lack of safety measures in this area can easily be perceived by tourists.

As those safety measures having the highest impact on trust in the case of their provision are predominately those that have the highest impact on mistrust in the case of their absence, a tourism community is well-advised to implement and put particular emphasis on these measures. However, the remaining safety measures and conditions should also not be neglected, since they all considerably contribute to an increase in trust in destination safety in the case of their provision (mean value between 4,65 and 5,71 on a 7-level scale) and likewise to mistrust in the case of their absence (mean value between 4,15 and 5,09 on a 7-level scale).

Finally, another research aim of the online survey was to test if the Asymmetry Principle of trust (Slovic 1993) holds true with respect to trust in the safety of a tourist destination. According to the Asymmetry Principle (Slovic 1993) it is much harder to build up trust than to loose trust. To verify this assumption, Slovic (1993) conducted a study with 103 college students who had to evaluate the trust and mistrust impact of 45 hypothetical events in a nuclear power plant on a 7-point rating scale. Several of the events were formulated in a way, which was supposed to lead to an increase in trust and others so that they were thought to create mistrust. Based on the notion of asymmetry, negative hypothetical events or news about the nuclear power plant should reduce trust in the management of the nuclear power plant to a much higher extent, than conversely positive hypothetical events or news should foster trust. The item "A potential safety problem was found to have been covered up by plant officials", for example should create mistrust to a much higher extent than in the opposite the item "There have been no reported safety problems at the plant during the past year" should create trust. The results from Slovic's (1993) study are illustrated in figure 6.2.

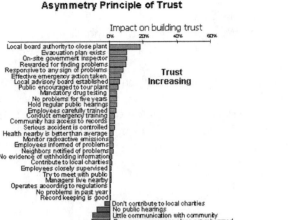

Fig. 6.2 Asymmetric Principle of Trust. Source: Slovic, P. (1993)

As can be seen in figure 6.2, the negative hypothetical events indeed led to a much higher extent of mistrust than the positive hypothetical events helped to build up trust in the safety of the nuclear power plant. Slovic's (1993) findings thus support his notion of trust asymmetry.

To test if Slovic's (1993) findings on the asymmetric principle are generalizable to tourists' trust in destination safety, participants of the online survey assessed the 15 safety measures and conditions regarding their impact on trust in the case of their provision as well as regarding their impact on mistrust in the case of their absence. T-tests for paired samples, which compare the trust impact of every single "trust item (positive information) with the distrust impact of the corresponding "distrust item (negative information), were used in data evaluation. The according results are presented in figure 6.3.

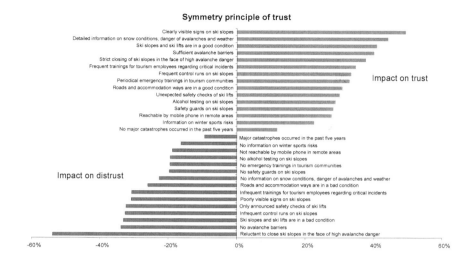

Fig. 6.3 Symmetric Principle of Trust

Due to the findings depicted in figure 6.3, results of the study at hand contradict Slovic's (1993) assumption of asymmetry. Based on the results of the t-tests, the provision of adequate safety measures and conditions has at least the same, in some cases even a significantly higher impact on trust as in contrast, the lack of these measures and conditions has on mistrust. These results suggest that the Asymmetry Principle of trust, confirmed in an industrial context (Slovic, 1993), cannot be generalised and applied within a tourism context.

6.3.3 Psychometric study: Qualitative dimensions of damage perception

The central aim of the psychometric study was to evaluate, if risk judgements, based on psychometric risk characteristics, provide the same overall picture of risk perception as the results of the interviews and the online survey.

In the questionnaire the risk were presented in the form of short damage scenarios. Below, the damage scenario for a cable car accident is shown examplarily.

Cable car accident
Just before dusk yesterday evening, the last cable car took 20 skiers on its last journey of the day up to the height of 2200m above sea-level. After 200m, the cable car suddenly crashed down into the gorge below and landed on a stream bed. 3 people were killed and 17 were injured, several of them seriously.

Each of the 15 damage scenarios had to be rated on the following 9 psychometric risk dimensions: (1) dreadfulness of the event of damage, (2) memorability of the event of damage, (3) media interest evoked by the event of damage, (4) search for perpetrators evoked by the event of damage, (5) decrease of trust in the safety management as a result of the event of damage, (6) economic consequences for the tourist destination, (7) negative effects on destination image, (8) fears induced through the event of damage and (9) crisis potential.

Damage perception was highest for the scenarios "breaking of an embankment dam" and "potable water poisoning". On all of the nine dimensions, these negative events reached scale values of >5 on a 7-point rating scale. Ratings were also high for the "food poisoning" scenario, with scale values of >4 on the nine psychometric scales. Moreover, the damage potential of the scenarios "cable car accident", "terrorist attack on the Europa-bridge" and "rockfall on a village" was perceived as very high: these events of damage received a scale value of >4 on eight of the nine risk characteristics. With a scale value of <4 on eight of the nine dimensions, the damage potential of the scenarios "thunder storm" and "mass movement" was perceived to be the lowest.

Ratings on the nine dimensions were aggregated and factor analyzed. In accordance with earlier studies, a two factorial solution was found. The extracted factors were labelled *emotional consequences* and *consequences for the community*. The factor *emotional consequences* explains 71.07%, the factor *consequences for the community* 14.92% of the variance in risk perception. Figure 6.4 depicts the localisation of the fifteen damage scenarios on these two factors.

6 Alpine tourist destinations– Perception of risks and events of damage 191

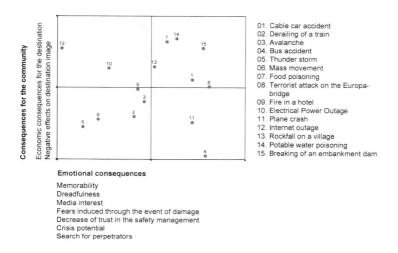

Fig. 6.4 Localisation of hypothetical damage events on the dimensions emotional consequences and consequences for the community

The factor labels *emotional consequences* and *consequences for the community* indicate that the factors extracted in the present study are different from those identified by Slovic, Fischhoff and Lichtenstein (1980) or Fischhoff, Slovic, Lichtensein, Read and Combs (1978).

Admittedly, the factor *emotional consequences* includes risk dimensions that are identical or comparable with those that shaped the "dread" or "severity" factor in previous studies, namely dreadfulness, crisis potential and fears induced through the event of damage. However, also the following risk characteristics, which were not considered in the risk perception studies mentioned above, show high factor loadings on the *emotional consequences* factor. These are memorability of the event of damage, media interest evoked by the event of damage, search for perpetrators evoked by the event of damage and decrease of trust in the safety management as a result of the event of damage. Due to the fact that all risk dimensions refer to emotional aspects and consequences of loss-incurring events and considering that the latter risk characteristics are not at all represented by the original factor labels dread risk or severity, the first factor was called *emotional consequences*. Damage events that received the highest ratings on the *emotional consequences* factor are "breaking of an embankment dam", "terrorist attack on the Europa-bridge", "bus accident", "plane crash" and "cable car accident". It is striking that with the exception of "bus accident" these events of damage all have a very low probability of occurrence. Insofar, the results are in line with Sandman's (1987) statement, according to which "the risks that kill you are not necessarily the risks that anger and frighten you".

In previous studies, characteristics referring to the scientific knowledge regarding a risk and the familiarity with it showed high loadings on the second factor. These dimensions were not included in the current study, which is why the second factor is completely different from the one identified by Slovic et al. (1980). The second factor consists of dimensions describing the consequences of loss-incurring events for the community, namely economic consequences for the destination and negative effects on destination image. Consequently, this factor was called *consequences for the community*.

Results of the factor analysis highlight the obvious fact that the nature of risks considered in the survey as well as the rating dimensions on which risks are assessed have a decisive influence on which factors are identified as the central ones for risk perception.

Risk management should give highest priority to damage events that show high ratings on both factors, i.e. on the *emotional consequences* and on the *consequences on a community level* factor. Such scenarios are the ones situated in the upper right quadrant of figure 6.4, namely "potable water poisoning", "food poisoning", "breaking of an embankment dam", "rock fall on a village", "cable car accident" and "terrorist attack on the Europa-bridge. Based on the interview findings as well as on the results of the online survey, these 'dreaded' risks are admittedly not those perceived to be typical or of high relevance for Tyrol.

Finally, findings of the psychometric study prove that the damage scenarios differ significantly from each other regarding their assessment on the nine rating dimensions. In other words, the hypothetical events of damage differ significantly in terms of their dreadfulness and memorability, regarding the media interest and the search for perpetrators they evoke. Also on the remaining dimensions, which are decrease of trust in the safety management, economic consequences for the destination, negative effects on destination image, fears induced through the event of damage and crisis potential, the damage scenarios receive significantly different ratings.

Regarding the dimension dreadfulness it is striking that events of damage with fatalities ("breaking of an embankment dam", "bus accident", "cable car accident") are seen as the most dreadful ones. This observation is in line with another result of the study that stresses the influence of the variable severity (fatalities/no fatalities) on perceived dreadfulness (see below). Scenarios that have been rated as high in terms of memorability ("terrorist attacks on the Europa-bridge", "breaking of an embankment dam") are consistently low probability – high consequence events. From this one could conclude that the more extraordinary loss-incurring events are, the higher their memorability is. Media interest is also judged to be high in the case of low probability – high consequence events ("terrorist attack on the Europa-bridge", "breaking of an embankment dam") and in the case of events of damage with fatalities ("cable car accident"). The high media interest with respect to these events might be due to the fact that the event characteristics unpredictability/surprise (low probability – high consequence events) and negativity (fatalities) are central news values influencing media interest (Galtung and Ruge 1965).

Generally, all of the 15 damage scenarios obtained comparably high ratings on the dimension media interest. This outcome corresponds with media reality, since it reflects the media's preference for bad news and the news value of negative events. In accordance with previous studies (Jungermann and Slovic 1993) and with other results from the present study, the search for perpetrators plays an inferior role in the case of natural hazards ("thunder storm", "mass movement", "avalanche", "rockfall on a village"). Decrease in trust in a destination's safety management is highest for "breaking of an embankment dam". Maybe safety expectancies towards technological protection measures and a false sense of security account for this finding. They are both impaired by the "breaking of an embankment dam".

Scenarios involving the most negative economic consequences and the most negative effects on destination image are according to the subjects' ratings "potable water poisoning", "food poisoning" and "breaking of an embankment dam". This might have to do with the fact that all of these events are likely to imply an at least temporal absence of tourists and consequently financial losses as well as a loss of reputation. Results further suggest that the most fear inducing loss-incurring events are "water poisoning", "food poisoning", and "terrorist attack on the Europa-bridge". Possibly these events of damage are perceived as particularly fear-provoking because they pose threats from which one would actually feel safe in Tyrol. If, in spite of this feeling of safety, such an unexpected event would really occur, the extent of fear will be even greater. Crisis potential is on the one hand perceived to be highest for extreme events of damage with a very low probability of occurrence ("terrorist attack on the Europabridge", "breaking of an embankment dam"). On the other hand, crisis potential is judged to be high for events that have a negative impact on recreational value because of illness ("water poisoning" and "food poisoning"). The highest crisis potential might have been attributed to these events because they promote effects that go beyond the event of damage itself, namely loss of reputation and the absence of tourists.

6.3.4 Experiment on Story Effects in damage perception

The influence of social context factors on damage perception was tested in two experimental studies based on the risk story approach (Wiedemann, Clauberg and Schütz 2003). In line with the risk story approach and with findings on framing effects (Tversky and Kahneman 1981; Sandman, Miller, Johnson and Weinstein 1993; Wang 1996; Levin, Schneider and Gaeth 1998) the slightly different description of one and the same loss-incurring event leads to differences in damage perception. In other words, the risk story approach postulates that damage perception is subject to story effects.

To test this hypothesis, three slightly differently framed damage scenarios were designed for each risk. The neutral version just provides information on what has happened and on who or what has suffered harm. Below, the neutral description or the baseline version for the scenario "avalanche accident" is cited exemplarily:

Avalanche Accident (Baseline/ Neutral Version)
After ongoing snowfall, an avalanche was set off last week in the mountain resort of Obergurgl, Tyrol. A house, in which a family of five lives as well as the two holiday apartments, where seven guests were staying, were buried under the snow. Two of the guests were killed.

The second damage story version (Framing Version I – mistakes and failures in risk management) contains, in addition to the baseline version, a further paragraph describing failures and mistakes in risk management. The objective information, that is information on what has happened and on who or what has suffered harm, remains the same.

Avalanche accident (Framing Version I – Failures and Mistakes in Risk Management

After ongoing snowfall, an avalanche was set off last week in the mountain resort of Obergurgl, Tyrol. A house in which a family of five lives as well as the two holiday apartments, where seven guests were staying, were buried under the snow. Two of the guests were killed. **Despite early warnings from weather experts, the occupants were not evacuated.**

Framing Version II still goes a step further and additionally talks about failures and mistakes not only in risk but also in crisis management.

Avalanche accident – (Framing Version II – Failures and Mistakes in Risk and Emergency Management)

After ongoing snowfall, an avalanche was set off last week in the mountain resort of Obergurgl, Tyrol. A house in which a family of five lives as well as the two holiday apartments, where seven guests were staying, were buried under the snow. Two of the guests were killed. Despite early warnings from weather experts, the occupants were not evacuated. **Complications during the rescue operation occurred, because responsibilities of the rescue services were not clear.**

As the three story versions for an avalanche accident demonstrate, the event of damage as such (avalanche accident) as well as its consequences (2 fatalities) remains the same across all the three scenarios. Only social context factors, namely mistakes and failures in risk management (Framing Version I) and failures and mistakes in emergency management (Framing Version II) are varied, to be able to analyze their influence on damage perception.

Story effects were analyzed by means of a one-way analysis of variance. Results prove that in the baseline version (M = 3.88) the extent of outrage evoked by the events of damage is perceived to be significantly lower (F = 12.48**) than in framing version I (M = 5.12) and framing version II (M = 5.50). Hence, in accordance with the assumptions of the risk story approach (Wiedemann, Clauberg and Schütz 2003), the description of mistakes and failures in risk and emergency management does indeed have an impact on the extent of outrage. Moreover, study findings provide evidence for story effects with respect to trust in the destination's safety management. Trust in destination safety was significantly lower (F = 13.65**) in framing version I (M = 3.48) and framing version II (M = 2.88) as compared to the baseline version (M = 4.23). Thus, trust after a loss-incurring event is eroded to a higher extent, if failures and mistakes in risk and crisis management are perceived.

These results suggest that perceived mistakes and failures in risk and emergency management have the potential to alter the appraisal of loss-incurring events, even though the objective extent of damage remains the same.

6.4 Summary and outlook

Taken together, study findings suggest that future tourism risk perception studies should not only examine the typicality or the existence or absence of holiday risks at a destination, but also consider psychometric risk characteristics. The latter should be taken into account, since a comparison of the interview results and the results of the online questionnaire with those of the psychometric study reveals that other risks are perceived to be of importance, if risk assessment is based on psychometric characteristics. In the interviews and in the online questionnaire, participants had to judge selected risks regarding their typicality for the tourist destination Tyrol and regarding their existence within the destination, respectively. Based on this procedure, winter sport related risks and natural hazards proved to be the most significant ones. If, in contrast, risks are assessed on psychometric characteristics, other risks – namely food risks (water poisoning, food poisoning) and industrial risks (breaking of an embankment dam) – prove to be of highest relevance.

Findings of the psychometric survey imply that a destination's safety management should not restrict itself to the management of winter sport related risks and natural hazards, but additionally should concentrate on "dreaded" risks. Managing "dreaded" risks, even if they do not typically occur in Tyrol, is of decisive importance, because if such dreaded risks result in events of damage, emotional reactions and public outrage are supposed to be high.

Safety measures and conditions pertaining to safety on ski slopes and protection against natural hazards have been found to have the highest impact on trust in destination safety in the case of their provision and on mistrust in the case of their absence. This result of the online survey also has implications for risk management. It highlights the necessity of putting special emphasis on such safety measures and conditions. Furthermore, results of the online survey prove that the asymmetry principle of trust (Slovic, 1993) does not hold true in a tourism context.

As perceived mistakes and failures in risk and emergency management induce outrage and negatively impact trust in destination safety, findings from the experiments stress the importance of being prepared to cope with potential loss-incurring events.

The reported results, especially those of the psychometric study and the experiments, indicate that the consideration of psychological aspects of risk and damage perception makes a valuable contribution to a merely quantitative risk assessment. Through the additional consideration of psychological characteristics other, new risks are attracting attention and can thus be identified. An additional advantage is the focus on the social context of risks, on aspects that often only become apparent after the occurrence of a negative event: human reactions, reaction of the public to risks or loss-incurring events. These reactions influence on the one hand the development and intensity of public risk debates. On the other hand, they determine whether an event of damage remains an event of damage or develops further into a crisis.

In the article title the question was raised if alpine destinations are safe havens in turbulent times? At first glance this appears to be the case. There are only a few risks that are perceived to be typical for the destination Tyrol and those risks that are crucial with respect to destination choice are perceived to be of no or only low relevance for the destination Tyrol. However, this picture changes drastically after the occurrence of an event of damage. If an event of damage and the underlying alpine risks are interpreted by the public as man made, especially in the case of failures and mistakes in risk and crisis management, then a tourism community is under pressure and faces the risk of plunging into a crisis.

References

Banse G (1996) Risikoforschung zwischen Disziplinarität und Interdisziplinarität – Von der Illusion der Sicherheit zum Umgang mit Unsicherheit. Berlin, Ed. Sigma

Bauer RA (1960) Consumer behavior as risk taking. In: RS Hancock (Ed.) Dynamic marketing for a changing world. Chicago: American Marketing Association, pp 389-398

Beck U (2007) Weltrisikogesellschaft. Auf der Suche nach der verlorenen Sicherheit. Frankfurt a. M., Suhrkamp

Bentley TA, Page SJ (2006) Tourist Injury. In: Wilks J Pendergast D, Leggat P (Hrsg.): Tourism in Turbulent Times- Towards Safe Experiences for Visitors. Advances in Tourism Research Series. Elsevier Ltd. Oxford, pp 155-168

Cheron E, Ritchie JRB (1982): Leisure Activities and Perceived Risk. *Journal of Leisure Research,* Vol. 2, No.4, pp 139-154

Coxon APM (1999) Sorting Data: Collection and Analysis – Quantitative Applications in the Social Sciences. Thousand Oaks: Sage Publications

Douglas M, Wildavsky A (1982) Risk and Culture. Berkely, CA: University of California

Dunwoody S (1992) The media and public perceptions of risk: How journalists frame risk stories. In: Bromley DW, Segerson K (Eds.) The social response to environmental risk. Boston: Kluwer

Finucane ML, Alhakami A, Slovic P, Johnson SM (2000) The affect heuristic in judgements of risks and benefits. Journal of Behavioral Decision Making, Vol. 13, No.1, pp 1-17

Fischhoff B, Slovic P, Lichtenstein S, Read S, Combs B (1978) How Safe is Safe enough? A Psychometric Study of Attitudes towards Technological Risks and Benefits. Policy Sciences, Vol. 29, No.9, pp 127-152

Fuchs M, Peters M (2005): Die Bedeutung von Schutz und Sicherheit im Tourismus: Implikationen für alpine Destinationen. In Pechlaner H, Glaeßer D (Hrsg.) (2005): Risiko und Gefahr im Tourismus – Erfolgreicher Umgang mit Krisen und Strukturbrüchen. Erich Schmidt Verlag, Berlin, pp 155-171

Galtung J, Ruge MH (1965) The Structure of Foreign News. Journal of Peace Research, Vol.2, No.1, pp 64-91.

Gardner GT, Gould LC (1989) Public Perceptions of the Risks and Benefits of Technology. Risk Analysis, Vol. 9, No.2, pp 225-242

George R (2003) Tourists' Perceptions of Safety and Security while visiting Cape Town. Tourism Management, Vol. 24, No.5, pp 575-585

Holzheu F, Wiedemann PM (1993) Perspektiven der Risikowahrnehmung. In Bayerische Rück (Hrsg.): Risiko ist ein Konstrukt – Wahrnehmungen zur Risikowahrnehmung. Knesebeck GmbH, Co. Verlags KG, München, pp 9-19

Hunter-Jones J (2000) Identifying the responsibility for risk at tourism destinations: the UK experience. Tourism Economics, Vol.6, No.2, pp 187-198

Ioannides D, Apostolopoulos Y (1999) Political Instability, War and Tourism in Cyprus: Effects, Managements and Prospects for Recovery. Journal of Travel Research, Vol 38, No. 1, pp 51-56

Jungermann H, Slovic P (1993) Charakteristika individueller Risikowahrnehmung. In Bayerische Rück (Hrsg.): Risiko ist ein Konstrukt – Wahrnehmungen zur Risikowahrnehmung. Knesebeck GmbH, Co. Verlags KG, München, pp 89-107

Kahneman D Slovic P, Twersky A (Eds.) (1982) Judgement under Uncertainty: Heuristics and Biases. Cambridge University Press

Lepp A, Gibson H (2003) Tourist Roles, Perceived Risk and International Tourism. Annals of Tourism Research, Vol.30, No.3, pp 606-624

Levin IP, Schneider SL, Gaeth GJ (1998) All Frames are not Created Equal: A Typology and Critical Analysis of Framing Effects. Organizational Behavior and Human Decision Processes, Vol. 76, No.2, Nov., pp 149-188

Loewenstein GF, Weber EU, Hsee CK, Welch N. (2001) Risk as feelings. Psychological Bulletin, Vol.127, pp 267-286

Luhmann N (1991) Soziologie des Risikos. Walter de Gruyter: Berlin/New York

Luhmann N (1993) Risiko und Gefahr. In Krohn W.G (Hrsg.): Riskante Technologien: Reflexion und Regulation. Einführung in die sozialwissenschaftliche Risikoforschung, Suhrkamp: Frankfurt. a. M, pp 138-185

Mawby RI (2000) Tourists' perception of security: the risk fear paradox. Tourism Economics, Vol. 6, No. 2. pp 109-121

Mitchell VW, Vassos V (1997) Perceived Risk and Risk Reduction in Holiday Purchases: A Cross-Cultural and Gender Analysis. Journal of Euromarketing, 6/3, pp 47-77

Page SJ, Bentley TA, Meyer D (2003) Evaluating the nature, scope and extent of tourist accidents – The New Zealand experience. In Wilks J., Page S.J. (Eds.) Managing Tourist Health and Safety in the New Millenium. Pergamon, Oxford

Page SJ, Meyer D (1996) Tourist accidents: An exploratory analysis. Annals of Tourism Research, Vol. 23, No.3, pp 666-690

Palmlund I (1992) Social Drama and risk evaluation. In: Krimsky S., Golding D (Hrsg.) Social theories of risk. Westport, CT: Praeger

Phillip R, Hodgkinson G (1994) The management of health and safety hazards in tourist resorts. International Journal of Occupational Medicine and Environmental Health, Vol.7, pp 207-219

Pikkemaat B, Weiermair K (2003) Safety and Security Issues – From a Tourist Destination Perspective. In Weber S, Tomljenovic R (Eds.): Reinventing a Tourism Destination. Scientific Edition Institute for Tourism: Zagreb, pp 271-281

Pizam A, Tarlow PE, Bloom J (1997) Making Tourists Feel Safe: Whose Responsibility Is It? *Journal of Travel Research*, Vol.36, No.1, pp 23-28.

Renn O (1992) Concepts of Risk: A Classification. In Krimsky S., Golding D (Hrsg.): Social Theories of Risk. Praeger: Westport/ London

Roehl WS, Fesenmaier DR (1992) Risk Perceptions and Pleasure Travel: An Exploratory Analysis. Journal of Travel Research, Vol. 2, pp 17-26

Rohrmann B. (1999) Risk Perception Research. Review and Documentation. Arbeiten zur Risikokommunikation, Volume 69, Forschungszentrum Jülich

Sandman P (1987) Risk Communication: Facing Public Outrage. *EPA Journal* 13, (9), pp 21-22

Sandman P, Miller PM, Johnson BB, Weinstein ND (1993) Agency communication, community outrage, and perception of risk: Three simulation experiments. Risk Analysis, Vol. 13, pp 585-598

Seddighi HR, Nutall MW, Theocharous, AL (2001) Does cultural background of tourists influence the destination choice? An empirical study with special reference to political instability. Tourism Management, Vol. 22, pp 181-191

Simon HA (1955) A behavioral model of rational choice. Quarterly Journal of Economics, Vol.69, pp 99-118
Sjöberg L (2000) Factors in Risk Perception. Risk Analysis, Vol.20 No.1,
Sjöberg L, Moen BE (2004) Explaining risk perception. An evaluation of the psychometric paradigm in risk perception research. Torbjorn Rundmo: Rotunde, No. 84, pp 1-12
Slovic P (1962) Convergent validation of risk-taking measures. Journal of Abnormal and Social Psychology, Vol.65, pp 68-71
Slovic P (1992) Perception of risk: Reflections on the psychometric paradigm. In Krimsky S, Golding D (Eds.): Social theories of risk. New York: Praeger, pp 117-152
Slovic (1993) Perceived Risk, Trust, and Democracy. Risk Analysis, Vol. 13, No. 6, pp 675-682
Slovic P (1999) Trust, Emotion, Sex, Politics, and Science: Surveying the Risk-Assessment Battlefield. Risk Analysis, Vol.19, No.4. pp 689-701.
Slovic P, Finucane ML, Peters E, Mac Gregor DG (2002) The affect heuristic. In: Gilovich T, Griffin, D, Kahneman D (Eds.): Heuristics and Biases. The Psychology of Intuitive Judgement. Cambridge University Press, pp 397-420.
Slovic P, Fischhoff B, Lichtenstein S (1977) Cognitive Processes and Societal Risk Taking. Reprinted (2003) in Slovic (Ed.) The Perception of Risk. Earthscan Publications Ltd., London
Slovic P, Fischhoff B, Lichtenstein S (1980) Facts and Fears: Understanding Perceived Risk. Societal Risk Assessment: How Safe is Safe enough?. New York, Plenum, pp 181-214
Slovic P, Kunreuther HC, White GF (1974) Decision processes, rationality and adjustment to natural hazards. In Natural Hazards: Local, National, Global. New York: Oxford University Press
Sönmez S, Graefe A (1998a) Determining Future Travel Behavior from Past Travel Experience and Perceptions of Risk and Safety. Journal of Travel Research, Vol.37, No.2, pp 171-177.
Sönmez S, Graefe A (1998b) Influence of Terrorism Risk on Foreign Tourism Decisions. Annals of Tourism Research, Vol. 25, No.1, pp 112-144
Starr C (1969) Social benefit versus technological risk. Science, 165, pp 1232-1238
Thompson M, Ellis R, Wildavsky A (1990) Cultural theory. Boulder, CO: Westview PressTversky A, Kahneman D (1981) The framing of decisions and the psychology of choice. Science, 211, pp 453-458
Wang XT (1996) Framing Effects: Dynamics and Task Domains. Organizational Behavior and Human Decision Processes, Vol.68, No.2, Nov., pp 145-157
Wiedemann PM, Clauberg M, Schütz H (2003) Understanding amplification of complex risk issues: The risk story model applied to the EMF case. In: Pidgeon N, Kasperson R, Slovic P (Eds.): The social amplification of risk. New York: Cambridge University Press, pp 286-301
Wilks J, Page SJ (2003): Managing Tourist Health and Safety in the New Millennium. Pergamon

7 Protective measures against natural hazards – are they worth their costs?

A. Leiter, M. Thöni, H. Weck-Hannemann

7.1 Introduction

Natural hazards such as debris flows, avalanches and floods can lead to intense damage when they hit society and its values. This social vulnerability is observable on different levels and ranges from human live and human health to buildings, settlements or infrastructure and from the loss of production in industry and tourism to restricted social mobility.

The widely acknowledged diagnosis of an increasing social exposure to natural hazards asks for an integral risk management that includes instruments for reducing, preventing and transferring risks. When talking about risk in the area of socio economic research, it can be defined as the product of the possible negative consequences for society and the probability of being affected. One concept that is able to meet these requirements can be found in the integral risk management approach. This approach combines permanent and temporary as well as active and passive protective measures. Accordingly, active strategies aim to prevent the risks at the origin, which means to reduce natural hazards from their probable occurrence (e.g. structural measures) whereas passive ones accept natural hazards as such and introduce prevention measures to decrease the damage potential (organisational measures such as evacuation). Adding the time dimension to this concept leads to permanent and temporary protection. Permanent strategies include measures with long dated impacts, such as land use planning activities, whereas temporary ones try to protect in the short run (e.g. early elimination of avalanches). Concluding, it can be stated that this integral concept affects natural processes as well as social values.[1]

Since social vulnerability can be seen as the centre of interests in natural hazard management, the described alternatives and following up the possible protection output should be compared. The aim is to provide an opti-

---

[1] See Kienholz (1994, 2004), PLANAT (1998), Heinimann et al. (1998), Wilhelm (1999), Ammann (2003) and Stötter et al. (2004). A comprehensive discussion of the term "risk" and the risk concept can be found in e.g. Fuchs (2004).

mal protection level for society and its values based on social acceptability. From an economic perspective this social acceptability can be stated by finding an answer to the question: Which mitigation measures are best suited to reduce collective risk in an adequate way? Answering this question identifies two main directions: the first one outlines the collective risk reduction, the societal level that reflects social preferences towards mitigation measures and therefore the demand side. The second one refers to the cost side of mitigation measures and reflects the supply within natural hazard management.

In order to prepare an answer toward this question, different decision support instruments like Cost Benefit Analysis (CBA), Cost Effectiveness Analysis (CEA), Multi Criteria Analysis (MCA) or Cost Utility Analysis (CUA) can be used. All these instruments are able to create a sound basis for decisions in natural hazard or risk management, nevertheless two different categories can be defined. Whereas MCA, CEA and CUA promote a cost efficient outcome by comparing costs on a monetary level with benefits on a "physical level", like saved lives, CBA works on both sides – costs and benefits – on a monetary level and therefore leads to an economic efficient outcome. To put more emphasis on this fundamental difference one instrument of each category was chosen, namely CEA and CBA, to allow an extended discussion.[2]

How these two instruments (CEA and CBA) work and how they can contribute to the integral risk management approach will be subject of section two of this chapter. This is followed by a conceptual discussion about the instrumental boundaries and possibilities, which will lead to the crucial difference between the two approaches: the monetary valuation of benefits. When talking about the evaluation of benefits on a monetary level, most discussions in the area of natural hazard management lead to one particular question: Can the protection of human life be monetarily assessed? Since the protection of human life is seen as a central benefit of protective measures it seams to be particularly interesting to focus on this topic. For this reason the third section of this paper outlines the main characteristics of monetary valuation of human life. In this context a study is presented that gives an example of a valuation method (contingent valuation method, CVM) which is able to examine a societal value for human life. Therewith, this chapter tries to draw the bow from the conceptual discussion to the effective implementation of economic values in decision support systems like CBA.

---

[2] For a comprehensive discussion concerning decision support systems within natural hazard management see Gamper et al. (2006).

7.2 Alternative Decision Support Instruments

7.2.1 Cost Benefit Analysis and Cost Effectiveness Analysis

Both methods, CBA and CEA serve as decision support instruments to evaluate either public or private projects based on the emerging costs and benefits. However, the comparison of costs and benefits is implemented differently in each instrument: in the case of CEA the costs are monetarily evaluated and are compared to benefits which are measured in non-monetary "real" values. In other words, negative impacts (costs) of a project are evaluated in monetary terms, such as material costs or wage costs whereas the positive impacts (benefits) such as saved human lives are physically quantified. Consequently, CEA is focusing on a cost efficient outcome, which makes use of restricted resources in order to achieve the envisioned objective in the best possible way and at lowest cost. The decision criteria of cost efficiency can be understood as an economic principle of managerial rentability and profitability which corresponds to the economic perspective of optimizing the supply side.

In contrast to CEA, CBA goes one step further and evaluates both indicators – costs and benefits – monetarily. This does not only lead to a cost efficient outcome but at large to economic (or allocative) efficiency. Allocative efficiency can be defined as the societal optimal outcome that is reachable when cost efficient production on the supply side (managerial optimum) meets the effective demand of individuals or society (economic optimum). Thus, economic efficiency is concerned with both, the production costs of suppliers and the resulting benefits on the demand side. It ensures that the output is produced at minimal costs whilst at the margin the consumer's willingness to pay just equals the cost of production.

In summary it is notable that CEA simply offers information about whether one project is more cost efficient than another, but it does not provide details if a project could offer a contribution to the increase of (social) welfare.[3] On contrast, the welfare valuation is explicitly included in the CBA as a central component and results in this analysis being one of the most highly developed decision support instruments within economic theory and welfare economics. Mishan (1998) summarises this differentiation

---

[3] See Groot et al. (2004).

as follows:[4] "... To be rather rude about it, the analysis of cost effectiveness can be described as a truncated form of cost benefit analysis...".

Based on this assessment this paper will focus on the discussion of the CBA to position this widely recognized instrument within natural hazard management and to explore its potential.[5]

7.2.2 Cost Benefit Analysis - a conceptual approach

Since the comparison of impacts of protective measures is one of the main focuses of socio economic research within natural hazard management, it is of great relevance to underline the conceptual approach of CBA. In order to be able to evaluate different alternatives CBA should fulfil the following sub tasks (figure 7.1).[6]

The first step is to define the project (e.g. maximizing social welfare by implementing protective measures against natural hazards and related risks). Therefore, the objectives and boundaries of the project should be determined both spatially and institutionally.[7] This definition includes two separate steps: on the one hand, the exact delimitation of the project, with which the content and territory related parameters are put in place. On the other hand, it is necessary to establish the group of affected people, whereby it is important to resolve the winners and losers. Within these boundaries project alternatives can be determined.

With regard to the delimitation of the project, distinctions are primarily based on whether the analysis is being made for the decision support of an individual or a private company (private sector CBA) or for a public institution such as a regional administrative body (community, region, state or federation) or a inter- or supranational organisation (social CBA). This classification results in a fundamental deviation with regard to the assessment criterions: whilst the decision of an individual is ruled by maximizing the own benefit, the private company focuses on the maximisation of profits. Considering society, it is imperative to maximise the sum of benefits on an aggregated societal level. This societal level can also be influenced

---

[4] Under the term "Cost Effectiveness Analysis" many such as Mishan (1998) understand that both the Cost Effectiveness (CEA) as well as Cost Utility Analysis (CUA) is subsumed.
[5] For a detailed comparison of the Cost Effectiveness and Cost Benefit Analysis see Weck-Hannemann and Thöni (2006).
[6] For a detailed survey of the steps compare Gamper et al. (2006).
[7] For the standard approach used in a Cost Benefit Analysis see amongst others Hanley and Spash (1993), Hackl and Pruckner (1994), Boardman et al. (2001), Tietenberg (2004) and Hansjürgens (2004).

by the consideration of spatial-institutional differentiation: it is necessary to establish, whether CBA will be used for a community or a region, or whether it will be more generally focused for the use of public funds on either a state, national or supranational level.

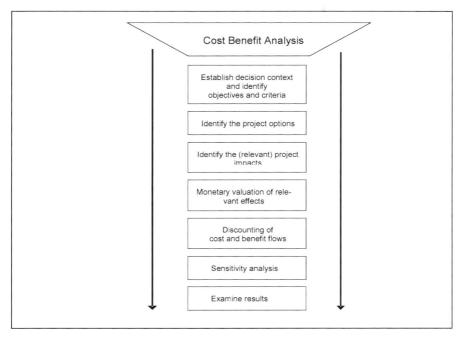

Fig. 7.1 The process of decision making with CBA

To achieve these above mentioned and occasionally diverse objectives alternative (project) options need to be identified and developed in a second step.

In a following third step it is necessary to qualify and filter out the project impacts that result from the alternative options and the connected economically relevant effects (e.g. resource use, occupational effects, income effects or the impact on transportation). As a matter of principle a distinction is made between positive and negative effects. Positive effects arise from an improved provision with consumer goods and services and lead to an increase in satisfaction in the sense of an increase in welfare. The negative effects of a project result from the suppression of alternative public or private activities that in turn would have led to an increase in welfare. Exemplarily project impacts can be real (technical) and pecuniary, direct and indirect or tangible and intangible.

In economic terms negative and positive impacts are understood as costs and benefits, respectively. The following table (table 7.1) exemplifies possible costs and benefits arising from the implementation of an alpine protective measure.

Table 7.1 Possible costs and benefits resulting from protective measures

Costs	Benefits
- building costs (material, explosives, personnel costs) - maintenance work - evacuation costs - change in landscape	- prevented material damage - modified possibilities of land use - prevention of injuries - protection of human life/ reduction of fatalities

However, in order to be considered as a relevant impact within CBA, there has to be a cause-and-effect relationship between some physical outcome of the project and utility of human beings notwithstanding.[8] A positive (negative) impact or benefit (cost) is either an increase (decrease) in the quantity or quality of goods that generate positive utility or a reduction (increase) in the price at which they are supplied. Thereby it is not relevant whether a market for the specific good exists or not. The priceless impacts are referred to as externalities which have to be included in a social CBA but not necessarily in a private sector CBA. On the other hand, transfer payments should be excluded from a social CBA as they constitute no using-up of real resources but are merely redistributions between individual members of society.

A fourth step in the process of a CBA is to physically quantify these relevant effects (impacts) and to transform them into costs and benefits representing an increase or decrease in social welfare. As previously mentioned, the Cost Benefit Analysis is characterised through the quantification of all occurring costs and benefits on a monetary level, through which the physical effects are transformed into a homogenous unit and are consequently also made comparable.

Primarily, these monetary values can be defined via market prices. But how should one proceed when such information is not available, e.g. when there is no market for goods such as landscape, ecology, recreation or human life? Basically there are a number of methods available in economics

---

[8] See Boardman et al. (2001) and Hanley and Spash (1993).

to assess such non-market goods monetarily. Either a specially developed questioning technique (CVM) is used to directly evaluate the appraisal of public goods or one uses the indirect methods of observing the individual behaviour on markets for substitutive or complementary goods (e.g. travel cost method or hedonic pricing).[9] The valuation of non-market goods does not necessarily have to arise from the benefit that is directly generated through consumption (use values), but can also include optional values, as well as the existence and legacy values of a good (non use values).

With regard to the temporal dimension not only the direct impacts of the project should be taken into account, but also the possible following costs (such as the maintenance of technical facilities) as well as future incomes. To be able to compare these temporally different accumulative costs and benefits, they must be accordingly homogenised. The temporal discounting can occur for private sector projects on the basis of market interest rates, which can also be used as a reference for the social discount rate if the capital market is functioning.[10]

Insecurities with regard to the expected events and the estimation of the involved damage variables can be depicted through the calculation of the according safety equivalent. In the case of risk compensation offered by an insurance company the corresponding insurance premiums can be used as a point of reference. In addition, it is possible to make a rough estimate of the effects of critical assumptions through sensitivity analyses. Since CBA is an instrument used ex ante it enables sensitivity analyses to discover which parameters (eg. discount rate, physical quantities) influence the final outcome. This procedure supports to test for the robustness of the findings and therefore helps to reduce one source of uncertainty..

The comparison of the alternative costs and benefits makes it possible to arrange the different options according to their welfare effects. Therefore the CBA results in a hierarchy which assesses the measures depending on their allocative efficiency and consequently also according to their socially optimal impact.

Results stemming from CBA allow to answer two questions: the first criterion demands that the (present value of) benefits (B or PV(B)) of a project prevail over the (present value of) costs (C or PV(C)). Those projects which display a positive net benefit (or a net present value NPV for

---

[9] An overview of the different valuation methods can be found amongst others in Pommerehne (1987), Hanley and Spash (1993), Hackl and Pruckner (1994) and Boardman et al (2001). – For further information on this topic compare section 7.3 of this chapter.

[10] To which extent the market interest rate is applicable as a social discount rate is discussed in e.g. Hanley and Spash (1993, chapter 8).

discounted benefits and costs) can be seen as economically profitable, in other words, through their realisation the welfare of society as a whole increases (Pareto-improvement) (see equation 7.1). To distinguish which project, under consideration of all the relevant alternatives, promises the highest increase in welfare, a second criterion must be met: the project that proves itself as the optimal strategy is the one out of the available options that maximizes the expected net benefit (see equation 7.2).

Cost Benefit Criterion I: Project acceptance if $NPV > 0$ (eq. 7.1)

with $NPV = PV(B) - PV(C)$

and $PV(B) = \sum_{t=0}^{n} \frac{B_t}{(1+s)^t}$ $PV(C) = \sum_{t=0}^{n} \frac{C_t}{(1+s)^t}$

Cost Benefit Criterion II: Optimal choice if $NPV=max$ (eq. 7.2)

with $\max NPV = \max[PV(B) - PV(C)]$

i.e. $\max NPV > NPV_1 > NPV_2 > NPV_i$

where s, t, i represent the interest rate, time indicator and available altenative, respectively.

The CBA consequently provides an answer to (a) whether a project should be realised at all, that means whether it is profitable from an economic point of view and (b) which project out of all the relevant alternatives is the optimal or Pareto-efficient decision in the sense that there are no other projects which lead to a higher net benefit or a project with which even one person is in a better position without someone else as a result being worse off.[11]

To clarify these considerations the illustration in the following graph (figure 7.2) has been divided into two parts: the above graph (figure 7.2a) represents the (present value of) total costs PV(C) and the (present value of) total benefits PV(B) (e.g. of one protective measure) and in the lower

---

[11] See Hackl and Pruckner (1994).

graph (figure 7.2b) the corresponding marginal costs MC and marginal benefits MB are shown.

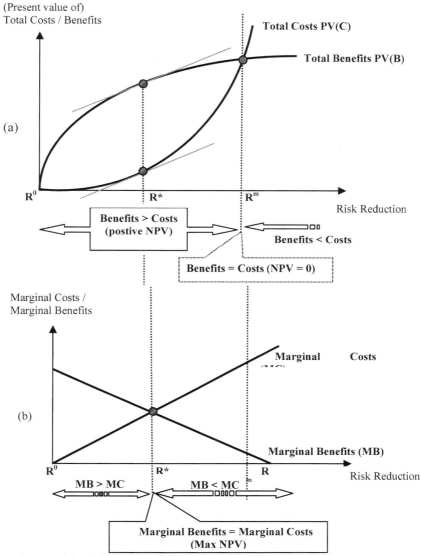

Fig. 7.2a and 7.2b: Cost Benefit Criteria I and II

The curves correspond with the assumptions of positive and increasing marginal costs as well as positive and decreasing marginal benefits which are assumptions commonly accepted in economic theory. According to the previously mentioned criterion I, all projects in which the total benefits exceed the total costs are economically profitable. Thus, all the projects between R^0 and R^m are economically justifiable seeing as up until the point of interception both total curves result in a positive net benefit or a positive net present value, respectively (NPV = PV(B) − PV(C) > 0 or PV(B) > PV(C)).

The optimal strategy out of all these economically profitable alternatives can be found by implementing the before mentioned criterion II. Consequently, out of all the possible project variants the optimal alternative is the one which has a maximum (positive) net benefit. This optimum can be found at the point of interception R* of marginal costs and marginal benefits (MC = MB). Formally this result can be derived from the optimality criterion (max NPV):

$$\text{Max NPV} = \max (PV(B) - PV(C)) \Rightarrow MB - MC = 0 \Rightarrow MB = MC.$$

In summary it is essential for the derivation of the optimum that all the relevant alternatives are taken into consideration and can be traded off against each other. Furthermore, all the relevant costs and benefits with regard to the examined objective should be integrated. Thus, all the internal costs and benefits have to be considered in a private sector CBA, but all the internal as well as the external costs and benefits have to be taken into account from a social perspective (social CBA). It is particularly difficult to depict the external effects monetarily due to the lack of market data. However, as previously mentioned it is possible to make use of innovative direct and indirect methods in order to acquire an estimation of the external effects of goods. This will be discussed in more detail in the following section 7.3.

In this context, however, it should be mentioned that it would be misleading to represent the benefit of a protective measure through the cost estimation of its implementation. The benefits of a protective measure should rather be determined through the valuation of the prevented damages. Costs and benefits have to be strictly separated and collected independently from each other.

7.3 The Valuation of Benefits

The monetary valuation of cost and benefit components associated with the realization of a project is a basic prerequisite for conducting a CBA. Only when one succeeds in expressing all relevant advantages and disadvantages of a project in one dimension (e.g. monetary value), it is then possible to compare these factors in the CBA. Whilst the good and service market offers information about the monetary value for some components (e.g. cost of protective barriers against natural hazards, prevention of damage to buildings), no market information exists for other variables (e.g. prevention of damages to the landscape, protection of human life). If the latter category is disregarded this leads to an incomplete and distorted evaluation of the consequences of a project.

The identification of the benefits together with the valuation of goods and services that belong to the group of non-market goods therefore requires special methods. The available procedures are used to determine the subjective valuation for a certain commodity or a certain service of involved individuals. This is done using the concept of willingness to pay (WTP) that examines how much each person would be willing to pay for the utilisation of specific services or goods. By means of WTP it is possible to estimate the benefits resulting from consumption. Thus, the valuation of the benefit in monetary units is made possible.

As mentioned before, there are direct and indirect methods available which are frequently used to calculate the WTP for public goods (e.g. travel cost method, hedonic pricing and contingent valuation method).[12] These methods are relevant when variables such as the protection of landscapes, animal or plant types need to be valuated. Regions that are used by humans as recreational areas are also an important component of the benefit valuation. By implementing protective measures these areas, which could otherwise only be used under high risk (e.g. skiing areas, hiking trails), can be secured.

A further beneficial component of protective measures, which is often not considered due to its challenging valuation, is the protection of human life. For this reason the focus of the following section is placed on this factor. The study of topical literature has shown that this central category of potential damages through (alpine) natural disasters is often factored out of economic valuations. Thus, this important output of protective measures, the protection of human life, is not or only marginally included in cost benefit considerations. If, however, the protection of human life is not included then this benefit implicitly receives a value of zero.

In many studies on the efficiency of measures against alpine natural hazards the particular advantage of protective measures is "only" portrayed

---

[12] The implications of each respective method, their advantages and disadvantages as well as their applicability in natural hazard areas are analysed in Leiter (2004); for a general discussion see also Hanley and Spash (1993).

in physical values, quoting the number of lives saved. Only few[13] assign a monetary value to saved human lives using the human capital approach. This approach, however, leads to problematic implications with regard to certain groups of people (e.g. unemployed, retirees).[14] An estimation of the number of protected human lives is not sufficient when the related benefit should be included in a CBA. Solely monetary values are taken into consideration in such a statement. Consequently, the question emerges how the protection of life should be valued monetarily. Is life not unpayable? Thus, an infinitely high amount should be appointed to human lives, which is not useful either for its inclusion as a benefit component.

One instrument that allows measuring the benefit of life saving investments monetarily is the "value of statistical life" (VSL). The VSL concept works with an observable phenomenon, namely that individuals accept certain fatal risks for certain advantages. For example: people do not nourish themselves properly, do not exercise enough, smoke even though they are aware of the negative effects on their health; employees accept risky work conditions for a higher wage; individuals take excessive speed to save time and money. These examples show that people are quite willing to take risks in return for advantages or money. The VSL makes use of these considerations: it mirrors that rate at which individuals themselves are willing to exchange income for risk changes keeping their benefit level constant. Information about the individual WTP for a risk reduction is obtainable from the job market (hedonic pricing) or through surveys (CVM).

The field of application for the VSL is many-sided. Amongst others, the VSL is used to evaluate the prevention or reduction of workplace risks, transportation risks, pollutant risks, natural risks or nutritional risks. As can be seen in several studies[15] not every type of risk is judged in the same way. Moreover, regional, cultural and individual characteristics can influence the WTP considerably. Consequently, the most exact information about the individual valuation of fatal accidents that were prevented through specific protective measures can be obtained through primary data collection which enables to consider the specific characteristics of the region which is being investigated. Such a study was conducted within the framework of the project "Socio-economic evaluation of protective measures" at the Centre for Natural Hazard Management - alpS. This primary survey is an example of evaluating non market goods: the prevention of life threatening incidents. The survey design will be described in more detail in the following section.

---

[13] See, e.g., Wilhelm (1997) and Fuchs (2005).
[14] For a critical discussion see Leiter (2004).
[15] See, e.g., Slovic et al. (2000) and Sunstein (2004).

7.3.1 Survey design – primary data collection

The study was conducted in Tyrol – based on the CVM – and determined the individual WTP for reduced avalanche risks. In the context of the survey people were asked about their WTP to prevent fatal avalanche accidents. The survey took place in September/October 2004 as well as February 2005. In autumn 992 people and in winter 1,005 people from all areas of Tyrol were interviewed. Additionally, the questionnaire collected information on attitudes towards risks, risk behaviour, risk perception, as well as socio demographic characteristics (e.g. sex, age, education) of each individual.

The central question focussed on the investigation of the WTP for preventing an increase in mortality risk. The respondents were given an initial risk level, which referred to the average number of fatal avalanche accidents over a ten-year period. As a next step, it was assumed that this level of risk would increase if routine maintenance work stopped being undertaken. The interviewed people were then asked whether they would be willing to pay a certain amount in Euros every month to ensure the upkeep of the hitherto existing protection. Three different initial values of € 2.5/5/10 were given to each person, whereby these were randomly assigned to each interviewee. Depending on the initial answer it was surveyed whether the interviewee would be willing to also pay € 5/10/20 (if the initial answer was positive) or at least € 1.3/2.5/5 (if the initial question was answered with "no").[16] If the second question was also answered with a no, then the interviewer would inquire whether a payment would generally be rejected, and if yes, for which reasons.

This type of questioning belongs to the closed question format, which is preferred[17] because it mirrors consumption decisions in practice (e.g. deciding between making a purchase or not) and as a result provides the interviewee with a common and known situation. In addition it has been argued that this type of questioning reduces strategic behaviour.[18] Using this format yields a range of payment offers which depict the dependent variable in the estimation model.

---

[16] Based on the dual question and the restriction of the answer possibilities to yes/no this type of questioning is called "double bounded dichotomous question format".

[17] See Arrow et al. (1993).

[18] The authors also mention that besides strategic intention there are other reasons for not expressing the true WTP (e.g. the need to give a "good" answer or to fulfil the interviewer's expectations).

Given that different studies[19] document that the communication of small risk variations cause problems in understanding or that it can be difficult for some individuals to envisage the good in question, the risk change was depicted in a logarithmic chart (see figure 7.3).[20] On the right hand side of the diagram information is given about other mortality rates of the Tyrolean population. On the left, human populations of different sizes are depicted to give a better idea of the number of affected people.

The majority of the interviewees were asked to evaluate the prevention of the doubling of the initial risk, while a third of the winter sample (333 interviewees) referred their answers to a quadruplication of the baseline risk. This subdivision of the winter sample enabled to test the sensitivity of WTP to the dimension of the risk change (scope test), i.e., the WTPs are examined with regard to whether or not they react to significant characteristics such as the extent of the risk variation. This is seen as an important indication for the validity of the results. As the questionnaire also collects information on risk attitude, risk behaviour, and risk assessment, as well as individual characteristics, further validity checks on the WTP results are possible.

---

[19] E.g. Hakes and Viscusi (1997), Shanteau and Ngui (1987).
[20] This is in accordance with Corso's et al. (2001) findings.

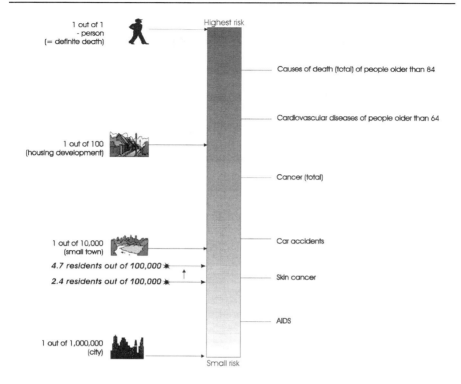

Fig. 7.3 Causes of deaths in Tyrol in the year 2002

7.3.2 Descriptive statistics

The aim of this section is to provide an overview of the underlying data. On the one hand the focus is on the presentation of the socio-demographic characteristics of the individuals. On the other hand, risk specific answers which are also of importance in the econometric analysis are explained in more detail.

Socio-demographic characteristics

The following table gives an insight into the individual characteristics of the interviewees and allows a comparison with the characteristics of the Tyrolean population.

Table 7.2 Sample and population characteristics

Variable	Sample average Obs.[I]	Sample average Mean	Census Mean
Female	1996	0.53	0.52[A]
age	1954	37.08	43.79[A]
alone	1958	0.39	0.35
housemembers	1982	2.82	2.56[B*]
children/capita[II]	1296	0.64	0.23[A]
birthaut	1997	0.84	0.88[A*]
smoking	1988	0.51	0.30[C]
income/month	1128	1,079	1,114[D]
healthy	1937	0.76	0.80[C]
moderate illness	1937	0.20	0.16[C]
bad illness/bad disability	1937	0.04	0.04[C]
elementary/junior high school	1967	0.22	0.37[A]
apprenticeship	1967	0.33	0.33[A]
vocational school	1967	0.16	0.13[A]
secondary school/post sec courses	1967	0.20	0.10[A]
college/university	1967	0.09	0.07[A]
employed fulltime	1961	0.53	0.48[A]
employed parttime	1961	0.10	0.07[A]
employed shorttime	1961	0.02	0.03[A]
retired	1961	0.12	0.22[A]
homemaker	1961	0.03	0.10[A]
student	1961	0.11	0.06[A]
unemployed	1961	0.02	0.03[A]
Others	1961	0.06	0.02[A]

Notes: [I] The differences between 1.997 and the data in the column "Obs." are a consequence of the missing observations in the corresponding category.
[II] The assumption that those who did not give an answer do not have children leads to a median/average of 0.42.
[A] Population 2001. Source: Statistics Austria. Statistical Yearbook 2005, Table 2.14.
[B] Source: Tiroler Landesregierung (Tyrolean State Government). Tyrolean Population – Findings of the Census 2001, Table 25.
[C] Population 1999 > 15. Source: Tiroler Landesregierung (Tyrolean State Government) 2003. Health Report 2002, Table 3.4.1.
[D] Monthly net income (= yearly income/14) of employees in 2003. Source: Statistics Austria, Statistical Yearbook 2005, Table 9.07.

The survey sample refers to Tyrolean people over 14-years of age, who were interviewed in Sept./Oct. 2004 and Feb. 2005. The public opinion poll includes the whole population (673,504) of 2002 (exclusions are mentioned). When possible, children under the age of 15 (123,855) were excluded to improve the comparability (* shows where no division was made). Nevertheless, the findings of this comparison have to be tentatively interpreted, as in several classes different definitions are used in the current study and the public opinion poll.

According to table 7.2, 53% of the interviewees were female, 39% single and 51% smokers. The average interviewee was 37 years old and lived in a household with 2.8 people. 84% of the interviewees were born in Austria. The average monthly income was 1,079 Euro. The health status (self-assessed), education level, and employment status were recorded in categories. A comparison with the data from the public opinion poll shows that the sample represents the Tyrolean population with regard to sex, country of birth, income, as well as health status quite well. Large variations were found with regard to age, the number of children per capita, and smoking behavior. Furthermore, the sample includes people with a higher level of completed education, which at least partially also affects the employment status (e.g. more students, less retirees).

This comparison allows an insight into the structure of the underlying data and can be seen as a brief description of the interviewed persons.

Risk perception

In order to receive information about risk perception the participants were asked to estimate the number of deaths caused by avalanches per year. It is evident from the statistics of the Avalanche Warning Service in Tyrol (2003) that over the last ten years an average of 16 people died per year in avalanche related accidents. The following graph shows how the individual answers corresponded with the statistical facts.

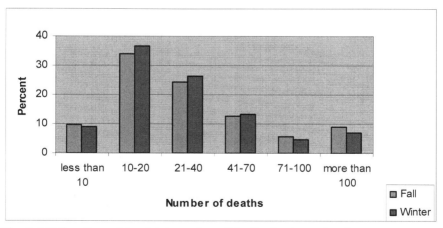

Fig. 7.4 Estimation – (absolute) number of deaths due to avalanches

Figure 7.4 reveals that over 30% of the interviewees correctly estimated the number of avalanche-induced deaths. However, the majority of people questioned overestimated the number of fatalities. At the same time, it is noticeable that the deviations in the relevant category are smaller in winter than in autumn.

Beside this "absolute" estimate of avalanche fatalities we also introduced a relative measure of perceived avalanche risk by using the information given in Figure 7.3. We asked respondents to draw in a line where they think the average risk of dying in an avalanche of the Tyrolean population is located. We interpreted the distance from the bottom of the graph to the self-plotted line - measured in millimetres - as another indicator for individual risk perception. It is this information which was included in the regression analysis to control for the influence of individual perception on the (monetary) valuation process. Beside the perceived dimension of the risk in question literature has shown that also perceived risk characteristics are important factors in valuation and hence, should be included in the analysis. The next paragraphs describe how the according information was collected.

Psychological studies deal with numerous risk characteristics. It has been revealed that involuntary, uncontrollable risks with high catastrophe potential lead to a higher level of rejection of the corresponding risk.[21] The acceptance of a risk is higher the greater the perceived benefits are that can be achieved through the risk. For example: several skiers are willing to accept a higher avalanche risk in order to be able to ski off-piste in areas that are not safe. According to Sunstein (1997), the question of (in)voluntariness can also be linked with the question of responsibility or blame for a fatal incidence. Thus, a risk is perceived as voluntary when the individual exposes himself/herself to the risk. These findings have been taken into account within the framework of the current study by asking a question with regard to the blame of fatal avalanche accidents. The survey participants were asked whether they thought that fatal avalanche occurrences were an anthropogenic event. Figure 7.5 depicts the given answers.

---

[21] See Slovic (1982), Slovic (2000) and Sunstein (1997).

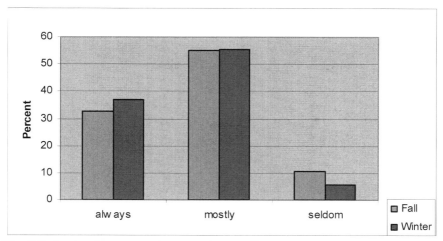

Fig. 7.5. Avalanches – an anthropogenic event?

The reason why a greater number of interviewees in winter thought that fatal avalanche accidents are anthropogenic could be linked to the fact that in the winter of 2005 all deaths that were recorded took place in ski areas where people put themselves at risk, as opposed to avalanche accidents on the road, or in residential areas.

Another survey question referred to alternative protective measures. The interviewees were asked to rate the importance (more/equally/less important) of the prevention of deaths either caused by traffic accidents, air pollution, food poisoning, rock fall/landslide/rock slide, flood or by radioactive radiation in comparison to the prevention of avalanche induced deaths, even if the number of prevented deaths is the same for all protective measures. Figure 7.6 reveals that for the majority of people only the prevention of traffic accidents is rated as more important than the prevention of avalanche induced deaths. A large number of interviewees rated the other categories as equally important (air, food, rock fall, flood) or as less important (radiation).[22]

---

[22] Figures are based on the total sample (autumn and winter).

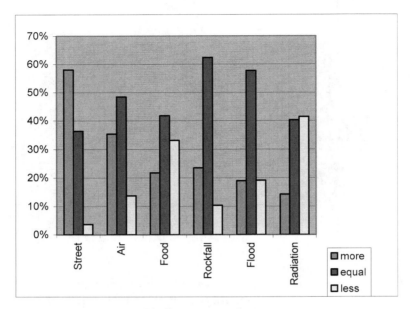

Fig. 7.6 Importance of alternative protective measures in comparison to protective measures against avalanches (total sample)

7.3.3 Econometric analysis

As previously mentioned, the format of the payment questions merely produces an interval within which the WTP can be found. This means that an exact Euro amount is not available, but has to be estimated from the yes/no answers. Only the position of the WTP is known, namely, if it is above or below the payment bid, based on whether the interviewee confirms a payment with yes or declines it with no. This results in the following formulae:

$$YY_i = 1 \text{ if } WTP_i^* \geq B_i^H; \ 0 \text{ otherwise;}$$
$$YN_i = 1 \text{ if } B_i^I \leq WTP_i^* < B_i^H; \ 0 \text{ otherwise;}$$
$$NY_i = 1 \text{ if } B_i^L \leq WTP_i^* < B_i^I; \ 0 \text{ otherwise or}$$
$$NN_i = 1 \text{ if } WTP_i^* < B_i^L; \ 0 \text{ otherwise.}$$

(eq. 7.3)

where YY, YN, NY and NN depict the answers (e.g. YY = two times yes) and B^H, B^I, B^L the higher, initial or lower payment bid, respectively. These equations are estimated with the help of the maximum likelihood approach, assuming a Weibull and log-normal distribution of the error term. Two different models are calculated: while model 2 takes the positive probability of a WTP of zero into account (= spike in the distribution at point 0), model 1 does not provide any separate treatment for any 0 WTP statements (i.e. WTP of 0 is subject to the Weibull or log-normal distribution).[23]

7.3.4 Results

Willingness to pay for risk prevention

The average WTP for the prevention of fatal avalanche accidents, which was estimated from the individual responses, is summarised in table 7.3. According to Carson (2000) the arithmetic mean is calculated as it depicts the adequate measurement if the values need to be incorporated into a CBA.

Table 7.3 further clarifies the surprising result: even though avalanches occur in winter and it can be assumed that current events influence perception, the WTP for protective measures against avalanches is lower in winter than in autumn. A study by Slovic et al. (2000) explains that certain risk characteristics, such as voluntariness or the controllability, can influence the individual risk assessment and the attitude towards risk regulation. Sunstein (1997) mentions that the question of (in)voluntariness of risks can also be linked to the person who is seen responsible or can be blamed for a fatal incident. The fact that all fatal avalanche accidents were off-piste on unsecured areas may have led the interviewees to rate avalanche risks as voluntary and manageable. Thus, if individuals are of the opinion that each person can easily reduce its risk exposure through a change of behaviour, then their WTP for protective measures against avalanches decreases.[24]

Table 7.3 depicts the monthly mean WTP based on a simple regression which only includes the bids and a constant. Monthly WTP varies between 3.6 and 6.2 Euro depending on the used distribution assumption and esti-

---

[23] A detailed explanation of the estimation method as well as the included explanatory variables can be found in Leiter and Pruckner (2005).
[24] Further influencing factors on the WTP are discussed in Leiter and Pruckner (2005).

mation approach. This amounts to an annual value between 43.2 and 74.4 Euro.

Table 7.3 Mean WTP/month (in Euro)

	Weibull			Log-normal		
	Autumn	Winter	Total	Autumn	Winter	Total
Observations	940	642	1582	940	642	1582
Mean/month (Model 1)*	4.82 (2.70)	4.48 (2.77)	4.69 (2.73)	6.17 (3.56)	5.58 (3.61)	5.93 (3.59)
Mean/month (Model 2)*	3.93 (1.07)	3.62 (1.06)	3.80 (1.08)	4.25 (1.11)	3.95 (1.06)	4.13 (1.11)

Notes: * standard deviation in parenthesis; Model 1 = base model; no special treatment of "0-WTP"; Model 2 = mixed approach; consideration of a positive probability of "0-WTP".

Value of statistical life

As previously mentioned, the VSL mirrors the rate at which the interviewees were willing to exchange income for risk alteration, keeping the individual utility level constant. The VSL can be calculated from the individual WTP for reduced mortality risks: if the annual WTP is divided by the risk evaluated change (in the Tyrolean example 1/42,500) then the quotient can be interpreted as the VSL. Using the data from table 7.3 the VSL lies between 1.8 and 3.1 million Euro. Table 7.4 contracts the Tyrolean figures with results found in comparable international surveys.

Table 7.4 The value of statistical life

Author/Study	Country	Method	Risk	VSL
Alberini, Hunt & Markandya 2004	France, Italy, Great Britain	CVM	general	2.26 Mio. €
Alberini, Scasny & Braun Kohlova 2005	Czech Republic	CVM	environmental/ climate risks	2.86 Mio. €
Baranzini & Ferro Luzzi 2001	Switzerland	Hedonic pricing	workplace risks	6.5 – 9.5 Mio. $
European Commission 2000	EU	different	different	0.65 – 2.5 Mio. €
Leiter & Pruckner 2005	**Austria**	**CVM**	**avalanche risks**	**1.8 – 3.1 Mio. €**
Maier, Gerking & Weiss 1989	Austria	CVM	transport risks	1.8 – 4.9 Mio. €[a]
Persson, Norinder, Hjalte & Gralen 2001	Sweden	CVM	transport risks	2.6 Mio. $
Spengler 2004	Germany	Hedonic pricing	workplace risks	1.65 Mio. €
Weiss, Maier & Gerking 1986	Austria	Hedonic pricing	workplace risks	4.4 – 7.4 Mio €[a]

Notes: [a] Values are converted into 2005 € using the domestic CPI series (Statistic Austria)

7.3.5 Application of the results and an outlook

The presented study enables a more precise estimation of individual benefits resulting from the prevention of avalanche accidents in comparison to the derivation of the VSL from other studies, which not only refer to different regions and/or cultures, but also vary in the type of risk that is evaluated. The data calculated for Tyrol sheds light on how advantageous the population estimates the prevention of fatal avalanche accidents. Using the instrument "VSL" enables the monetary valuation of the most important output of protective measures against avalanches, namely the protection of human life. This value can now be used in future CBA which allow for the implementation of a comprehensive analysis of the social welfare effects of protective measures. This additional information increases in importance when only a restricted number of projects can be realised, due to a limited budget. These relevant projects had to be previously chosen from available alternatives. For example, a decision could be made whether a certain section of road, a housing estate or a business establishment should be protected from avalanches. By answering this question it is possible that, due to the inclusion of the monetarily estimated utilities for protecting human life, a change occurs in the order of the favoured projects.

In addition, the study at hand provides a building block for the examination of a benefit transfer. "Benefit transfer" is understood as the transferability of primary results to other regions, cultures and goods or service types. A corresponding analysis could primarily focus on the transferability within a region of different types of risks (other natural risks, health risks) and subsequently examine the transferability to other regions. This type of generalisation enables to considerably reduce costs and time expenditure involved in a CBA, which would be advantageous for the society as well as single institutions and companies. Furthermore, the EU also articulates interest in studies which deal with the (monetary) value of the prevention of fatal accidents: „ ... *DG Environment notes the need to carry out further research to refine them [interim values] for future use, and will update them as and when evidence become available ... ".*[25]

7.4 Conclusion

Considering the given resource shortage and the increasingly restricted public budget, there is an urgent demand to allocate goods efficiently, in other words to move towards a more demand oriented use of scarce resources. The CBA which aims to reveal the social evaluation of the costs and benefits associated with alternative protective measures can contribute to place the collective decisions on a solid foundation. A basic requirement to include all the advantages and disadvantages of a specific measure in a CBA is the monetary valuation of the relevant components. One approach for monetary valuation of non market goods is the CVM which was discussed and illustrated by means of a current study.

The information created through CBA can make a valuable contribution to support decisions with regard to an efficient allocation of resources. Even if the appropriateness of the approach as well as the information gained are a matter of ongoing discussions, CBA can be taken as a transparent, flexible and comprehensive instrument which is of particular importance when scarce resources have to be allocated. In other words, CBA not only offers an answer to the question if protective measures are worth their costs but also gives information about their societal demand. Furthermore, CBA allows – especially notable in the context of integral risk

---

[25] http://europa.eu.int/comm/environment/enveco/others/recommended_interim_values.pdf

management - to simultaneously address questions of risk analysis, risk assessment and risk management and to provide appropriate solutions.

References

Alberini A, Hunt A, Markandya A (2004) Willingness to Pay to Reduce Mortality Risks: Evidence from A Three-Country Contingent Valuation Study, Nota di Lavoro 111.2004, http://www.feem.it/NR/rdonlyres/8904A715-57A3-4FDD-A7E9-52318537EEFF/1258/11104.pdf (15.8.2005)

Alberini A, Scasny M, Braun Kohlova M (2005) The Value of Statistical Life in the Czech Republic: Evidence from a Contingent Valuation Study, http://www.webmeets.com/files/papers/EAERE/2005/24/alberini%20scasny%20EAERE%202005.pdf

Amann W (2003) Integrales Risikomanagement von Naturgefahren, in Jeanneret F et al. (eds.): Welt der Alpen – Gebirge der Welt. Haupt, Bern/Stuttgart/Wien

Amann W (2006) Risk Concept, Integral Risk Management and Risk Governance, in Dannemann S, Ammann W, Vulliet L (eds): Risk 21 – Coping with Risks Due to Natural Hazards in the 21st Century, Taylor & Francis, London, pp 3-23

Amt der Tiroler Landesregierung, Lawinenwarndienst Tirol (2003) Schnee und Lawine 2002-2003, Innsbruck

Arrow K, Solow R, Portney PR, Leamer EE, Radner R, Schuman H (1993) Report of the NOAA Panel on Contingent Valuation, http://www.darp.noaa.gov/library/pdf/cvblue.pdf (05.05.2005)

Baranzini A, Luzzi GF (2001) The Economic Value of Risks to Life: Evidence from the Swiss Labour Market, Swiss Journal of Economics and Statistics 137(2), pp 149-170

Boardman AE, Greenberg DH, Vining AR, Weimer DL (2001) Cost Benefit Analysis – Concepts and Practice, Prentice Hall, Upper Saddle River

Carson RT (2000) Contingent Valuation: A User's Guide, Environmental Science & Technology 34 (8), pp1413-1418

Corso PS, Hammitt JK, Graham JD (2001) Valuing Mortality-Risk Reduction: Using Visual Aids to Improve the Validity of Contingent Valuation, The Journal of Risk and Uncertainty 23 (2), pp165-184

European Commission. Recommended Interim Values for the Value of Preventing a Fatality in DG Environment Cost Benefit Analysis, http://europa.eu.int/comm/environment/enveco/others/recommended_interim_values.pdf (17.10.2005)

Fuchs S (2004) Development of Avalanche Risk in Settlements - Comparative Studies in Davos, Grisons,Switzerland, Dissertation, University of Innsbruck

Fuchs S, Alpin MC (2005) The net benefit of public expenditures on avalanche defence structures in the municipality of Davos, Switzerland, Natural Hazards and Earth System Sciences 5, pp319-330

Gamper CD, Thöni M, Weck-Hannemann H (2006) A conceptual approach to the use of Cost Benefit and Multi Criteria Analysis in natural hazard management, Natural Hazard and Earth System Sciences 6, pp 293-302 (http://www.copernicus.org/EGU/nhess/6/2/293.htm)

Groot W, Maassen van den Brink H, Plug E (2004) Money for Health: the equivalent variation of cardiovascular diseases, Health Economics, www.interscience.wiley.com

Hackl F, Pruckner G (1994) Die Kosten/Nutzen-Analyse als Bewertungsinstrument der Umweltpolitik, in Bartel R, Hackl F (eds) Einführung in die Umweltpolitik, Vahlen, München, pp 81-100

Hakes JK, Viscusi WK (1997) Mortality Risk Perceptions: A Bayesian Reassessment, Journal of Risk and Uncertainty 15, pp 135-150

Hanley N, Spash CL (1993) Cost Benefit Analysis and the environment, Cheltenham

Hansjürgens B (2004) Economic Valuation through Cost Benefit Analysis – Possibilities and Limitations, Toxicology 205, pp 241-252

Heinimann HR, Hollenstein K, Kienholz H, Krummenmacher, Mani B (1998) Methoden zur Analyse und Bewertung von Naturgefahren, Umweltmaterialien 85, Naturgefahren, BUWAL, Bern

Kienholz H (1994) Naturgefahren – Naturrisiken im Gebirge, Schweizerische Zeitschrift für Forstwesen 145(1), pp 1-25

Kienholz H (2004) Alpine Naturgefahren und -risiken – Analyse und Bewertung, in Gamerith W, Messerli P, Meusburger P, Wanner H (eds) Alpenwelt – Gebirgswelten: Inseln, Brücken, Grenzen, Deutsche Gesellschaft für Geographie, Heidelberg/Bern, pp 249-258

Leiter A (2004) Meilenstein I – Zusammenstellung und Auswertung der verfügbaren Methoden zur Risikobewertung bzw. Erfassung der WTP für eine Risikoverringerung, alpS working paper

Leiter AM, Pruckner GJ (2005) Dying in an Avalanche: Current Risk and Valuation, Working Paper, School of Economics, University of Adelaide, http://www.economics.adelaide.edu.au/research/wpapers/doc/econwp05-16.pdf

Maier G, Gerking S, Weiss P (1989) The economics of traffic accidents on Austrian roads: Risk lovers or policy deficit? Empirica 16 (2), pp 177-192

Mishan E (1998) Cost Benefit Analysis; an informal introduction, London

Persson U, Norinder A, Hjalte K, Gralen K (2001) The Value of a Statistical Life in Transport: Findings from a New Contingent Valuation Study in Sweden, The Journal of Risk and Uncertainty 23 (2), pp 121-134

PLANAT (1998) Von der Gefahrenabwehr zur Risikokultur. Bern

Pommerehne WW (1987) Präferenzen für öffentliche Güter, Mohr/Siebeck, Tübingen

Shanteau J, Ngui ML (1989) Decision Making Under Risk – The Psychology of Crop Incurance Decisions, http://www.ksu.edu/psych/cws/pdf/insurance_paper90.PDF

Slovic P, Fischhoff B, Lichtenstein S (1982) Facts versus fears: Understanding perceived risk, in Kahneman D, Slovic P, Tversky A (eds.) Judgment under uncertainty - Heuristics and biases, Cambridge, pp 463-489

Slovic P, Fischhoff B, Lichtenstein S (2000) Facts and Fears: Understanding Perceived Risk, in Slovic P (ed) The Perception of Risk, London, 137-153

Spengler H (2004) Kompensatorische Lohndifferenziale und der Wert eines statistischen Lebens in Deutschland, Zeitschrift für Arbeitsmarktforschung 37 (3), pp 269-305

Stötter J, Meissl G, Ploner A, Sönser T (2004): Developments in Natural Hazard Management in Alpine Countries Facing Global Environmental Change, in Steininger KW, Weck-Hanmennan H (eds) Global Environmental Change in Alpine Regions. Recognition, Impact, Adaptation and Mitigation. Edward Elgar, Cheltenham, pp 113-130

Sunstein CR (1997) Bad Deaths, Journal of Risk and Uncertainty 14, pp 259-282

Sunstein CR (2004) Are Poor People Worth Less Than Rich People? Disaggregating the Value of Statistical Lives, AEI-Brookings Joint Center for Regulatory Studies, Working Paper 04-5, http://ssrn.com/abstract=506142

Tietenberg T (2004) Environmental Economics and Policy, Pearson, Boston et al., 4th edition

Weck-Hannemann H, Thöni M (2006) Kosten-Nutzen-Analyse als Entscheidungsgrundlage im Naturgefahrenmanagement: Verfahren – Vorzüge – Vorbehalte, alpS-Working paper WP - 09 - 2006

Weiss P, Maier G, Gerking S (1986) The Economic Evaluation of Job Safety – A Methodological Survey and Some Estimates for Austria, Empirica 13 (1), pp 53-67

Wilhelm C (1997) Wirtschaftlichkeit im Lawinenschutz, Eidgenössisches Institut für Schnee- und Lawinenforschung, Davos

Wilhelm C (1999) Naturgefahren und Sicherheit der Bevölkerung im Gebirge oder: Von der Schicksalsgemeinschaft zur Risikogesellschaft. Forum für Wissen 2, pp 1-9

8 Analysing Decision Mechanisms for Natural Hazard Management

C.D. Gamper, P.A. Raschky, H. Weck-Hannemann

8.1 Introduction

The traditional economic approach to natural hazard management follows the idea of internalizing external effects of the public good, so-called "protective measures". From an economic perspective protective measures can be classified as a public good. Such a good is characterized by non-rivalry, meaning that public goods are those for which consumption by one individual does not detract from the ability of others to consume them. Furthermore, it is characterized by non-exclusivity, in other words by indivisibility ('all or nothing' provision) and collective consumption (individuals cannot be excluded from consumption) (Edwards-Jones et al. 2000). In contrast to private goods, there is no market which takes over the decision process for the allocation of the goods. This so-called market failure is compensated by the state taking over the decision responsibility, whereby a choice has to be made how to decide collectively.

The individual that installs a protective wall does not receive compensation for all (i.e. for the internal but not for the external) benefits that the wall generates. By internalizing the external effects, a pareto-optimal situation similar to that of the market could be achieved. However, the analysis of market failure in natural hazard management cannot be solely reduced to the public good problem and externalities. The whole set of reasons for market imperfections (Fritsch et al. 2007) might apply to the different protective measures (e.g. see Raschky and Weck-Hannemann 2006/2007a for imperfections on the market for natural hazard insurances).

The suggestion of traditional (neo-classical and welfare) economics to correct market imperfections is that the government intervenes. In the case of externalities, welfare economics focuses on the internalization of the external effects but ignores the problems of collective-decision making and therefore the decision-making process itself (Buchanan 1991a). This process takes place within the existing institutional framework that determines the behaviour of individuals involved in the political process (e.g. voters, politicians, bureaucrats). The welfare-economic approach implicitly considers the state as a benevolent identity that tries to increase the welfare of its citizens (Pappenheim 2000, Blankart 2006). The research focus is usually limited to the effect of the state intervention and neglects the aspect of the political decision-making process, its relevant actors, and potential imperfections within the given setting. These imperfections, however, could also be part of the problem or increase inefficiencies. For example, the introduction of a (Pigovian) tax aiming at internalizing negative external effects demands, from the very beginning, the direct and active intervention of the state through taxes, without testing whether an alternative decision rule (e.g. market or negotiation) would lead to a more efficient result (Bonus 1996). It ignores the possibility that state intervention results in further transaction costs and could thus even increase the negative effects and worsen the situation. Traditional economics assumes that the institutional framework is given and cannot be the subject of change (Buchannan 1991b). Coase (1988) argues that "economic policy involves a choice among alternative institutions [...]", an essential part of the analysis that is faded out by the majority of economists. Consequently, the analysis of state interventions for coping with market imperfections as well as cost-benefit analysis (CBA) as a decision-support system for deciding about the provision of public projects (Leiter et al. 2007) should incorporate alternative institutional arrangements in their calculation.

In regard to the instrument of cost-benefit analysis, Hanusch (1987) argues that it sets distributional aspects aside. Cost-benefit analysis does not primarily focus on the question of who bears the costs and who benefits from the installation of a protective construction. However, distribution is an essential factor in the allocation of public resources and the acceptance of a political decision (Blankart 2006).

A further criticism of cost-benefit analysis as it is applied nowadays stems from the methods used to disclose individual preferences for public goods. The welfare-economic approach demands cost-benefit analysis for the efficient allocation of public goods. As individual preferences for public goods are not reflected on markets, special methods are made use of to disclose these preferences. A recent book by Lackner-Frey (2004) suggests that the existing instruments used to reveal individual preferences do not pay any attention to the insufficient description of human cognitive processes. The author argues that the aspect of limited individual rationality and the concept of endogenous preferences (i.e. the idea that individual preferences can emerge and change during the dynamic decision-making process) have not been incorporated into the theory of preference disclosure. Natural hazards and their impacts are highly complex issues that even experts sometimes are not able to fully embrace. The complexity of the situation, the uncertainty, and the long-term character of natural hazards and protective measures pose an immense challenge to the cognitive skills of individuals. How can the average citizen obtain and process all the relevant information and thus state "true" preferences? This might result in biased evaluation outcomes that can distort the cost-benefit analysis. Therefore, it is hard to formulate precise recommendations for economic or political activities. To counteract these problems, Lackner-Frey (2004) demands the application of a process-oriented approach to the evaluation of individual preferences for public goods. In her opinion, the analysis of the decision-making rules and the institutions that affect individual behaviour could provide a clearer picture of the development of preferences themselves and the preferences of public goods in particular.

Beside these theoretical points of criticism (for additional arguments see, e.g. Hansjürgens 2004), problems arise within political practice. Boardman et al. (2001) point out that the perception of what benefits and costs actually are depends on the politicians and bureaucrats involved in the process as well as their position and agency. This perception is based on whether they are "spenders", "analysts" or "guardians". Blankart (2006) argues that certain actors within the political process might abuse cost-benefit analyses for the justification of their decision instead of determining the most efficient alternative, a point that is certainly true for all decision support systems. A study by Bauer (1986) summarizes his experiences of the Austrian audit control collected from different practical cost-benefit studies as well as cost-effectiveness studies in Austria, Germany and Switzerland. He describes how these instruments have been manipulated by politicians as well as experts in order to justify a desired option. Bauer (1986) lists the following possibilities of how the instruments might have been influenced:

- omitting undesired alternatives,
- disadvantageous design of undesired alternatives,
- leaving out project aims that might be disadvantageous for the desired project alternative,
- one-sided weightings by decision-makers and/or experts,
- systematic over- or under-weighting of sub-objectives in order to reach the desired project alternative,
- biased or selective communication of analysis results,
- no attention paid to the results of the analysis.

Despite the theoretical points of criticism and the potential of misuse within the political practice, cost-benefit analysis is a well accepted and often used instrument within the political decision-making process. It builds an important basis of information for the political process and can contribute to an objective dialogue on the allocation of public funds (Blankart 2006). Cost-benefit analysis is a useful tool to reduce all possible impacts of a project to one criterion and therefore simplify the amount of complex information. This is essential for the decision-making process, but should not be misused to justify outcomes, desired by only a few participants.

The focus on the decision-making process itself could help to avoid these problems and to overcome certain theoretical points of criticism. Therefore, a process-oriented approach may be seen as an extension of the existing approach rather than a substitute.

8.2 A process-oriented economic approach

The standard economic approach basically defines efficiency in natural hazard management with the marginal costs equal to marginal benefit theorem. This view is output-oriented: basically the output or project is chosen that delivers the highest net-benefit. An alternative economic approach focuses on the decision-making process itself and tries to overcome the disadvantages of the standard economic approach. This process-oriented view tries to find which decision rules or decision system build the frame under which individual preferences for natural hazard management and protective measures are reflected in a most efficient way (Weck-Hannemann 2006).

This approach receives its theoretical background from new institutional economics (NIE) in general and the related fields of constitutional economics and public choice theory. In the following, focus will be placed especially on aspects of institutional economics, in particular the process-analytic aspects of NIE. This economic theory (with its major scholars Ronald Coase, Douglass North and Oliver Williamson) forms the foundation for any further analysis and models relevant for a process-oriented analysis. Nevertheless, in order to provide a more comprehensive overview comparisons are drawn to the approach of standard economic theory.

In his fundamental work, North (1990) wants the reader to think about making the same transaction in a different country such as Bangladesh. Even without knowing the institutional setting of Bangladesh, one could imagine that certain transactions or economic activities (e.g. setting up a shop for cars) are performed differently in Bangladesh than for example in a European country. Institutions define and constrain the individual's set of choices. If this applies to economic activities such as setting up a car shop, this should also apply to decisions related to protection against natural hazards.

Decisions linked to protection measures and natural hazard mitigation strategies take place in a complex situation. A vast amount of different environmental, economic and social variables need to be taken into account.

Such a complex situation creates uncertainty for the individual. It is the purpose of institutions to reduce individual uncertainty (Richter and Furubotn 2003).

According to North (1990), "(I)nstitutions are the rules of the game in a society or, more formally, are the humanly devised constraints that shape human interaction". This rather general definition of institutions is more specified by Frey (1990), who classifies three types of institutions: Firstly, institutions are rules or procedures that clarify how decisions are made within society. These decision-making systems can be the market, democracy, hierarchical decision-making or negotiating systems. In relation to natural hazard management the market and possible imperfections and distortions as well as drawbacks of the existing political mechanisms are of particular importance. Secondly, institutions can make formal or informal rules that influence individual behaviour such as laws, social norms, traditions or informal rules defining the behaviour within a group or the family. Regarding natural hazards this type of institution could range from the general legal setting (regulating e.g. preventive measures, financial support in the case of emergency or the distribution of competencies) to the behavioural standards (e.g. within a ski-hiking group). Thirdly, institutions can be groups of individuals that share a common aim e.g. organizations. These could be, for example, political parties, authorities, companies, NGOs or clubs. Beside politicians, companies or interest groups, organizations such as insurance companies, experts or tourist associations might be of special interests in the area of natural hazards.

8.2.1 How can a process-oriented analysis be designed?

A process-oriented approach to the question of efficiency in the area of natural hazard management tries to find answers to the following questions: Who, where and how are decisions made? What are the weaknesses of the existing decision-making mechanisms? Which alternative decision-mechanisms reflect the individual preferences in an optimal way? Based on these questions, the analysis can be structured in a similar manner. Firstly, the relevant decision-making systems and decision makers are identified and the interactions within these existing structures are observed. Secondly, each decision-making system will be checked for possible imperfections and distortions. Thirdly, after the weaknesses of each system have been highlighted, better or best-practice models as a reference standard can be developed and suggestions for the implication of models in accordance with the existing institutional setting can be made.

Structuring the relevant decision-making systems and the different decision-makers

In a first step the status quo of the decision-making mechanisms for natural hazard management and protective measures is examined. As Frey and Kirchgässner (2002) argue, the characteristics of the different decision-making systems and their institutional arrangements need to be described in order to analyse the alternatives and to identify the most productive setting. This analysis mainly applies economic theories from public choice and new institutional economics. The process-oriented approach is based on the economic model of human behaviour (homo oeconomicus) with the assumption that individuals behave rationally and try to maximize their utility under given constraints. They systematically react to the incentives provided by the institutional framework (Frey 1990, Kirchgässner 1991, Weck-Hannemann 1994). Therefore different categories of decision-makers can be identified (e.g. consumers, politicians, voters or bureaucrats) by taking into consideration that for all members of each group the same institutional setting applies. The action of each individual is thus predictable in accordance to the group of decision-makers he or she belongs to and the behaviour of one representative member of the alternative category can be derived.

The first step at this part of the analysis is a rough categorization of the mechanisms of decision-making and stakeholders involved in the area of natural hazard management and protective measures. We can basically identify four decision-making systems or decision rules based on the structure by Frey (1990): the political process, the system of bureaucracy and administration, the market or price-system and finally decisions within groups, which are referred to as the negotiation procedure.

In an attempt to structure these different decision mechanisms for natural hazard management, the following figure was created to give an overview of the relevant forms of decision alternatives for protective measures and thus, the corresponding desired level of risk among society's actors (figure 8.1).

8 Analysing Decision Mechanisms for Natural Hazard Management

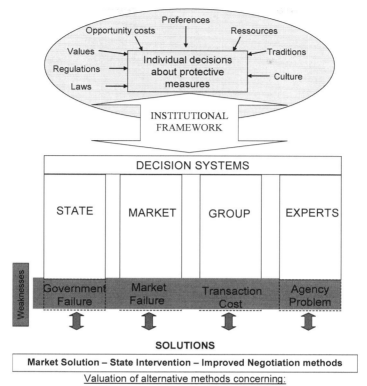

Fig 8.1 Alternative Decision Mechanisms for Natural Hazard Management

Within these four systems a set of sub-rules for decision-making apply. For example, the political decision (i.e. category "state" in figure 8.1) can either be made through direct democracy or representative democracy. Decisions about the installation of a protective wall against avalanches or a dam within a municipality could either demand a simple majority or follow the unanimity rule. Natural hazard management can either be centralized or decentralized in its decision structure.

Concerning the market, it is of general interest which parts of natural hazard management and what sort of protective measures are actually allocated by the price-system. An example for decision-making within a group in the area of natural hazard management can be given for extreme sports, e.g. ski-hiking tours in an avalanche-endangered back-country.

Within these systems a number of decision-makers act:
- the citizen, the voter, the tax-payer or the consumer (depending on the decision-making system), demanding and finally financing protection against natural hazards and support/relief once a natural catastrophe occurred;
- the politician, who is responsible (depending on the constitutional and legal frame) for protective measures;
- bureaucrats in public administration, who are involved in the preparation and implementation of protective measures;
- experts, delivering information and opinions on the risk and available protective possibilities;
- organisations, associations and economic identities (particularly the insurance industry).

Besides the rules applying within the different decision-making systems, institutions such as informal rules, social norms and traditions still have to be taken into account.

Imperfections and weaknesses of the existing decision-making mechanisms

After describing the status quo, the next step is the analysis of the weaknesses of the existing decision-mechanisms. Prior to choosing a decision-making system for the allocation of natural hazard management and protective measures, it is essential to compare the existing alternatives (Frey and Kirchgässner 2002). If the market fails to efficiently allocate resources for protective measures, one should not simply call for state intervention, as this could result in even greater inefficiencies. This part of the analysis is once again structured by the different decision systems, e.g. the political process, the market and the group and the different actors within the systems.

Based on the theory of public choice, individuals within the political process act according to the general assumption of rationality: their behaviour is determined by their wishes (preferences) and the constraints they face. The politicians and the public administration - the "suppliers" of natural hazard management and protective measures - are basically constrained by the institutional design of the political decision-making process (Weck-Hannemann 1994): firstly, politicians are interested in being re-elected in order to stay in power. Facing an up-coming election, politicians will try to initiate programs that reflect the preferences of the voters in the best way. Furthermore, they might incorporate the demands of specific associations or interest groups that have a great impact on the outcome of the elections. However, they always have to weigh up the benefits of their decision (the support of special interest groups) against its costs (possible loss of voters, because of privileges for the special interest groups). Secondly, politicians have to deal with budget constraints. Public budgets are limited and so are the public financial resources available for preventive measures and post-catastrophic relief action. In order to implement their programs, politicians only have limited budgets. For example, the federal government of Germany (including federal states and local authorities) provided almost € 10 billion emergency relief after the catastrophic flooding in Germany in 2000. This sum was financed by postponing a planned tax reform (Schwarze and Wagner 2004). Thirdly, politicians depend on public administration when they want to implement their program (principal-agent relationship). Additionally, in the case of natural hazard management and in particular the planning process, politicians as well as bureaucrats often rely on information and recommendations by natural hazard experts and scientists. As both public bureaucrats and experts do not face a re-election constraint, they may use their informational advantage for their own interest which can either be a budget-intensive and/or a discretionary policy.

Testable hypotheses on the adequacy (efficiency) of alternative decision-making processes can now be derived from the above assumptions on the actors' behaviour and the incentives resulting from alternative institutional frameworks (Frey und Kirchgässner 2002). In addition, the empirical examination of these hypotheses is challenging. Economic theory does not strive for a precise calculation of a system's inefficiencies per se, but rather focuses on a comparative assessment of different institutional settings and their abilities to provide goods and services in an efficient way, i.e. with minimal cost and as favoured by consumers or voters. Such a comparative institutional analysis demands that the specific goods and services are provided by different systems. Frey and Kirchgässner (2002) present two ways of comparing the public and the private provision of a certain good: on the one hand, if the publicly and privately provided goods are identical, a direct comparison of the costs is possible. On the other hand, if the public and the private sector provide disparate goods, the differences have to be calculated with statistical methods such as multiple regression analyses capturing the differences in size, type or quality in detail. Additionally, differences in the provision (e.g. by central state or municipalities or by a voluntary or compulsory insurance system) also have to be incorporated into the multiple regression analysis.

Suggestions for the development of new and the implementation of alternative models

Both neo-classical (welfare) economics and Neo Institutional Economics (NIE) have been applied to the area of environmental economics. One of the main issues in this field is the analysis of public goods and corresponding external effects and the development of models and methods to correct this market imperfection. Bonus (1996) compares a neo-classical model – the Pigovian tax – with a NIE-model – the Coase-theorem - and shows how they both approach the issue of external effects. The neo-classical model suggests that external effects should be internalized either by Pigovian taxes if the effects are negative or by Pigovian subsidies in the case of positive external effects. This solution always demands active state-intervention, which can sometimes even increase the market imperfections. Bonus argues that the problem is setting the right tax level. However, the government needs information about the external costs, which are not directly available because of the market failure itself. In Bonus' opinion the costs of gaining this information (transaction costs) are not taken into account by the neo-classical approach. The NIE-approach incorporates this problem into its model. It emphasises the installation of useful institutions instead of internalising external effects. Institutions can be used to reduce individual uncertainty as long as there is a general agreement or consensus on the existing rules (Buchanan and Tullock 1962). This links the institutional economic approach, used so far, with the ideas of constitutional economics. On the constitutional level institutional settings for certain constraints such as the acceptable risk for society (and based on that the optimal level of protection) are determined. Additionally, rules, systems and processes will be constituted to decide about future allocation of resources in natural hazard management and on protection strategies.

A comprehensive process-oriented view includes the whole set of alternative decision-making systems, such as the market, democratic systems, bureaucracy or group decisions. This means that the optimal model could either be alternative political rules, such as direct or representative democracy, or market solutions, such as markets for formerly public goods (e.g. emission trading schemes), or group decisions, as long as the decision-making rule fulfils the criterion of being best-suited to reflect individuals' preferences.

Consider, for example, the imperfections and distortions on the natural disaster insurance markets. As shown by Raschky and Weck-Hannemann (2007a/2007b) state intervention through post-catastrophic financial relief provided by the government could even cause a further distortion to the insurance market. The introduction of a mandatory insurance system could increase efficiency. However, a comparative analysis of institutional settings is necessary in order to define the optimal arrangement of decision-making systems and institutional constraints.

An array of alternative decision-making rules and institutional arrangements exist that can reduce existing imperfections and increase efficiency in natural hazard management. Nevertheless, depending on the existing institutional arrangement, the complexity of the situation and the variety of imperfections a comprehensive analysis as well as a profound change of the institutional setting is demanded. Such a broad analysis and general models can only be provided by a process-oriented approach.

A decision support technique that builds on the idea of traditional economic approaches such as cost-benefit analysis does, however also take a process-oriented approach, is called multi-criteria analysis. The tool itself originates from the science of operations research but has been applied to public good decisions, especially about the environment, in the field of environmental economics for the past two decades and extensively during the last decade. Even though it is not declared as a process-oriented economic instrument in the literature, it will be demonstrated in the following why this could nonetheless be argued for.

8.3 Multi Criteria Decision Analysis – combining output- and process-orientation for an efficient decision-making process

Decisions made in natural hazard management in general, and for protective measures in specific, can be characterised as environmental decisions not only because the issue in question is nature, but also because of the characteristics generally known from environmental decisions. In the case of environmental issues, and therefore protective measures against natural hazards, politicians (as state representatives and therefore responsible for the allocation decision) are confronted with a complex choice. Environmental decisions, as might be argued for other political issues too, affect a multiplicity of parties, from individuals to organisations of producers and consumers. Moreover, the modes of environmental decision-making, as well as the underlying assumptions, tools and criteria are highly diverse and add on to this multiplicity in decision-making. Taking protection measures as an example: for the proper installation of avalanche protection measures the civic, geological, legal or meteorological experts' opinions are needed as well as environmental, economic, ecological and social criteria have to be taken into consideration.

On the one hand, economics can provide decision-makers with an assessment of these issues providing results and advice in the form of decision-support systems, like cost-benefit analysis (CBA). This outcome-oriented tool, as mentioned earlier in this paper, represents people's preferences indirectly via valuation methods that are incorporated into the analysis. On the other hand, economics can provide a process-oriented approach (constitutional economics) where people directly decide themselves and preferences are aggregated in the political decision process. The first approach offers a decision-support system for politicians indirectly representing preferences as well as leaving final decisions to politicians. The second approach is one that directly asks people to state their preferences and to take part in the decision process (whether via chosen decision rules or directly in the process of a prevailing matter).

A third approach, namely multi-criteria analysis (MCA), offers a new instrument to combine both elements of the economics' task so far. In itself it is, similar to cost-benefit analysis (CBA), a decision-support tool. On the other hand, it offers a way to incorporate participation in its analytical process, thus the direct representation of preferences as well as their intensities.

The conventional environmental economic tool usually applied to the kinds of decision sets mentioned above is CBA that tries to facilitate a more efficient allocation of society's resources (Boardman et al. 2001). The method seeks to translate all relevant considerations into monetary terms and can therefore select the most efficient project from a portfolio of alternatives (Hanley and Spash 1993). MCA, which can be defined as "the study of methods and procedures by which concerns about multiple conflicting criteria can be formally incorporated in a decision making process" (International Society on MCDM 2004) seems to complement CBA and moreover help overcome some of CBA's weaknesses. MCA includes techniques for comparing impacts in ways which do not involve giving all inputs and outputs explicit monetary values but could also include for example other numerical or qualitative measurements.

Similar to CBA, MCA builds on the idea of aggregating information flows of costs and benefits of different alternate projects. Like CBA, MCA considers the wide range of preferences and valuations arising when a group decision is faced. However, the way preferences are incorporated into the analyses is quite different. CBA's preference revelation depends on the respective assessment method; consequently preferences are gained through direct or indirect procedures of observing individual behaviour. MCA considers preferences directly through stakeholders' (who act as representatives of individuals' preferences) involvement in the process of the analysis. The main difference of the methods in that respect lies in the level of preference disclosure. While CBA can also generate a value that is applicable to a certain number of project areas, like i.e. the value of statistical life in the area of natural hazards (Leiter at al. 2007), MCA can only gain preferences for specific project alternatives (for a more detailed discussion of the two methods applied to natural hazard management refer to Gamper et al. 2006).

In the following, a brief outline of the procedure of MCA will be given in order to then show MCA's strength in its application to a wide range of political areas.

8.3.1 MCA – the procedure

In short, MCA can be drafted as follows: O is a finite set of n feasible options and G is a set of m evaluation criteria, it is possible to build a $n \times m$ matrix *(P)*, whose elements $p_{i,j}=g_j(o_i)$ *(i=1,2,...n; j=1,2,...m)* represent the evaluation of option i by means of criterion j. An option o_1 is evaluated to be better than option o_2 (both belonging to the set O) according to the jth criterion if $g_j(o_1)>g_j(o_2)$ (Munda 1995).

In more detail, the procedure follows the subsequent path: in a first step the decision context is established which involves outlining the aims as well as identifying decision makers and other stakeholders. An institutional compilation, performed mainly on historical, legislative and administrative documents, can be used to achieve this. The map of actors resulting from this analysis might find adjustments during the process of the MCA (De Marchi et al. 2000). A shared understanding of the decision context that is the political, social, economic and administrative structure, is essential because the impacts can be manifold and a lot of people may be affected whose preferences and perceptions need to be recognised (Omann 2004).

Subsequently, in a second step, objectives can be created either using a top-down approach, in the case of having a larger project where objectives need to be set by a central body, or a bottom-up approach, where various stakeholders participate in generating objectives (Edwards-Jones et al. 2000). An overall objective can be broken down into a subset of objectives, thus higher level goals are dependent on lower level ones: in natural hazard management, a higher level of security might be the main aim when installing protective measures against avalanches, but at the same time a sub-goal could be to minimise the environmental impact (i.e. the loss of biodiversity incurred through the building of high altitude feeder roads). Goals need to be clear, specific, measurable, agreed, realistic and time-dependent, that means being classified into long-term or immediate goals.

Only MCA identifies the criteria that allow measuring the strength of the options in fulfilling the objectives in this step. To meet this requirement, a criterion needs to be measurable, if not quantitatively, then qualitatively, to show how well an option performs in relation to that criterion (Dogson et al. 2001). In the example above, a possible criterion could be the number of species lost through human induced behaviour. The finalisation of the chosen criteria requires assessing them against a range of qualities: criteria should be complete, operational, decomposable (two factors should not be in opposition in a single criterion), non-redundant, minimal, and defined in terms of time. The time-definition (temporary consequences or permanent ones) causes difficulties when aggregating and comparing the results. With monetary techniques, discounting is a reasonably well established procedure for aggregation. Apart from the fact that this might not always be plausible, this does not solve the problem with criteria measured in non monetary terms .

In MCA, criteria should be developed through participation of the stakeholders, in order to make sure all interests are represented and can then be regarded when conducting the analysis.

The third step involves identifying all the relevant options for achieving the objectives. The number of options may vary between 2 (e.g. should a certain project be undertaken or not?), any discrete number (i.e. 10 different ways for building protection measures against avalanches) and infinity (Fandel and Spronk 1983). In the first case, we speak of a 0-1 choice system where one chooses between the current and a new situation (Munda 1995); in the second case we have a finite number of options available. Fandel and Spronk (1983) suggest creating a subset of alternatives in case of a high number of options. Given the complexity of decision making problems, it is not always possible to define the set 'A' a priori. In a continuous situation, the set of options 'O' is progressively elaborated (Munda 1995) such as MCA is used to specify the best option under given constraints, like i.e. costs (Dogson et al. 2001).

There are different MCA techniques available; some require complete information from decision makers about their objectives, others are more participative and work with decision makers to clarify their priorities. Again others analyse problems without relying on preference information at all (Edwards-Jones et al. 2000). Nonetheless, there are steps that all MCA's have in common. (A comprehensive overview of the available MCA techniques is given in Omann (2004) and Gamper (2004)).

The creation of the performance matrix requires the analysts to determine relationships between options and their impacts on criteria. More precisely, the fourth step follows in two parts: scoring and weighting.

So far, the different options have different measures attached to them (i.e. biodiversity loss is measured in numbers, direct costs are measured in monetary units). Because one cannot compare different measures straight away, a common one in terms of a scale needs to be found. If we face multiple attributive decisions, a numerical scale is created to deal with this complexity. For more simplified decision structures this is not necessary, and an easy dominant or outranking relation can be evaluated (for further details see Fandel and Spork, 1983 or Vincke et al. 1989). Usually a scale is chosen between 0 and 100. So, for example, if there are 3 different technical solutions to build avalanche barriers and the first one costs 30 Million Euros, the second 40 and the third 50, then the first solution will be allotted with a score of 100 as it is the cheapest option and the last one with a score of 0 as it is the most expensive one. Now, the way the analyst processes this information (what is the score for technical solution 2?) depends on the particular MCA method she is using. Basically, one can go about this scoring process by building a value function and simply reading off values in a function given the lowest and highest scores. Another approach is to use an interval scale through the integration of expert judgments (direct rating). In a third alternative, one can choose an indirect method by letting experts make verbal pair-wise assessments. Analytical Hierarchy Process or Rembrandt and Macbeth are examples for conducting these steps. Once the alternatives and the relevant criteria are defined, the criterion scores can be determined and consequently an impact matrix can be constructed (second part) taking the following example (table 8.1).

Alternatives still can not be compared as a unit of preference since the values do not as yet reflect preferences as such. In taking the exemplary matrix from above, this could mean that a local governor might find the subsidisation criterion more important than the biodiversity one, as opposed to the ecologist's expertise who finds biodiversity utmost important. As a consequence, weights can be seen as trade-off values, indicating how much of one criterion you would be prepared to give up in return for an improvement on another criterion (Belton and Vickers 1990). Again, these weights are taken into the impact matrix, this time an option's score on a criterion is multiplied with its weights. After assigning this to all criteria the sum of these products gives the overall preference for that

Table 8.1 Example of an impact matrix of a protection measure

Criteria/Options	No Avalanche Barrier	Technical Solution 1	Technical Solution 2	Technical Solution 3
Economic Direct (financial) costs	0 Mil. Euros	10 Mil. Euros	20 Mil. Euros	30 Mil Euros
Environmental Environmental Impact Assessment – loss of biodiversity	No loss	Loss of 50 species	Loss of 25 species	Loss of 10 species
Social Number of People being protected	No people saved	200 people saved	100 people saved	300 people saved
Institutional Subsidies at national level	None	70%	50%	40%

option. This process is repeated for each of the options and finally a rank can be given to each option's score (Dogson et al. 2001). These scores can be re-built into the adjusted impact matrix. With these results it will be possible to rank the options (or in the case of simpler models, one option can be recommended).

MCA seeks to support the selection of projects and policies which are most efficient and optimal. The interpretation of MCA results can be illustrated in costs and benefits graphs where a relative value-for-money comparison can be shown (Dogson et al. 2001) and also dominating options drawn out (being cheaper and more beneficial). In all ex ante as well as ex post cases of MCA, the analyst must make predictions concerning future physical flows, future relative values, as well as a number of criteria and stakeholders, therefore a sensitivity analysis, as an essential final stage, is conducted (Hanley and Spash 1993).

8.3.2 MCA's potential for natural hazard management

In situations of decision processes in natural hazard management it has been shown that the realisation of plans for protective measures could not be put forward because of a lack of acceptance among the affected interest groups. In order for such projects to be successful, decision makers need instruments that enable them to assess all possible alternatives according to the necessary criteria while at the same time allowing for the inclusion of stakeholders' preferences. MCA values qualitatively and quantitatively all positive and negative impacts of measures against natural hazards. Based on this generated information, the final decision maker is in possession of a ranking of alternatives according to the level of acceptance among the relevant parties. Through participation processes a higher level of acceptance can be achieved since the affected parties do not only take part in the decision making process but also tend to increase their willingness to compromise through the increased level of their information basis.

On the downside of MCA, sometimes considerable, transaction costs can arise: the elicitation of preferences through discussion processes among stakeholders can carry high information and time costs. Not going into detail of this discussion for the scope of this paper is limited, it needs to be stated that trade-offs between all the described alternatives have to be made and the costs and benefits of each of the options have to be considered carefully given the decision at hand.

8.4 Conclusions

In natural hazard and integral risk management, economics is asked to contribute – besides the estimation of potential damages as well as the identification of cost efficient means – to the identification of optimal protection measures and optimal (remaining) risk. In the case of private goods, the 'invisible hand' of the price system results in the optimal accordance of demand and supply in a competitive market. With protective measures being classified as public goods, however, the market fails. The decision about optimal prevention measures is therefore a collective or social choice.

Cost-benefit analysis (CBA) as the standard approach in economics based on welfare economics is geared to identify the optimal strategy for alternative measures and remaining risk by balancing social benefits against social costs. As an outcome-oriented approach taking a public perspective it considers and compares the net benefits of all relevant alternatives, i.e. in the case of integral risk management the set of alternative prevention measures and the benefits and costs from all individuals involved. As a decision-support system it addresses the question of optimal (remaining) risk and the efficiency of alternative risk management strategies by selecting the alternative that has the highest net benefit.

Apart from this output-oriented approach of standard welfare economics also a process-oriented approach, based on new institutional and constitutional economics, is concerned with efficiency, but it does so as a meta-decision problem. The basic question of this alternative politico-economic approach is the same as in cost-benefit analysis, namely: "what is the optimal or acceptable risk and efficient prevention for society?". However, instead of focusing on the socially optimal outcome it is basically concerned with the rules the members of society themselves would consider optimal to decide about acceptable remaining risk and prevention measures. According to this approach those decision rules are considered to be optimal that all members of the society - or more specifically, all stakeholders or all those benefiting from such measures and those bearing the costs - could agree and decide upon (Olson 1969, Blankart 2006). In principle, a wide range of alternative decision rules are at their disposal: the mechanism chosen might be the market or price system. Alternatively, bargaining could be the preferred decision mechanism or the political system with democratic participation rights and specific voting rules. Moreover, it could be decided to refer to a bureaucratic mechanism, or even the advice of experts might be decisive.

While cost-benefit analysis offers a decision-support mechanism for politicians indirectly representing individual preferences as well as leaving final decisions to political representatives, the process-oriented approach focuses on political decision processes explicitly where people directly decide themselves and thus their preferences are expressed and aggregated via the voting process. The politico-economic focus therefore is to establish those fundamental institutions which make politicians and bureaucrats most responsive to individuals' interests and which finally lead to the best possible fulfilment of individual preferences (Frey and Stutzer 2007).

Conducted in a participative manner, multi-criteria analysis (MCA) is argued to offer a new instrument which combines elements of both, the outcome-oriented and the process-oriented approach. On the one hand, it is a decision-support tool for politicians similar to cost-benefit analysis. On the other hand, it aims to incorporate participation of relevant stakeholders in its analytical process and thus realises the direct representation of preferences as well as their intensities.

The focus on a process-oriented approach and multi-criteria analysis as a decision-support system in natural hazard management is to be considered as an appeal to take individual preferences seriously but also to pay special regard to their allowance in public policy. Through participation processes, individual preferences are put forward directly into the political decision-making process. Moreover, through participation, a higher level of individual knowledge may be achieved as well as higher acceptance of public policies can be expected. More direct involvement and participation in political decision-making processes thus helps to overcome market failure on the one hand, and in addition it helps to cope with political failure in natural hazard management on the other hand.

References

Bauer, C (1986) Nutzen-Kosten-Untersuchungen – Fruchtbringende Entscheidungshilfen oder Tarnkappen für politische Willkür? Der Öffentliche Sektor 11(1), pp 36-48

Belton V, Vickers, S (1990) Use of simple Multi-Attribute Value Function incorporating visual interactive sensitivity analysis for Multiple Criteria Decision Making. In: Bana e Costa, CA (ed). Readings in Multiple Criteria Decision Aid, Berlin, Springer, pp 320-334

Blankart, CB (2006) Öffentliche Finanzen in der Demokratie: Eine Einführung in die Finanzwissenschaft, 6th edition, Vahlen, Munich

Boardman, AE, Greenberg, DH, Vining, AR, Weimer, DL (2001) Cost-Benefit Analysis: Concepts and Practice, 2nd edition, Prentice Hall, Upper Saddle River

Bonus, H (1996) Institutionen und Institutionelle Ökonomik: Anwendungen für die Umweltpolitik. In: Gawel, E. (ed) Institutionelle Probleme der Umweltpolitik, Zeitschrift für Angewandte Umweltforschung, Sonderheft 8. Berlin, pp 26-41

Buchanan JM, Tullock, G (1962) The Calculus of Consent. Logical Foundations of Constitutional Democracy, Ann Arbor, Michigan

Buchanan, JM (1991a) Constitutional Economics, IEA masters of modern economics, Oxford

Buchanan, JM (1991b) The Economics and the Ethics of Constitutional Order, Ann Arbor, Michigan

Coase, RH (1988) The Firm, the Market and the Law, The University of Chicago Press, Chicago

De Marchi B, Funtowicz SO, Lo Casciao S, Munda, G (2000) Combining Participative and Institutional Approaches with Multicriteria Evaluation. An Empirical Study for Water Issues in Troina, Sicily. Ecological Economics 34, pp 267-282

Dogson J, Spackmann M, Pearman A, Philips, L (2001) Multi-Criteria Analysis: A Manual. Office of the Deputy Prime Minister, London, United Kingdom

Edwards-Jones G, Davies B, Hussain S (2000) Ecological Economics. An Introduction, Blackwell, London

Fandel G, Spronk J (1983) Multiple Criteria-Decision Methods and Applications, Springer, Berlin

Frey, BS (1990) Ökonomie ist Sozialwissenschaft, Vahlen, Munich

Frey BS, Kirchgässner G (2002) Demokratische Wirtschaftspolitik. Theorie und Anwendung, 3rd edition, Vahlen, Munich

Frey BS, Stutzer A (2007) Should National Happiness be Maximized? Institute for Empirical Research in Economics, University of Zürich, Working Paper No. 306

Fritsch M, Wein T, Ewers HJ (2007) Marktversagen und Wirtschaftspolitik, 7th edition, Vahlen, Munich

Gamper CD (2004) Building Energy Scenarios for Eastern Styria (Austria) using Multi Criteria Mapping with the Inclusion of Participatory Methods. Master of Science Thesis, University of Edinburgh

Gamper C (2005) Can Public Participation Lead to an Increase in Efficiency and Effectiveness of Decision Making? A Preliminary Assessment of the Relevant Concepts. alpS Working Paper Series, WP 06-05

Gamper C, Thöni M, Weck-Hannemann H (2006) A Conceptual Approach to the Use of Cost-Benefit and Multi Criteria Analysis in Natural Hazard Management. Natural Hazards and Earth System Sciences 6, pp 293-302

Hanley N, Spash CJ (1993) Cost-Benefit Analysis and the Environment, Edward Elgar, Cheltenham

Hansjürgens B (2004) Economic Valuation through Cost-Benefit Analysis – Possibilities and Limitations. Toxicology 205, pp 241-252

Hanusch H (1987) Nutzen-Kosten-Analyse, Vahlen, Munich

International Society on MCDM (2004) International Society on Multi-Criteria Decision Making http://www.terry.uga.edu/mcdm/ (accessed date 17.6.2004)

Kirchgässner G (1991) Homo oeconomicus, Mohr/Siebeck, Tübingen

Lackner-Frey E (2004) Öffentliche Güter im individuellen Entscheidungskalkül. Möglichkeiten und Grenzen verschiedener Präferenzenthüllungsverfahren, Dr. Kovač, Hamburg

Leiter A, Thöni M, Weck-Hannemann H (2007) Protective Measures against Avanlanches – Are they Worth their Costs? (forthcoming in this volume)

Mueller D (1996) Constitutional Democracy, Oxford University Press, New York

Munda G (1995) Multicriteria Evaluation in a Fuzzy Environment. Theory and Applications in Ecological Economics, Physica, Heidelberg

North DC (1990) Institutions, Institutional Change and Economic Performance, Cambridge University Press, New York

Olson M (1969) The Principle of 'Fiscal Equivalence': The Division of Responsibilities Among Different Levels of Government, American Economic Review 59, pp 479-487

Omann I (2004) Multi-Criteria Decision Aid as an Approach for Sustainable Development Analysis and Implementation, PhD Thesis, Karl-Franzens University Graz, Graz

Pappenheim R (2000) Neue Institutionenökonomik und politische Institutionen, Peter Lang, Frankfurt

Raschky PA, Weck-Hannemann H (2006) El "riesgo de caridad": Análisis económico de la ayuda gubernamental tras las catástrofes naturales.Gerencia de riesgos y seguros 93, pp 17-31

Raschky PA, Weck-Hannemann H (2007a) Charity Hazard – a Real Hazard to Natural Disaster Insurance? Environmental Hazards (forthcoming)

Raschky PA, Weck-Hannemann H (2007b) Vor- oder Nachsorge? Ökonomische Perspektiven. In: Glade T, Felgentreff C (eds) Naturrisiken und Sozialkatastrophen, Elsevier, Heidelberg (forthcoming)

Richter R, Furubotn EG (2003) Neue Institutionenökonomik, 3rd edition, Mohr Siebeck, Tuebingen

Schwarze R, Wagner G (2004) In the Aftermath of Dresden: New Directions in German Flood Insurance, The Geneva Papers on Risk and Insurance 29(2), pp 154-168

Vincke P, Gassner M, Roy B (1989) Multicriteria Decision-Aid, Wiley & Sons, West Sussex

Weck-Hannemann H (1994) Die politische Ökonomie der Umweltpolitik. In: Bartl R, Hackl F (eds) Einführung in die Umweltpolitik, Vahlen, Munich, pp 101-117

Weck-Hannemann H (2006) The Efficiency of Protective Measures. In: Ammann, WJ, Dannemann S, Vulliet L (eds) Risk 21 - Coping with Risks due to Natural Hazards in the 21st Century, Taylor & Francis, London, pp 147-154

9 Alternative Risk Transfer and Alternative Risk Financing

M. Gruber, R. Wiesner

9.1 Introduction

There are, in general, several ways of managing risks. In case risks cannot be avoided or minimised, companies also dispose of certain mechanisms in order to retain, fund or transfer them. Because of the constantly growing complexity and volume of (actuarial) risks and the development of new risk classes over the last decades, the insurance market has often proved to be unable to provide coverage for certain risks. Consequently, these risks are deemed to be "uninsurable". In addition to traditional instruments alternative solutions have been developed, in particular for high risk layers such as risks related to natural catastrophes.

The widely used term "ART" comprises various forms of alternative risk transfer mechanisms and alternative risk financing mechanisms ("ARF"). In the first case, actuarial as well as other risks are transferred from one business entity to another, using new structures. Additionally, not only financial and operational risks but also actuarial risks can be funded by innovative financial instruments. These mechanisms are said to be "alternative" as conventional risk transfer and financing structures are amended, combined and finally used as alternatives to traditional mechanisms, whenever additional cover and capacity is needed.

9.2 Self-insurance, Captives and Risk Retention Groups

Self-insurance programmes, Risk Retention Groups and Captive Insurance Companies are mainly US phenomena, originating from hardened insurance market conditions and crucial insurance liability crises, in particular during the 1970s and 1980s. Besides, additional forms of pooling arrangements have emerged (see Pollner 2001). An overview of various-self-insurance mechanisms is given in Figure 9.1.

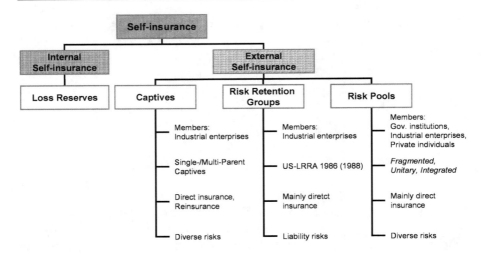

Fig. 9.1 Self-insurance Mechanisms

Self-insurance in its proper meaning comprises no risk transfer, as this would be the case for traditional insurance contracts. Instead, identified risks are retained by the company, which provides cover for future economic losses by building up loss reserves on its balance sheet. These reserves are earmarked, which means, that the funds shall not be used for any other purpose than the coverage of certain economic losses, as for example claims arising from natural catastrophes or product liability. As the company provides funds prior to the realisation of actual losses, self-insurance is often termed as pre-loss finance (see Banks 2005).

One problem in relation to **pre-funded retention**, which is another term used for self-insurance, may be the extent to which the company is able to get outsiders to recognise the actual earmarking of these funds. A lack of credibility associated with loss reserves on the balance sheet of a company can become a serious problem for risk retaining firms (see Culp 2006).

Beside this "original" form of self-insurance, which in legal terms is also called "**internal self-insurance**", several other forms have emerged. Captive Insurance Companies and Risk Retention Groups are the most wide-spread forms of the so-called "**external self-insurance**" (see Mueller 1988). Actually, external self-insurance mechanisms all base upon the combination of risk transfer and risk financing in the way, that special (separate) entities take responsibility for the transformation of risk (transfer or financing), instead of accounting for expected future losses by building earmarked reserves appearing on the balance sheet of the company. The special entity may be a separate corporate body, but this is not an absolute requirement for the qualification of some contract as external self-insurance. In general, internal and external self-insurance only make sense for a company, if identified risks are homogeneous, properly measurable and ratable.

In the following paragraphs, various structures of Captive Insurance Companies are presented. Thereafter, the notion of Risk Retention Groups will be explained.

As already mentioned above, **Captive Insurance Companies** ("captives") constitute a special concept of external self-insurance. Thereby, some industrial enterprise establishes a legally separate corporate entity in order to transfer or finance certain types of risks that are not insurable – or only insurable against very high premiums - on the traditional insurance market. Traditional insurance premiums are to a much larger extent dependent on insurance market cycles than captive insurance premiums, as the captive insurance contract is set up between two familiar entities (see European Commission 2000).

There are two criteria by which one can **distinguish** between several forms of captives: the number of parent companies and the intended field of application. On the one hand, captives can be set up by either one single company (Single-Parent Captives) or by multiple companies (Multi-Parent Captives). On the other hand, companies may implement a captive structure for means of either direct insurance or reinsurance. In any case, corporate risks are transferred from parent companies to the captive against the payment of a premium. The arrangement itself can be compared with a traditional insurance contract (see Borch 1990; Farny 1995).

In the case of **Single-Parent Captives** an industrial enterprise provides the captive with 100% of share capital, which can comprise internal funds, financial claims and other assets of the parent company (see Figure 9.2). The required guarantee fund varies among state regulations. The European Union, for example, passed a directive regulating the minimum capital requirements for reinsurance companies as well as for reinsurance captives domiciled in the member states. Whereas for traditional reinsurance companies the guarantee fund shall not be less than EUR 3 mio., the home member state is allowed to set the minimum guarantee fund required for captive reinsurance undertakings at a lower amount (but at least at EUR 1 mio.) (see Art 40 Directive 2005/68/EC). For direct (captive) insurers the guarantee fund, for example for the class of indemnity insurance, is set at the amount of at least EUR 300.000 (see Art 17 (2) First Council Directive 73/239/EEC).

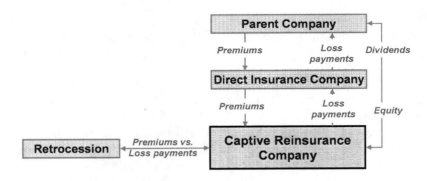

Fig. 9.2 Exemplary Structure of Single-Parent Reinsurance Captives

This capital as well as all insurance premiums paid to the captive by the parent company are placed in the market for rather short-term low-risk investments, in order to guarantee liquidity at the event of loss payments. Single-Parent Captives are applied to obtain a flexible and tailor-made risk coverage, including participation in positive investment developments.

Multi-Parent Captives are established by several companies, which provide the captive with the required amount of capital and with premium payments. A distinction can be made between the so-called Group or Association Captives, Protected Cell Captives (also called Segregated Cell Captives) and Rent-a-captive structures (see Culp 2006). A **Group or Association Captive** consists of a number of companies that are typically operating in the same branch of business and, consequently, facing similar classes of risk. Due to this homogeneous risk exposure and the joint experience with typical risks in this particular branch of business, the captive solution enables companies to directly apply their knowledge, generate additional benefits from tailor-making their insurance contracts and participating in positive loss experiences over the years. In the case of an average Group or Association Captive, the companies' contributions to the core capital are intermixed, which means that loss payments are served by the joint capital stock (including premium payments) of the captive.

This is not the case for **Protected Cell Companies**, where every participating company is given its own insurance account ("cell"). Payments into and out of this account are transacted independently from other cells. If the capital in the cell does not suffice, parts of the core capital may be liquidated to pay for the claims, if the other participants agree. Later on, the particular company will normally have to replenish the capital fund of the Protected Cell Company (see e.g. Jersey Companies Law 1991 and Amendment No. 8, 2006; Guernsey Companies Law 1996).

During the last few years a new captive concept has emerged in several legislations. These captives are called **Incorporated Cell Companies**. In contrast to the Protected Cell solution, the "incorporated cells" are treated as separate legal entities, which have to report separately to the insurance authorities, and which are legally enabled to enter into reinsurance contracts – for the single cell. Only the core capital and the core administrative body are shared by the cell owners. The balance sheet of the cell needs to be consolidated with that of the cell owner. Each share holder enjoys participation in positive loss developments of the incorporated cell (see e.g. The Incorporated Cell CompaniesOrdinance 2006, Guernsey).

Rent-a-captive solutions are predominantly established by (re-)insurers or financial intermediaries as Rent-a-captive holding companies. These parties provide an administrative structure, the required capital stock and know-how in the field of risk transformation. Then, industrial enterprises can rent a customer account, which is similar to a protected cell and managed by the Rent-a-captive holding company. This concept may typically attract smaller industrial enterprises with little experience in risk management and risk transfer. In particular for these companies, the knowledge provided by financial intermediaries and (re-)insurance companies may be very valuable (see Culp 2006).

As these structures are all more complex than traditional insurance solutions the question arises, to what extent Captive Insurance Companies can generate additional benefits for industrial enterprises. One explanation can be found in the quite **compact definition** of Captive Insurance Companies by Christopher L. **Culp**, who defines the notion of captive as "a risk management structure or solution involving the participation of insurance and/or reinsurance companies that enable firms either to finance or transfer some of the risks to which they are exposed in a non-traditional way (…) thereby functioning as synthetic debt or equity in a corporate customer's **capital structure**" (Culp 2002). The possibility of **smoothing** a company's **balance sheet** by transferring risks, and consequently loss reserves, to a captive, which is owned by the insured company itself, is considered as a very attractive alternative to traditional insurance contracts. Beside the benefit of balance sheet protection, the insured company, at the same time being a share holder of the captive, **participates** in the positive as well as negative development of the Captive Insurance Company. Thus, a certain incentive for **risk preventive measures** within the parent company is given. By this means, captives can contribute to the depletion of **moral hazard**, which is a major problem faced by traditional insurers.

As captives are primarily established in low-tax domiciles the parent companies can additionally benefit from tax savings. In the1970s, when the captive industry started to grow constantly, there were two main reasons for companies to establish a captive: a hard insurance market with high premiums for certain risks, and tax advantages in specialised domiciles. At present, other **advantageous characteristics** of captive structures prevail, as for example their high degree of flexibility and adaptability to the companies specific needs, their contribution to an improved and target-oriented risk management, their potential for balance sheet protection and their weak dependence of the traditional insurance market cycles (see Lee and Ligon 2001; Eisenhauer 2004; Booth 2006).

Typical **domiciles** are, for example, Bermuda (~1400 captives), Cayman Islands (~700), Vermont (~ 500), Guernsey (~400), British Virgin Islands (~350), Luxembourg (~300), Ireland (~200), Isle of Man (~170) and Switzerland (~50) (see Aon 2005). As can be seen, the **European market** for captives – as for all alternative risk transfer mechanisms – is still in the early stages, but is expected to grow considerably within the next decade. In order to make the captive concept more attractive for a broader range of companies in Europe, additional emphasis needs to be placed on the harmonisation of insurance and capital markets, in particular on a closer collaboration and a mutual acknowledgement of legal regulations among the different EU member states in the field of corporate law, insurance law and fiscal law.

Another form of external self-insurance is the establishment of **Risk Retention Groups (RRGs)**. These self-insurance companies can solely be set up in the application area of the US Liability Risk Retention Act (LRRA 1986). The LRRA 1986 determines that, a Risk Retention Group is a corporation or other limited liability association, functioning as a captive insurance company and organized for the "primary purpose of assuming and spreading all, or any portion, of the liability risk exposure of its group members" (15 USC Sec. 3901-3902, Liability Risk Retention Act 1986) (member-owners). It must be chartered and licensed as a liability insurance company in one US state. It can also charter as an industrial or association captive under special state captive laws such as Vermont, Delaware, Colorado, Illinois.

According to the definitions of the US Government Accountability Office (GAO) "the Liability Risk Retention Act permits RRGs to provide commercial liability insurance and largely exempts them from regulatory oversight other than that performed by their chartering state" (see GAO 2005). It is to be emphasised that arising from their evolutionary history – liability crises in the 1970s and 1980s - RRGs can exclusively be applied to liability risks. The range of risks that could be insured by RRGs has never been extended thereafter. The LRRA 1986 does not specify characteristics of ownership, control and capital requirements, or establish governance safeguards.

This freedom of state regulation has led to widely varying regulatory standards and consequently to sophisticated domicile migration of RRGs. In 2005 the US GAO conducted a study to find out more about this issue. In the ensuing report, the GAO (see GAO 2005) calls for a strengthening of overall regulations and recommends that state insurance regulators adopt consistent regulatory standards for RRGs. Furthermore, partial preemption shall only be granted to states that adopt these standards, which also include certain minimum corporate governance rules.

9.3 Securitisation

Securitisation can be explained as the transformation of assets or liabilities from the balance sheet of a company into tradable assets. Primarily, illiquid financial assets with the same characteristics in terms of yield, maturity and geographical spread are pooled and re-packaged into marketable securities that can be sold to investors on the capital markets.

The **notion of securitisation** comprises various **financial instruments**, such as **contingent capital solutions, derivatives and bonds**. Whereas in contingent capital solutions emphasis is put on risk financing, derivatives and bonds are risk transfer mechanisms. Risks, and thus liabilities, that are identified by companies can be pooled and transformed into a securitised instrument in order to be transferred to the capital markets. The sector of securitisation is growing constantly on a worldwide level. To give an example, the European securisation issuance is expected to rise up to EUR 400 billion in 2006. Companies acting in different industries apply this mechanism in order to cope with a wide range of risks. Insurance derivatives and the so-called catastrophe bonds ("Cat bonds", "Act-of-God Bonds") being a specific field of application are often referred to as Insurance-Linked Securities (ILS) (see Albrecht and Schradin 1998; Bodie et al 2002). In the following, these two instruments will be focussed on.

Since the 1990s securitisation has increasingly been deployed for **catastrophic risks**, such as liabilities related to damages arising from hurricanes, earthquakes, floods, extreme temperatures and strong winds. In particular, reinsurance companies and large corporations issued **cat bonds** in order to reinsure or retrocede these low frequency-high severity risks appearing in their balance sheet (see Froot 2001).

Several parties are involved in these transactions, what makes this quite a complex deal (as shown in Figure 9.3). A "sponsor" that might be directly affected by natural hazards cedes liabilities to a special purpose vehicle (SPV), which will then carry out the actual securitisation by transforming the liabilities into a marketable catbond. This bond is issued by the SPV and traded on the capital market, where investors will hold shares of this bond and will be paid a coupon above average. The coupon payments in past transactions ranged approximately from LIBOR plus 100 bps to LIBOR plus 1400 bps (see Lane and Beckwith 2006). Then the sponsor enters into a reinsurance contract with the SPV. In the following, the cash payments collected by the SPV are handed over to a trust, which invests this capital into secure short-term assets (e.g. T-Notes with a maturity between one and ten years, other fixed-income instruments). The generated investment returns (floating rates) are usually bundled and swapped against fixed rates (e.g. LIBOR) in order to guarantee regular coupon payments to the catbond investors.

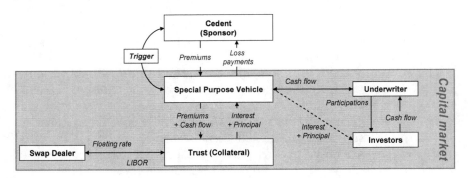

Fig. 9.3 Exemplary Cat Bond Structure

The time to **maturity** of cat bonds varies from one to five years. If no catastrophic event occurs during this period, the catbond pays out the agreed upon coupon. But if the catbond is triggered – in the case of a certain catastrophic event – the sponsor will demand for compensation payments, consequently leading the SPV to retain the coupon ("**coupon-at-risk**") or even parts of the principal of the bond ("**principal-at-risk**"). From an investor's perspective cat bonds are generally attractive, as they are relatively independent of other investments and offer above-average returns. On the other hand, the coupon and even the principal can be at risk.

There are three types of **triggers** that have been used for cat bonds: indemnity triggers, index-based triggers and parametric or technical triggers. An **indemnity-based** catbond is triggered, if the actually reported damages in a specified region or company within a specified time horizon will exceed an ex ante specified level. For the second type of trigger an **index** is calculated using data from particular insurance markets together with complex simulation methods. For example, companies such as Risk Management Solutions (RMS), EQECAT, AIR Worldwide Corp., PCS or Guy Carpenter accomplish these index calculations. In the past, particularly the Property Claims Service Index (PCSI) and the Guy Carpenter Catastrophe Index (GCCI) (see Guy Carpenter 2005b) were used as a basis for cat bonds and also as underlyings for insurance derivatives. If some marginal index level is exceeded, the catbond is triggered. If a catbond is related to a **parametric** or **technical trigger**, this means, that the payout pattern of the bond is linked to the development of a specific technical parameter, such as temperature, storm force, magnitude of earthquakes or rainfall.

Cat bonds are mainly used for the coverage of high-risk layers. These transactions typically include high retention levels, which counteracts moral hazard. In the first three quarters of 2006, cat bonds were issued worldwide amounting to approximately USD 3.3 billion (see Lane and Beckwith 2006).

In addition to cat bonds, **insurance derivatives**, such as options, swaps, futures and forward agreements, are deployed for the insurance of catastrophic risks. On the one hand, these transactions can be related to indemnity or modelled indices, such as the GCCI and the PCSI ("PCS Options", "PCS Futures"), and are mainly traded over-the-counter (see CBoT 1995). On the other hand, parametric or technical indices and triggers can be used as underlyings. In this case, the transactions are called **weather derivatives**. This special form of insurance derivatives will be emphasised in a separate section.

In recent years, the CBoT as well as other exchanges have made efforts to establish a liquid market for insurance derivatives, for example by improving index calculation methods, by reducing the option tick size and easing participation requirements. The CBoT amended its ISO-Option and ISO-Futures contracts, which had been traded from 1992 until 1995 with little success (only about 5.700 transactions in total). Thereafter, the CBoT concentrated on the newly set up PCS Catastrophe Options basing on the PSCI (see Abrecht and Schradin 1998). However, whereas the market for weather derivatives is growing constantly, the market for indemnity index based derivatives lags behind. The PCS Futures were delisted in 1996, the PCS Options in 2000 due to the lack of trading volume – only 6 contracts in 2000 (see also www.cbot.com).

A major problem related to all index-linked securities is the **basis risk**, which needs to be considered when closing these transactions (see also Cummins et al 2000). Basis risk can be explained as the risk of incomplete coverage due to discrepancies between the development of the underlying index and the actual losses. This discrepancy is created by the uncertainty surrounding the applicability of coverage for losses yet to occur (see Kerr 2006) Thus, the more accurately potential losses and indices are modelled, the more the provided cover will be attuned to the actual damages.

Though basis risk is a major concern, insurance derivatives and in particular cat bonds offer specific advantages to the sponsors as well as to investors on the capital markets. As these instruments are almost independent of insurance market cycles and developments on the capital markets, they can be used by insurers, reinsurers and also industrial enterprises for the purpose of portfolio **diversification**. Furthermore, the capacity of the insurance market is amplified by the additional **capacity** of the capital market. Another advantageous trait of securitised products is that they help to smooth the **balance sheet** of any company which intends to protect itself against major liabilities.

9.4 Financial Reinsurance and Finite Risk

These two terms have caused a number of controversies, investigations and new regulations within the last years. Financial reinsurance as well as finite risk contracts combine risk transfer and risk financing in a way that authorities were doubtful concerning their proper legal treatment. Some questions are still to be declared.

Early "finite" contracts were called **time and distance** policies offered by Lloyd's of London. These treaties involved the payment of a large one-off premium by the cedent to the reinsurer, as well as a fixed schedule for the repayment of funds to the cedent at maturity. This structure was used in order to smooth insurers' balance sheets and almost solely included the transfer of timing risk. Because of regulatory problems, these structures were abandoned (see Culp 2002).

A recently published **EU Directive** defines the term "finite reinsurance" as reinsurance under which the explicit maximum loss potential, expressed as the maximum economic risk transferred, arising both from a significant **underwriting risk and timing risk transfer**, exceeds the premium over the lifetime of the contract by a limited but significant amount, together with at least one of the following two features:

(i) explicit and material consideration of the time value of money,

(ii) contractual provisions to moderate the balance of economic experience between the parties over time to achieve the target risk transfer (see Art 2 (q) Directive 2005/68/EC).

Finite contracts originate in financial reinsurance that was used by insurance and reinsurance companies in order to transfer not only actuarial but also financial risks. Over the years, industrial enterprises entered the market for **finite covers**, which means, that a certain limited layer of risk is transferred to another business entity, such as insurance companies.

At present, in principal, finite contracts can be concluded either between an insurer and a reinsurer, or between a corporate end user and an insurer. In the first case, the **term** "financial reinsurance" is used, whereas the second case is referred to as "finite risk" contracts (see European Commission 2000). Both mechanisms can be deployed in order to manage, transfer and/or finance **various types of risk**, such as financial and operational risks, whereas traditional insurance contracts only provide cover for actuarial risks. The contracts may also feature the consideration of positive loss histories, particular **participation** clauses and **multi-year** contract periods in order to spread insured risks over time. Furthermore, the cedent (or insured) has the possibility to **regain** cash payments deposited in an own "loss experience account" of the insurer, if no event occurred during the contract period. In this **experience account** profits and losses on the actual underlying deal are tracked for the whole period. Repayments to the cedent may also include investment returns.

Finite covers are available in various forms. On the one hand, **retrospective** contracts facilitate the transfer of risks which were already realised in a past period, but though have not been reported to the cedent (IBNR losses, which stands for "incurred but not reported losses"). Loss Portfolio Transfers (LPT), Adverse Development Covers (ADC), funded Excess of Loss contracts (XOL) and Retrospective Aggregate Loss Covers (RAL) are among these retrospective finite forms. Also **Run-o**ff solutions are related to retrospective covers, as liabilities and loss reserves are "forwarded" to (re-)insurers against the payment of a premium. But in the case of Run-off solutions the underwritten risk by the insurer is not limited, thus not finite (see Swiss Re 2003; Banks 2005; Culp 2002).

On the other hand, **prospective** contracts, such as Financial Quota Share treaties (FQS) and Spread Loss Covers, are offered by (re-)insurers. Thereby, present and future losses are covered, similar to traditional insurance contracts. The different forms of finite covers are shown in Figure 9.4.

Another **distinction** between finite risk solutions is the point in time of providing funds by the involved parties. In the majority of cases, the insured pays a regular premium into his/her experience account managed by the insurer, who also makes for an appropriate capital investment. In the case of a compensation payment from the insurer to the insured after the conclusion of the finite contract, firstly the funds of this account are exhausted. If additional cash payments are required, the insurer acts as provider of supplemental capital (comparable to traditional excess-of-loss covers). Subsequently, the funds need to be replenished by the insured. This kind of contract design is also referred to as **post-loss financing**, because the insured will have to fill up the experience account after the occurrence of a specified event ("ex post") by paying adapted premiums into the experience account. These premiums are calculated regarding the present value of the actual loss payments (see Culp 2006).

The contract might also be shaped in a different way. The cedent pays a certain amount of cash into the experience account at the conclusion of the finite insurance contract. Thereafter, no premium payments are required. The insurer again makes for an appropriate capital investment, and if the contractually specified event occurs, the policy is triggered. Compensation payments are covered by the funds of the experience account. Additional cover is provided by the insurer, if stipulated in the initial contract. This type of finite cover can be seen as **pre-loss financing**.

The main advantages, but at the same time also the crucial legal question marks arising from finite risk contracts are the transfer of both timing risk and underwriting risk with flexible weights in each part, as well as the possibility of smoothing balance sheets and improving insurers' ratings (see Culp and Heaton 2005; Guy Carpenter 2005a; Metzler et al 2005).

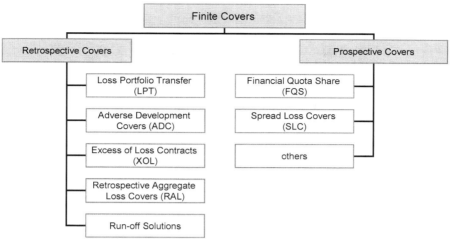

Fig. 9.4 Overview of Finite Risk Structures

9.5 Contingent Capital Structures

Contingent capital instruments provide the buyer with the right to issue and sell securities at a fixed price during a specific period of time if a predefined trigger event occurs. These securities may be debt, equity or some hybrid form. Contingent capital structures link insurance and financial markets closely by raising funds from capital markets upon the trigger of an insurance-related event.

Contingent capital instruments are not directed at pure risk transfer but rather at risk financing. The advantage they offer is to have access to less expensive capital after the occurrence of a financially stressful event (defined as a trigger). The price of regular funding in the capital market or bank loans would be much higher after such an event. Contingent capital bridges the gap between insurance and self-insurance, since it provides capital for a price, not full coverage for a loss.

The importance of contingent capital solutions in the market is constantly growing. In 2000 the volume of transactions was roughly USD 1.5 billion. This was followed by a decline in 2001, but the market improved in 2002 (Swiss Re 2003).

Contingent capital instruments can be divided into two broad classes: contingent debt and contingent equity. Contingent debt refers to debt financing, whereas contingent equity corresponds to equity financing after a trigger event. Within these two classes we can further subdivide into different contingent capital structures. Three main types will discussed here: standby credit facilities (debt financing), contingent surplus notes and catastrophe equity puts (equity financing).

a) In the case of standby credit facilities, a bank agrees to provide a certain amount of credit to a corporate borrower for a specific period of time. The corporation can draw on the credit line at any time in the case of a trigger event. Then the standby credit becomes an actual loan and must be repaid at a predetermined interest rate. In return for this option, the borrower owes the bank a commitment fee on any undrawn balances.

b) Accounting regulations in the USA enable insurers to issue surplus notes. A surplus note is a type of debt instrument or preferred stock that is treated as capital for regulatory purposes. Typically in a contingent surplus note structure the insurers and investors are linked by the establishment of a special purpose trust, which is capitalized on by investors through trust-issued notes paying an enhanced yield. The proceeds from the sale of contingent notes to investors are then used to finance the purchase of high quality securities, such as for example US Treasury securities. Upon the occurrence of a predefined trigger event the insurer has the right to sell contingent surplus notes to the trust to raise capital at fixed conditions. This provides immediate access to capital after the occurrence of a catastrophic underwriting loss for the insurer. In exchange for providing the initial commitment, contingent capital and taking the credit risk of the surplus notes, the investor achieves a yield that is greater than similarly rated high-quality, marketable securities. The option price on contingent surplus notes equals the difference of the yield paid to investors and the earned interest rates on the collateral assets which are held in the trust.

If the option is exercised the insurer has to repay the received capital to the trust at a predetermined interest rate during a specific time period. In this way the trust receives the capital necessary to buy back contingent surplus notes or to pay the interest to the investors. The advantage of the use of contingent surplus note structures for the insurer is to obtain a post-loss funding commitment in advance for fixed conditions that might be difficult to realise after a loss event (Froot 1999; Elliott 2001).

c) Another alternative of a contingent capital structure is known as catastrophe equity put. This kind of option entitles the holder to issue equity, e.g. new shares at a pre-negotiated fixed price, if a trigger (natural disaster) is breached. In a typical structure a firm purchases a put option from a counterparty that gives it the right to issue and sell new stock directly to the counterparty if the trigger is activated. In exchange the option writer receives an option premium. Since share prices after a catastrophic event normally decline, an equity put instrument offers the benefit of post-loss funding at predetermined levels. A disadvantage of catastrophe equity puts is, however, that the ownership structure can be affected if the option is exercised and new shares are issued. To avoid dilution that arises from the issuance of common shares, preferred equity can be issued (Froot 1999, p 15; Elliott 2001; ISO 1999).

Contingent capital instruments are usually structured like American put options. The price of this instrument is the option premium plus a loading that the capital provider adds. In addition, the price may also include a financing rate on the underlying asset if the option is exercised. This financing rate can be fixed or floating and may be based on the credit rating of the company (Culp 2002; Shimpi 2001).

Contingent capital facilities can supplement existing capital resources and complement bank facilities. They offer great structural flexibility and can be used to finance high-severity events, providing a company with post-event solvency and liquidity relief. Additionally, contingent capital can address risks that cannot be adequately hedged with traditional insurance tools. Contingent capital instruments offer diverse applications for insurance and reinsurance companies, financial institutions, manufacturing companies or municipalities. In general, it is a supplemental capital source that should be used in co-ordination with other financing products.

There are several reasons why providers of contingent capital instruments are usually insurance companies. Firstly, they have core competencies in risk analysis, pricing and assessment of high severity risks. Insurance companies also have the skills to structure contingent capital transactions flexibly and according to the needs of the customers. Additionally, contingent capital offers important diversification possibilities by broadening the product scope. Finite risk programmes, which are related to contingent capital, have already been structured and underwritten by insures for many years (Chubb Financial Solutions 2002).

Figure 9.5 demonstrates the general mechanics of a contingent capital instrument.

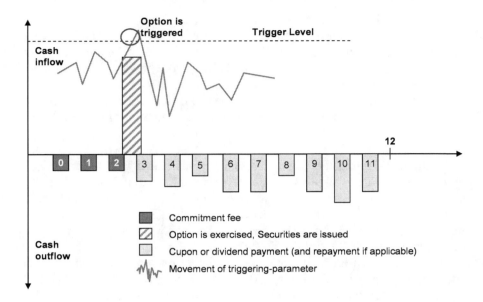

Fig. 9.5 Mechanics of contingent capital instrument (Source: Chubb Financial Solutions 2002)

9.6 Multi-line/Multi-year und Multi-trigger Products

Enterprise-wide risk management enables firms to conduct comprehensive risk identification, measurement, control and monitoring. Based on a detailed analysis of risks it is possible to develop problem- and industry-oriented coverage and to assess the retained risks. In response to the increasing interest in enterprise-wide risk management, many reinsurance companies began to offer integrated risk management (IRM) solutions. The possibility of simultaneous coverage of multiple risks and the ability of firms to participate in ex post profits constitute these structures as alternative risk transfer solutions (Culp 2002).

In this chapter two primary IRM products will be reviewed: multi-line/multi-year and multi-trigger solutions.

Multi-line/multi-year products (MMP) combine uncorrelated risks into an insured portfolio which allows efficient risk transfer and avoids overinsurance. Traditional insurance risk can be combined, e.g. fire, business interruption and liability. Therefore, traditional single insurance risks can be managed comprehensively so that the companies take advantage of the risk consolidation within their own portfolio of risks. MMPs can also include special risks currently covered by banks, e.g. exchange rates or commodity prices, as well as risks which are considered uninsurable if isolated. The liability limit and the deductible are aggregated across all different categories of risk and contract terms, rather than being calculated individually and on a yearly basis. So MMPs allow a distribution of risks within a risk portfolio and over time. Because of the substantial risk transfer capacity of MMP solutions, they present a challenge to the capitalisation of insurance providers. (Swiss Re 1999; Swiss Re 2003).

The key benefits of MMPs for policyholders are efficiency gains, stabilisation of risk costs, administrative efficiency and flexibility. Firstly, the volatility of the loss experience in the deductible on this integrated portfolio is usually less compared to individual insurance policies. Additionally, the business specific diversification potential for the deductibles offers cost advantages and reduces overinsurance. Secondly, premium rates are fixed for several years, so the policyholder takes advantage of stable risk costs. Thirdly, MMPs can reduce administrative costs, as the number of participating insurers and brokers decline and yearly contract renewal procedures are unnecessary. Fourthly, MMPs are tailored to customer needs and hence offer greater flexibility in contract design depending on the clients' risk portfolio and risk tolerance. (Swiss Re 1999; Herold and Paetzmann 1999).

Similar to finite risk products most MMPs include a feature which allows the policyholder to participate alongside the insurer in certain profit circumstances. If the policy remains unused, a refund of some of the premium may be possible. If the policy is used, joint participation in excess investment income may be possible. (Culp 2002)

In contrast to single-line products the aggregated deductible level is set on an integrated view of all risks and should cover the desired retention level on a portfolio basis. For the calculation of the optimal retention level and aggregated deductible the clients' actual loss experiences are analyzed and financial risks are modelled. (Culp 2002)

To address customer needs further, some insurance providers also combine MMPs with traditional products. One example of such a solution is Swiss Re's *Multi-Line Aggregated and Combined Risk Optimization* program. The product is a multi-line/multi-year structure that has a single annual aggregate deductible, a single aggregate exposure limit, and occurrence specific catastrophic excess of loss cover. (Culp 2002).

The fundamental difference between single-line insurance and MMPs in regard to their basis-structures is illustrated in figure 9.6.

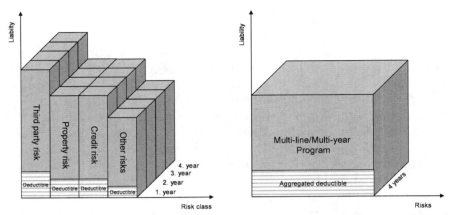

Fig. 9.6 Single policies vs. multi-line/multi-year policies (Source: Herold and Paetzman 1999; Romeike and Finke 2003)

In the case of MMPs a payment becomes due if the aggregate losses of different risk categories exceed the deductible. This means that MMPs are based on a single trigger event. Like MMPs, **multi-trigger products (MTP)** are also based on a holistic risk approach, but at least two triggers must be breached. In addition to an insurance event (first trigger), a non-insurance event (second trigger) also has to occur to receive a payment. One benefit of this structure is that it reduces the probability of a claims payment, as more conditions must be met, which allows insurance cover to be offered at a cheaper premium than with traditional products. Another primary reason for the use of MTPs is the mitigation of moral hazard problems. The second trigger is usually based on events whose outcomes the policyholder cannot influence, so the insurer can be sure that losses have not been deliberately caused. Common triggers are variables, such as commodity prices, exchange rates, interest rates or GNP. Nevertheless, there should be a high correlation of the trigger with policyholders' hedging interests so as to create an effective cover.

However, in practice MMPs and MTPs must overcome many obstacles and disadvantages. Traditional organisation structures in risk management and uncertainties in their treatment regarding tax laws and accounting principles decelerate their adoption as risk transfer instruments. Despite the theory MMPs and MTPs still carry high transaction costs which have hindered their breakthrough. (Swiss Re 1999; Culp 2002)

9.7 Weather Derivatives

Many industries and accordingly their sales and income are at risk due to weather fluctuations. It is estimated that nearly 20-30 percent of the U.S. economy is directly affected by the weather (CME 2006). Other estimations assume that 70 percent of all businesses face weather risk in some way (Jain and Foster 2000). Weather fluctuations still cannot be controlled but with weather derivates a relatively new financial tool has been created which offers protection against weather-related risks. The use of weather derivates enables firms to transfer a portion of their weather risk to capital markets or counterparties.

Weather risk and weather exposure

Weather risks refer to weather events which negatively affect the revenues and earnings of a company. The type and quantity of relevant weather parameters depends on the business area and can include such events as temperature, precipitation or humidity (Schirm 2001). These kinds of weather characteristics are called "noncatastrophic events" which typically have a high-frequency but single events have low severity. They are contrasted with catastrophe-related low frequency – high severity risks, e.g. hurricanes, tornados, etc. Weather derivates were developed to facilitate protection against profit impacts given adverse weather conditions but not for property or catastrophic risk protection (Clemmons 2002).

Weather risks belong to the category of operational risks. In particular they refer to the danger of losses due to external events, seeing as weather conditions cannot be influenced by an enterprise. Furthermore, weather risks are considered to be volumetric risks which can affect both supplies as well as sales.

The sensitivity to weather conditions can be defined as weather exposure. It quantifies the dependency of operating figures to adverse weather conditions and subsequently allows a comprehensive weather risk management. The determination of the weather exposure usually requires a detailed analysis of a company's data and potential influencing factors on a company's business success. Weather derivatives then serve as an efficient instrument to reduce weather risks and respectively weather exposure, and hence minimize their potential negative influence on company's success. A systematic weather risk management process therefore assures competitiveness and presumably causes positive effects and after (Jewson and Brix 2005):

- Reduces the year-to-year volatility of profits
- Profit smoothing leads to constant year-to-year tax burdens
- Low volatility in profits often reduces refinancing costs
- In a listed company low volatility in profits usually translates into low share price volatility, and less volatile shares are valued more highly
- Low volatility in profits reduces the risk of bankruptcy and financial distress

A sample of weather risk and corresponding financial risks faced by various industries is given in Table 9.1

Table 9.1 Illustrative links between industries, weather type and financial risks (Source: Climetrix 2006)

Risk Holder	Weather Type	Risk
Energy Industry	Temperature	Lower sales during warm winters or cool summers
Energy Consumers	Temperature	Higher heating/cooling costs during cold winters and hot summers
Beverage Producers	Temperature	Lower sales during cool summers
Building Material Companies	Temperature/Snowfall	Lower sales during severe winters (construction sites shut down)
Construction Companies	Temperature/Snowfall	Delays in meeting schedules during periods of poor weather
Ski Resorts	Snowfall	Lower revenue during winters with below-average snowfall
Agricultural Industry	Temperature/Snowfall	Significant crop losses due to extreme temperatures or rainfall
Municipal Governments	Snowfall	Higher snow removal costs during winters with above-average snowfall
Road Salt Companies	Snowfall	Lower revenues during low snowfall winters
Hydro-electric power generation	Precipitation	Lower revenue during periods of drought

Characteristics of weather derivatives

A weather derivative is a financial contract between two parties that has a payoff derived from the development of an underlying index. They differ from common financial derivatives as their underlying is not a tradable asset but meteorological variables.

As we already noted weather affects different companies in different ways. In order to hedge these different types of risk, weather derivatives can be based on a variety of different weather variables. These include temperature, wind, rain, snow or sunshine hours. The only condition is that the weather variables can be transformed into an index and that they are measured objectively without the influence of any counterparties.

The majority of weather derivatives use temperature data as underlying, which is transformed into an index via degree days (DD). Degree days are calculated as absolute difference of the daily average temperature and a reference temperature. Subsequently, the DDs are summed up constantly over the contract time to compute weather index values. The temperature is most commonly used as underlying as it offers some advantages in comparison to other weather variables; temperature can be measured reliably, it is continuous and shows relatively high correlation over larger areas (Dischel 1999).

The pay-off of weather derivates is derived from the weather index, not on the actual amount of money lost due to adverse weather conditions. Therefore, it is unlikely that the pay-off will compensate the exact amount of money lost. The potential difference between an actual loss and the received pay-off is known as basis risk. Generally, basis risk is reduced when the company's financial loss is highly correlated with the weather and when contracts of optimum size, structure and location are used for hedging (Jewson and Brix 2005).

Typically, a standard weather derivative contract is defined by following attributes:
- Contract period: defines a start date and an end date, usually a month or a season
- Measurement station: preferably close to the company's location to reduce geographical basis risk
- Weather variable: corresponding to weather exposure and hedging needs
- Underlying: index which aggregates the weather variable over the contract period
- Pay-off function: determines the cash-flows of the derivative
- Strike Level: value of the index at which the pay-off changes from zero to a non-zero value
- Tick and Tick Value: defines how much the pay-off changes per unit of the index

The pay-off functions of weather derivates rely on the pay-off functions of traditional derivative instruments, e.g. options, collars, straddles or swaps. Because weather is not a tradable asset, exercise of weather derivatives always results in cash settlement. In addition, contracts may involve financial limits in the maximum pay-off.

Weather derivatives in comparison to weather insurance contracts

Traditional insurance companies already offer protection against weather risks. So, is there really a need for weather derivatives or can weather risks be hedged by traditional insurance contracts? To answer this question the key characteristics of weather derivatives and weather contracts have to be contrasted.

The main difference between the two instruments is that the holder of an insurance contract has to prove that he actually suffered a financial loss due to adverse weather conditions in order to be compensated. Hence, weather insurance contracts consist of two triggers: a weather event plus a verifiable financial loss. If the insured is not able to show this, he will not receive payments from the contract (Becker and Bracht 1999). In contrast, the payments of weather derivatives solely rely on the weather index value, an actual loss does not have to be demonstrated. Weather derivates offer the option holder the advantage of receiving prompt compensation without the risk of potentially needing long lasting proof of the actual loss (Raspé 2002).

In conjunction with this feature stands another distinctive criterion. The buyer of a weather derivate does not need to have any weather sensitivity or intention to protect himself against adverse weather conditions, i.e. he does not need to show an insurable interest. On the other hand, insurance contracts are based on the concept of insurable interest, which means a direct relationship between the insured risk and the insurance purchaser has to exist, i.e. a natural exposure is required (Culp 2002). Weather derivatives can be bought for mere speculation (Alaton et al. 2002).

An additional important feature of weather derivates compared to insurance is that two counterparties with opposed risk exposures can enter into a contract to hedge each other's risk, e.g. via a swap structure. This is usually not possible in the insurance market (Alaton et al. 2002). Furthermore, swap contracts allow protection at no upfront costs, whereas a premium must always be paid for insurance contracts (Becker and Bracht 1999).

The typical usage of weather derivates and insurance contracts can be seen as another distinction between the two. As stated before, weather derivates are mainly constructed for protection against high-frequency/low-severity risks. Insurance contracts usually refer to risks of extreme or catastrophic nature but with low occurrence probability (Alaton et al. 2002). A possible explanation for these typical application fields can be found in the instrument specific basis risk. The pay-off of weather derivates shows no dependency on the actual loss which results in a natural basis risk and implies the risk of an insufficient compensation in extreme events. Insurance solutions are preferred in this situation as they are indemnity based and do not contain basis risk. In contrast, insurance contracts do not function well with normal, frequent weather risk as all losses have to be proven which is time-consuming and costly. The straightforward compensation process of weather derivates justifies the acceptance of basis risk. Furthermore, the existence of basis risk has the beneficial effect that weather derivates are lower priced than insurance contracts.

Other aspects in which derivatives and insurance differ include legal, tax and regulatory issues. A comprehensive discussion of these topics is given in Raspé 2002; Edwards 2002; Kramer 2006.

To summarize, weather derivates and insurance contracts show significant differences but are not to be considered as exclusively competing concepts. Rather they should be used as complementary instruments in weather risk management as both offer application specific advantages.

References

Albrecht P, Schradin HR (1998) Alternativer Risikotransfer: Verbriefung von Versicherungsrisiken. Zeitschrift fuer die gesamte Versicherungswissenschaft, 84. Jg, pp 633-682

Aon (2005) Aon maintains position as leading captive manager. Alternative Views 40, April 2005, pp 1-3

Becker HA, Bracht A (1999) Katastrophen- und Wetterderivate: Finanzinnovationen auf der Basis von Naturkatastrophen und Wettererscheinungen. Wien

Business Insurance - Data Joe. World's Captive Domiciles2006. http://businessinsurance.datajoe.com/app/ecom/pub_products.php, 22.09.06

Banks E (2005) Alternative risk transfer. Wiley & Sons, New York

Bodie Z, Kane A, Marcus AJ (2002) Investments. 4th edn, McGraw-Hill / Irwin, New York

Booth G (2006) A captive audience. *Reactions* - Captive Survey 2006, pp 30-33.

Borch KH (1990) Economics of insurance. North Holland, Amsterdam New York Oxford Tokyo.

CBoT (1995) The Chicago Board of Trade, catastrophe insurance: A user's guide.

Chubb Financial Solutions (2002) Enhancing financial flexibility using contingent capital. Risk Magazine Special Report June 2002, 15 (6)

Clemmons L (2002): Introduction to Weather Risk Management. In: Banks E (ed) Weather risk management: markets, products, and applications, Basingstoke, pp 3-13

Climetrix (2006) Climetrix – Weather Derivatives Software. http://www.climetrix.com, 15.08.2006

CME (2006) CME Weather Products, http://www.cme.com/trading/prd/weather/index14270.html, 09.08.2006.

Culp CL (2002) The ART of risk management. Wiley Finance, New York.

Culp CL, Heaton JB (2005) The uses and abuses of finite risk reinsurance. Journal of Applied Corporate Finance 17 (3)

Culp CL (2006) Structured Finance & Insurance. The ART of managing capital and risk. Wiley Finance, New York

Cummins JD, Lalonde D, Phillips RD (2000) "The Basis Risk of Catastrophic-Loss Index Securities" Working paper of the Wharton School of Pennsylvania. Available at SSRN: http://ssrn.com/abstract=230044

Dischel RS (1999) A Weather Risk Managment Choice. In: Geman H (ed) Insurance and weather derivatives from exotic options to exotic underlyings, London, pp 183-196

Edwards S (2002) Accounting and Tax Treatment. In: Banks E (ed) Weather risk management: markets, products, and applications, Basingstoke, pp 246-261

Eisenhauer, J G (2004) Risk pooling in the presence of moral hazard. Bulletin of Economic Research, 56 (1), pp 107-111

Elliott MW (2001) Contingent Capital Arrangements. Risk Management Section Quarterly 18 (2), pp 1-2

European Commission (2000) ART market study final report. Study Contract ETD/99/B5-3000/C/51

Farny D (1995) Versicherungsbetriebslehre. VVW, Karlsruhe

Froot K (1999) The evolving market for catastrophic event risk. In: Figlewski S and Levich RM (eds), 2002, Risk Management - The State of the Art, Massachusetts, pp 37-65

Froot KA (2001) The market for catastrophic risk: a clinical examination. NBER Working Paper 8110

GAO (2005) Risk Retention Groups – Common regulatory standards and greater member protections are needed. GAO-05-536

Guy Carpenter (2005a) Finite reinsurance and risk transfer: will concern in the United States shape the global debate? Guy Carpenter & Company, Inc

Guy Carpenter (2005b) The growing appetite for catastrophic risk. The catastrophe bond market at year-end 2004. Guy Carpenter & Company, Inc

Herold B, Paetzmann K (1999) Alternativer Risikotransfer. München

ISO (1999) ISO „Financing catastrophe risk - Capital market solutions". www.iso.com/studies_analyses/docs/study013.html, 28.10.2005

Jain G, Foster D (2000) Weather Risk – Beyond Energy. Weather Risk supplement to Risk magazine August 2000, http://www.financewise.com/public/edit/energy/weather00/wthr00-beyondenergy.htm, 10.08.2006

Jewson S, Brix A (2005) Weather derivative valuation: the meteorological, statistical, financial and mathematical foundations. Cambridge

Kerr DA (2006) Understanding basis risk in insurance contracts. Risk Management and Insurance Review 9 (1), pp 37-51

Kramer AS (2006) Critical Distinctions between Weather Derivatives and Insurance. In: Culp CL (ed) Structured finance and insurance: the ART of managing capital and risk, Hoboken, pp 639-652
Lane MN, Beckwith R (2006) The 2006 Review of the insurance securitization market. Lane Financial L.L.C. Trade Notes
Lee, W and Ligon, JA (2001) Moral hazard in risk pooling arrangements. The Journal of Risk and Insurance, 68 (1), pp175-190
Metzler M, Ockenga T, Kuehner C (2005) Deutsche Lebensversicherer: Nach dem großen Fressen – stabile Seitenlage. Fitch Ratings Special Report (10 October 2005)
Mueller H (1988) Selbstversicherung. In: Farny D et al. (ed) Handwoerterbuch der Versicherung – HdV. VVW, Karlsruhe, pp 781-784
Pollner J D (2001) Catastrophic risk management - using alternative risk financing and insurance pooling mechanisms. World Bank Policy Research Working Paper no. 2560. Available at SSRN: http://ssrn.com/abstract=632627
Raspé A (2002) Legal and Regulatory Issues. Banks E (ed) Weather risk management: markets, products, and applications, Basingstoke, pp 224-245
Romeike F, Finke R (2003) Erfolgsfaktor Risikomanagement: Chance für Industrie und Handel. Wiesbaden
Schirm A (2001) Wetterderivate: Einsatzmöglichkeiten und Bewertung, Working Paper, Universität Mannheim
Shimpi P (2001) Integrating coporate risk manatement. New York
Swiss Re (1999) Alternativer Risikotransfer (ART) für Unternehmen: Modeerscheinung oder Risikomanagement des 21. Jhdts.? Sigma 2/1999
Swiss Re (2003) The picture of ART. Sigma 1/2003

Law:

15 USC Sec. 3901-3902, Liability Risk Retention Act 1986 (1988)
Directive 2005/68/EC of the European Parliament and of the Council of 16 November 2005 on reinsurance and amending Council Directives 73/239/EEC, 92/49/EEC as well as Directives 98/78/EC and 2002/83/EC
First Council Directive 73/239/EEC of 24 July 1973 on the coordination of laws, regulations and administrative provisions relating to the taking-up and pursuit of the business of direct insurance other than life assurance
Jersey Companies Law 1991 and Amendment No. 8, Jersey, 2006
Guernsey Companies Law, Guernsey, 1996
The Incorporated Cell Companies Ordinance, Guernsey, 2006

10 Risk management

S. Ortner, J. Lammel, M. Pöckl, A.P. Moran

10.1 Introduction

The intense socio-economic development of the alpine region in the past 50 years has led to an increase in damage potential associated with natural hazards. The therewith addressed vulnerability demands a new handling of natural hazards and emphasises the necessary implementation of integral risk concepts. The aim of this contribution is to deal with this topic strategically and to enhance awareness of the overall alpS concept and its transdisciplinary approach.

The article elucidates the different phases of risk management and begins with one of the most essential aspects of the approach – the definition of "risk". Subsequently, on the basis of a universally valid risk definition, a risk management concept is presented. In the final section this model is then applied to practical examples on both a communal as well as on a corporate level.

10.2 Risk

The term risk originates from the Italian word "risco" and means cliff. In a scientific sense the term is defined differently as can be seen in the following examples:

Risk in Natural Hazard Sciences is defined as the product of:
- *The frequency or probability of a "catastrophic" event/disaster; and*
- *The scale of damage, as measured by the number of people involved and the value of material damage caused at the moment of the actual event. Hence, it accounts for the vulnerability of the affected people and their assets. The damage is therefore the product of the assets exposed to hazards and their vulnerability.*
(Amman et al. 2006)

Risk in psychology:

Risk represents the uncertain consequence of an event or an activity, i.e. risk always refers to the likelihood or chance of potential consequences. (Kates et al. 1985)

In common use the term risk is usually associated with negative consequences which arise from the uncertainty of the future. If one speaks about "risking something in the future" one assumes that one could lose something. However the future does not only entail dangers but also offers opportunities. To be able to determine the term risk in more detail it is necessary to consider which situations lead to the development of risks.

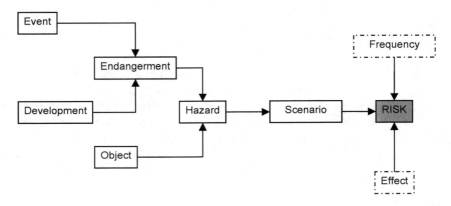

Fig. 10.1 Definition of Risk (based on ONR 49000 2004)

An event can occur suddenly (e.g. as an event causing damage) or can slowly and gradually appear (e.g. as an unexpected development). Such an event or development often leads to an endangerment. It can pose a potential source of damage or harm and can trigger a hazard. A hazard is only considered as such when a human, an object or an organisation is exposed to such an endangerment. Hence, firstly, a hazard can be defined as a potential threat of the targets that an organisation wishes to reach. Secondly it can pose a potential threat to the functions that ensure the safety of a system. Thirdly, it can be a consequence of negligence. Thus, the expectations that are placed externally on organisations or systems are not fulfilled. The observation of scenarios helps to better understand the future. Scenarios are specific and illustrative and show both opportunities and hazards in an organisation or system (ONR 49000 2004). In the scenarios both negative as well as positive aspects are incorporated.

The following figure exemplifies the risk approach upon which the risk management model is based.

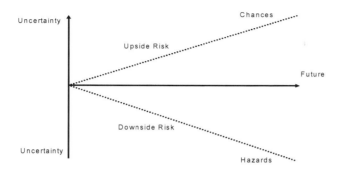

Fig. 10.2 The future

Figure 10.2 illustrates that risks can potentially include chances/opportunities as well as hazards. To deal with risks in a structured manner, it is necessary that risk management meets these and other requirements. In the following section a risk management model is presented that serves as the foundation for further applications.

10.3 Risk management model

To successfully implement risk management, a systematic approach in the sense of structured process phases with a clear assignment of tasks is needed. The subsequent figure depicts the risk management model that was developed throughout the course of the alpS research project "Contingency plans for companies".

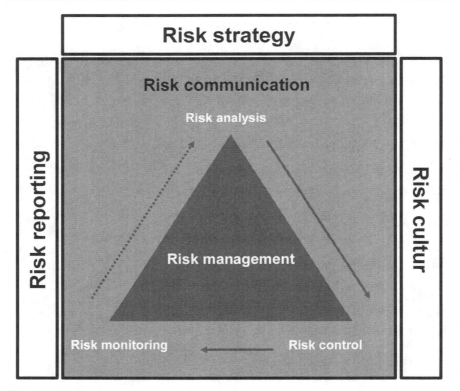

Fig. 10.3 Risk management model

Depicted in the form of the alpS triangle, the figure illustrates the inner process of the risk management model which consists of three general risk management steps: risk analysis, risk control and risk monitoring.

This process requires a system framework, defining the system that comprises strategic planning, risk reporting, and risk culture. Risk communication is responsible for the flow of information between the system framework and the process stages and also for the communication within the individual phases. Thereby strategic planning focuses on the essential processes of risk management in advance. Risk reporting is an instrument to ensure that the already obtained information can be easily and quickly retrieved at any time. Risk culture is a deeply important component of the model and influences all the other elements. It aims at establishing risk awareness, and a transparent information transfer. Moreover it supports a strategic way of thinking and acting in terms of risks.

10.3.1 System Framework - Definitions

Risk strategy

The term strategy is diverse in its definition. *In* Top Management Strategy *[7], Benjamin Tregoe and John Zimmerman, define strategy as the framework which guides those choices that determine the nature and direction of an organization.* The strategic core can consist of a vision or the definition of specific aims that risk management should achieve, for example:

- Provision of preventive risk information
- Strategy based on the motto – acting instead of reacting
- Improvement of risk awareness in the population
- Protection of the population
- Protection of companies

Hence risk strategies can be derived that determine the specifications and guidelines for the identification, controlling and monitoring of risks (PWC Deutsche Revision 1999).

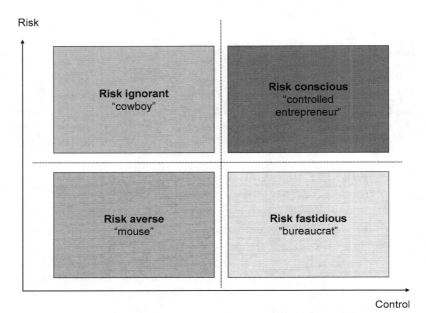

Fig. 10.4 Styles of risk management (based on KPMG 1998)

A good balance between risks and control is certainly the key to a successful handling of risks (KPMG 1998).

Risk culture

Risk culture leads to an elevated consciousness and willingness of the employees, population, etc., to perceive risks and consequently affects their sensitivity of risks. In short: only integrative risk management can be successful. This requires defining a risk strategy, setting the course for an open risk culture and implementing an information policy. Only then can risks be dealt with successfully.

By the term "risk culture" we refer to the way a society handles questions of safety and security. Risk culture emphasizes that insecurity can only be controlled by risk-oriented thinking. On the one hand we fear natural hazards and on the other hand there are practical limits to safety. A unified basis to describe risks due to natural hazards has to be defined.
(Ammann 2003b, 2003d, Malzahn and Plapp 2004)

The development and maintenance of a risk culture can be achieved through the following measures:

Risk awareness
Risk affinity
Risk accountability
Communication
Transparency
Integration

Risk reporting

Besides strategic planning and risk culture, risk reporting is a determining factor that is necessary for the successful implementation of risk management. Consequently, a lack of communication and information policies reduces the transparency of the risk management process (Gleißner 2001). Thus, risk reporting is a core competence of risk management. It is necessary to integrate adequate documentation in order to guarantee an appropriate reaction in the sense of a successful handling of risks and to achieve a sustainable risk management independent of specific individuals. Decision makers can be supported by a timely, appropriate and complete information policy that can be integrated in risk management. Thus, it is necessary to determine criteria for the processing of information which should serve as a general standard. For example, the following criteria could be used:

- essentiality
- punctuality
- preciseness
- completeness
- homogeneity

10.3.2 Process stages of risk management

Risk analysis

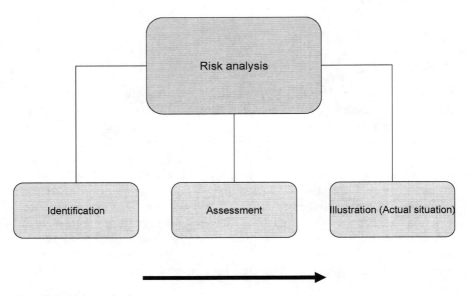

Fig. 10.5 Risk analysis

Risk analysis, the first step of risk management, can be represented as in figure 10.5, and subdivided into identification, assessment and illustration. Risk identification determines the "essential" risks. Risk assessment valuates the collected risks according to frequency and effects. At the end of a risk analysis the determined and valuated risks are illustrated, thus showing the actual risk situation. The obtained information is then the basis for the next step of risk control.

The goal of risk analysis is the most objective identification of the risk factors for a specific damage event, object or area.
(Amman et al, 2006)

All fundamental risks associated with the field of activity are collected. Thus, it is important to make use of all potential information whereby the aim of risk identification must be to guarantee a complete coverage of the essential risks. The result of the identification can also be described as "risk landscape". In a next step this risk landscape is upgraded by assessing the risks. Risk assessment is a component of risk reporting and a helpful instrument for risk control.

Following risk identification the next step in the risk analysis process is risk assessment. *Risk assessment by definition aims for an explicitly subjective answer to the question: What of that which can happen is acceptable?* (Ammann et al., 2006). After the definition of the term risk, an assessment in regard to frequency and impact is carried out.

Risks can be depicted in many ways. To provide a quick and exact overview, it seems necessary to combine the risk description with an illustration; hereby a risk matrix (risk landscape) can help. The risks are illustrated in a matrix according to their frequency (probability of occurrence) and their impacts.

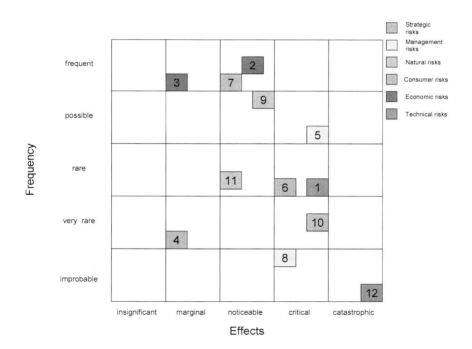

Fig. 10.6 Risk matrix: actual risk landscape

Risk control

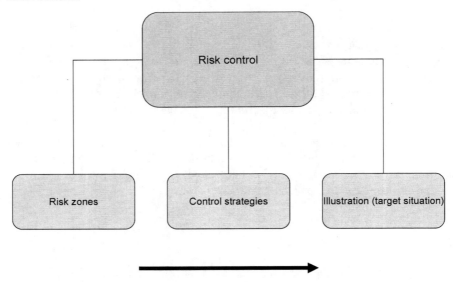

Fig. 10.7 Risk control

Figure 10.7 shows the essential steps of risk control. After the actual risk landscape is established, enquiries are made into the priority of risks. When this question has been answered, control strategies and measures can be developed. The assessment of the risk situation, after the determination of measures, is illustrated in the target situation of the risk landscape.

The first step questions whether risk coverage is necessary or not. The risks are assessed according to their priority for the client. The risk zones determine the different levels of priority. Subsequently, specific considerations are made regarding individual risk strategies.

There are various possibilities for the implementation of measures and control strategies – see figure 10.8.

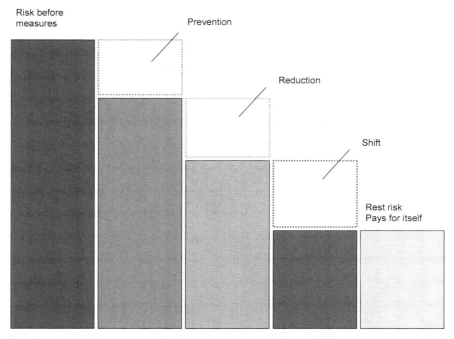

Fig. 10.8 Control strategies (based on Buderath/Amling 2000)

The target situation of the risk landscape represents the results of the provisional process. This is the case only after the actual state was shown in the risk analysis, the risk zones were determined and control strategies with measures were designed.

Risk monitoring

Following risk analysis and risk control, risk monitoring is the final step in the risk monitoring process.

Risk monitoring is an important stage in the context of risk management. In nearly all risk management models the monitoring and controlling of risks are fixed aspects. The identification of risks, the assessment of these and the placement of adequate measures alone is not sufficient for dealing with risks successfully. A holistic risk understanding, and a constant examination and monitoring of risks is necessary. Thereby the question must be asked whether the implemented measures are really as effective as intended by risk control or risk strategy.

The hitherto described risk management model is the basis for further applications e.g. risk management on a communal and corporate level. The following sections give an insight into the formation of risk management for different requirements and responsibilities.

10.4 Application of risk management

Examples of implemented risk management are given below, whereby the focus is placed particularly on the prevention of future damage. The importance of integrating risk management both on a communal as well as on a corporate level becomes especially apparent when adequately answering the question "what can maximally happen?".

10.4.1 Communal level – the city of Innsbruck

In the project "Innsbruck" a risk platform was created in which a 3D flood model, risk analysis, as well as risk assessment are joined.

3D Flood modelling

In order to conduct a three dimensional flood simulation in the chosen area, calculated by the Department for Hydraulic Engineering (IWI) at the University of Innsbruck, it is necessary to firstly generate an adequate terrain model. The software package for the simulation of free surface flow requires the so-called STL format as a data basis. This format represents the surfaces of 3D-bodies by triangulation. Each triangle is characterised by three vertices and the corresponding surface normal. Curving surfaces are approximated by the triangles.

The 3D terrain model for the test area (see figure 10.9) is created based on the following data:

- Surveying data (CAD format)
- Topographical map of the area
- Laser scanning data with a resolution of one meter in an ASCII format (XYZ data)
- Inn-cross section in an ASCII format (XYZ data)
- Analogue bridge plans

Generating a terrain model

The largest part of the terrain model is based on laser scanning data. Along the revetment walls and other relevant areas (bridge heads, dams) the accuracy of the model was improved with the help of surveying data (CAD format) and the topographical map of the area.

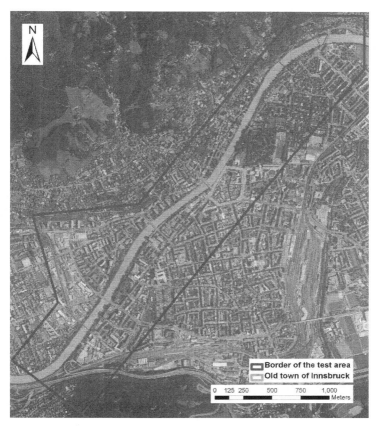

Fig. 10.9 Location of the test area

The construction of the river bed is rather complex. The laser scanning data only provides information on the actual water surface at the time of the survey flight. Therefore, the few existing cross sections of the Inn River provide the only source of data. Thus, with the help of CAD software a polygon mesh was constructed as the river bed between the existing cross sections. Due to reasons of accuracy, a finer mesh was applied, which integrated the bridge bearings and bridge piers (see figure 10.10).

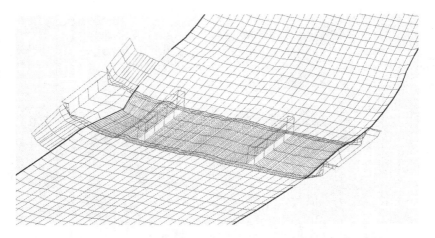

Fig. 10.10 intersection of the water surface with the embankment (blue) and three-dimensional polygon mesh (red) in the area of the Inn Bridge

On the basis of the data listed above (foremost laser scanning data) and the constructed river bed, a so-called TIN (Triangulated Irregular Network) was generated with a GIS application (see figure 10.11).

Fig. 10.11 Excerpt of the TIN in the area of the Inn Bridge

The buildings in the test area are given a height of 10 meters each and are included in the TIN. This simplification is possible because the height of the objects is hardly relevant in flood modelling. The final TIN can be seen in the following figure:

10 Risk management 291

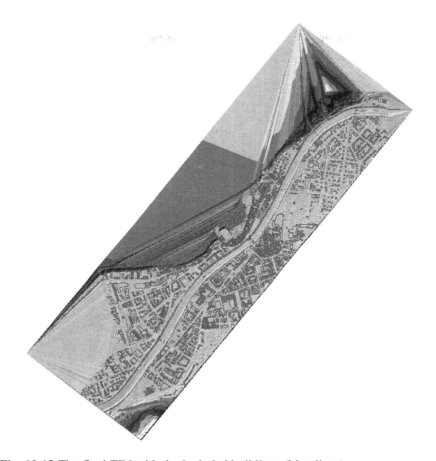

Fig. 10.12 The final TIN with the included buildings (blue lines)

The bridges are brought into a further workable form on the basis of analogue bridge plans by applying CAD software (see figure 10.13).

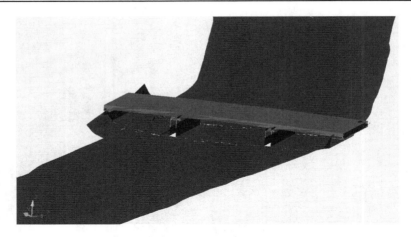

Fig. 10.13 The Inn Bridge (red) and the river bed (blue)

All data is assembled in a binary STL file, which is the basis for the flood simulation. For this purpose the bridges and the TIN are exported into an ASCII-STL file and subsequently united. The result can be seen in figure 10.14.

Fig. 10.14 Final terrain model

Evaluation of the flood simulation

The results of the flood simulation can be exported as an ASCII file (XYZ format).

Table10.1 Excerpt of an ASCII file

X	Y	Z
243237.6250	55383.6250	575.5327
243238.8750	55383.6250	575.5322
243240.1250	55383.6250	575.5319
243241.3750	55383.6250	575.5316
243242.6250	55383.6250	575.5317
243243.8750	55383.6250	575.5321
243245.1250	55383.6250	575.5327
243246.3750	55383.6250	575.5324
243247.6250	55383.6250	575.5322
243248.8750	55383.6250	575.5325
243250.1250	55383.6250	575.5330
243251.3750	55383.6250	575.5329

After the transformation of the point data into the right coordinate system, these can be converted to raster format and visualised in a GIS. The TIN is also converted into raster format. The water depth is obtained by subtracting this newly created raster layer from the flood simulation results. (see figure 10.15).

Fig. 10.15 The results of the computer based simulation for a discharge of 1900 m^3/s within the study area (green border)

10.4.2 Risk assessment

The aim of risk management for the city of Innsbruck is to achieve the best possible protection against risks (in phase 1 of the project flood risk was chosen as a test risk). This calls for the cooperation and collaboration of all those involved in the wide range of possible measures. The analysis of risks, the measures for hazard and damage mitigation, disaster control and emergencies pose the foundation for the implementation of flood protection measures.

Integral risk management in the project comprises the steps of risk analysis, risk assessment and risk control. Risk assessment is described as an essential component of risk management, which serves as the further basis for the valuation of risks. It is necessary to guarantee decisions that take the possible interaction of risks into consideration. Thereby, on the one hand the frequency and on the other hand the extent of the effects on the infrastructure is assessed.

The assessment system for the city of Innsbruck bundles extensive information, which is available for buildings, to a manageable size. Data was collected, with which individual objects are precisely described and on the basis of these the potential damage effects and their probability can be defined. Detailed information on all objects in flood endangered areas is the basis for strategic flood management as well as for efficient decisions and measures.

The reconditioning of the data and the creation of an information pool achieves a better flow of information and supports the cooperation between the different administrative positions in the case of an operation. All data such as real estate as well as electric cables, traffic systems and communication posts are ascertained, shedding light upon their vulnerabilities. It becomes apparent how the city of Innsbruck both preventively, as well as in the case of emergency, can ensure and optimise the resilience of the city, its stability and the existing network structures. In the case of disaster the high standard of knowledge concerning property and the infrastructure can minimise damage and ensure better functionality and effectiveness of the complete system.

Procedure of risk assessment

1. Choice of a test area

The test area reaches from the Karwendel Bridge to the Freiburger Bridge and has an area of ca. 3 km². This area was chosen because most emergency operations took place there during the flood event of 2005.

2. Establishment of building categories

For application planning it is decisive to be able to quickly and clearly recognise the function of a building and to be able to prioritise in the case of an operation. In addition, building classes are defined. Most buildings in the test area belong to two categories.

The following categories were established:

1. Residential building
2. Administration
    2.1. Public administration
    2.2. BOS facility (police, fire brigade etc.)
3. Basic provisions (e.g. groceries)
4. Resource, supply (e.g. petrol station)
5. Medical facilities
    5.1 Hospitals
    5.2 Clinical practice
    5.3 Pharmaceutical facilities
    5.4 Nursing homes
6. Infrastructure
    6.1 Infrastructure traffic
    6.2 Infrastructure communication
    6.3 Infrastructure power supply
    6.4 Infrastructure sanitation
    6.5 Infrastructure recreational facilities
7. Commercial companies
    7.1. Building and transport industry
    7.2. Sensitive infrastructure (e.g. chemical storage)
    7.3. Gastronomy
    7.4. Commercial industry (Production/service)
8. Educational insitutions
9. Churches
10. Special buildings (bridges, pumping stations, sewerage treatment plants, support buildings)

3. Building data

After the categorisation of the buildings is completed, more exact building data for the individual categories must be determined for the further procedure. The completed database with the total information includes all the building types in the test area.

Firstly data is determined that is to be collected for all buildings. This includes:
- address
- building category
- proprietors' spokesperson

- responsible person on-site
- level of the entrance
- basement and usage
- number of floors
- elevator
- roof type (flat)
- gas supply
- heating and fuel

Subsequently data was determined which could only be applied to individual categories and had to be collected. In the following table the specific building data is listed:

Table 10.2 Building categories and required building data

Building category	Building data
Residential building	Number of residents
Public administration	Type of facility
	Emergency power supply (Type, durability, power)
	Analogue telephone connection
BOS-facitilies	Type of facility
	Emergency power supply (Type, durability, power)
	Analogue telephone connection
Resource maintenance	Type of facility
Doctors' practices	Type
	X-ray
	Small operation room
Pharmaceutical facilities	Laboratory
	Toxic room
Nursing homes	Type
	Number of residents
	Number of nursing staff
	Emergency power supply (type durability, power)
Infrastructure traffic	Type
	Description of the facility
Infrastructure communication	Type
	Supply area
	Emergency power supply (type, durability, power)
Infrastructure energy	Type
	Supply area

Infrastructure sanitation	Type
	Description of the facility
Infrastructure Recreational facility	Type
	Visitor capacity
	Adequacy as emergency accommodation
	Capacity as emergency accommodation
	Emergency power supply (type, durability, power)
Building/Transport industry	Type
	Machinery
Risk enterprise	Type
	Description of the facility
Gastronomy	Type
	Restaurant
	Number of seats
	Accommodation
	Number of beds
Other commercial enterprises	Type of enterprise
Educational establishment	Type of facility
Churches	
Special buildings	Bridges:
	Type of construction
	Piping
	Trafficability
	Sewerage treatment plant:
	Description of the facility
	Pumping station:
	Description of the facility
	Support buildings:
	Construction (lengthwise, crosswise)
	Storage capacity

4. Collection of data

In the next step a standardised list of data with all necessary building data should be created. This data should be available locally and updateable at anytime. The data should be collected for every building and should provide a complete level of information for the test area.

5. Digitalisation of the data in an Excel table
6. Transferral of the data into a GIS

10.4.3 Corporate level – TILAK (The Tyrolean Provincial Hospital Company)

Risk management is gaining more and more importance for hospitals. An effective risk provision, e.g. for flooding, is achieved both by strategic risk management on a communal level (the city of Innsbruck), and on an individual company level. These should act self-responsibly and develop and implement a functioning risk management with appropriate protective measures. Hereby unforeseeable future developments should be taken into consideration and current decisions should not restrict future possibilities in the economic, construction and ecological field.

TILAK Innsbruck is one of the best examples of a company to shift from common flood protection to the implementation of an integral risk management. Thus, different flood scenarios are developed in which probabilities and effects of extreme events on the operation of the hospital are taken into consideration and solutions sought.

A minimisation of the flood risk by means of optimal construction measures and a complementary mobile protective system was implemented by the TILAK after the flood event on August 23, 2005. However areas a residual risk, which can trigger a crisis and endanger people, must always be taken into consideration. Facilities in these areas are dependant on the reliable and high quality supply of electricity and water, consistently good traffic accessibility and other services and factors. Besides the endangerment of patients and staff through flooding there can be a considerable financial loss.

Due to the problem of remaining risks, the TILAK risk management aims at reducing the potential extent of damages. This means that critical health care infrastructure must be described to obtain an insight to the network. In the case of flooding no or only little damage should be caused to the buildings of the Landeskrankenhaus Innsbruck (State Hospital Innsbruck). To achieve this, the complete hospital area (13 buildings and an extensive underground corridor system) was assessed in regard to risk. This assessment was developed throughout the course of the risk management process as an effective strategy. It is based on the presumption, that in the case of a flood, the damage and the protective measures are dependent on building use.

The focus of the project was on risk assessment, which was carried out for the infrastructure of the TILAK both in a written and graphic form. For the flood risk the focus was placed on all the basements and ground floors of the hospital premises. Based on existing room information a classification in three risk zones was made. The most important rooms – that must be protected and are indispensable for the operation of the hospital – were classified with the risk priority 1 and the colour red. Rooms with an elevated hazard for the running operation of the hospital were allocated with priority 2 and the colour orange. Rooms with priority 3 and the colour yellow are less endangered. Generally, the classification of risk priorities was based on the assumption of water leakage – possible different water levels were however included beforehand in the considerations of damage extent.

The existing room information is summed up in the so-called room book. This in combination with the object-related flood plans provide information on the probability of each room being affected. Additionally, for those in responsibility, this sheds light on which consequences should be expected. The respective hazard potential of individual rooms, and subsequently which ones are to be particularly observed, is made clear. Possible relocations or strengthened protective measures of particularly critical infrastructure can be considered.

The plans should provide a quick orientation to the individual floors. Furthermore, they give an overview of the procedures and provide controls for guaranteeing the functionality of each individual room in the application plan. In addition, it was made possible to add special annotations to the rooms that allude to the particularities of the room or give instructions.

The flood protection plans serve as an aid for the operation controllers in the case of a flood as they visualise and localise the endangered rooms in the TILAK area. They provide a quick, general overview of the individual floors and can be used as a check list. The plans were created with AutoCAD, whereby the 3 risk priorities (yellow, orange, red) are conform with those of the official room data.

10 Risk management   301

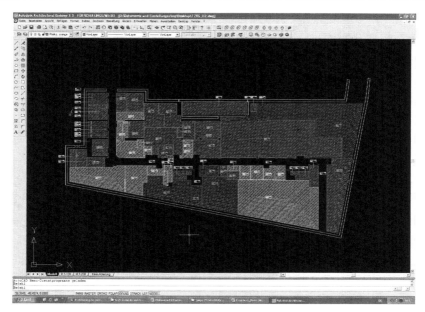

Fig. 10.16 Flood protection plan TILAK

The plan depicted above shows clearly the different risk priorities of the individual rooms. The red coloured rooms would be the first where safety measures and controls would be carried out in the case of water leakage. Also relocation into another section of the building, if technically possible, can be considered.

10.4.4 Corporate level – Ski lift operator – Schlick 2000 Schizentrum AG, Bergbahn AG Kitzbühel

After having concentrated on the visualisation of risk information (e.g. plans, maps etc.) for the TILAK, the second example shows a typical risk management instrument on a corporate level – the risk report. In a field study for two ski lift operators - Schlick 2000 Schizentrum AG and Bergbahn AG Kitzbühel – a risk assessment was conducted and a risk report compiled.

The individual components of the risk report: general considerations, strategies/results, definitions, risk analysis, risk details and risk monitoring, structure the essential information in regard to the risk situation. As such the basis for a transparent risk communication is formed. The aspect of risk awareness as well as the perception of risks plays a decisive role hereby. Only when it is possible to implement all of these features in an organisation/company can risk communication be improved in the future.

General Considerations

The section General Considerations reflects a part of the system framework. In advance the necessary information regarding the organisation, the project team for risk management, the analysis unit as well as general information concerning risk analysis must be gathered.

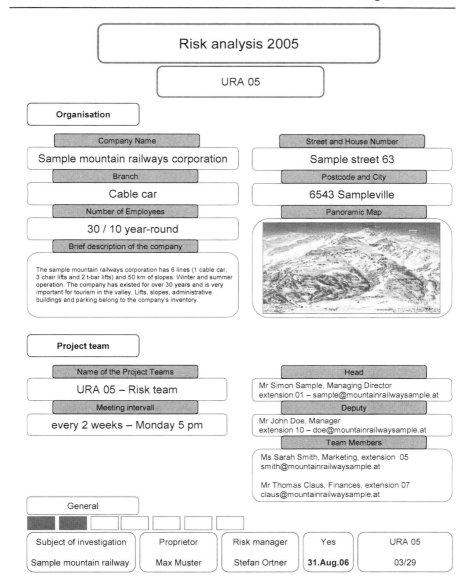

Fig.10.17 Risk handbook – general considerations

Strategies/results

The risk strategy and is an essential component of risk management. A risk strategy reflects the aim that is to be achieved.

Definitions

Further "organisational" details are clarified by stating precise definitions. An organisational and factual field of application is to be determined, the categories frequency and effects are to be defined and the risk zones (priority zones) are to be established.

Risk examination

The risk examination is a summary of the risk analysis and risk control. The main focus of the observation is to depict the actual and the target risk landscape. In order to obtain more accurate information on individual risks, each risk is dealt with separately in the section "Risk details".

Risk details

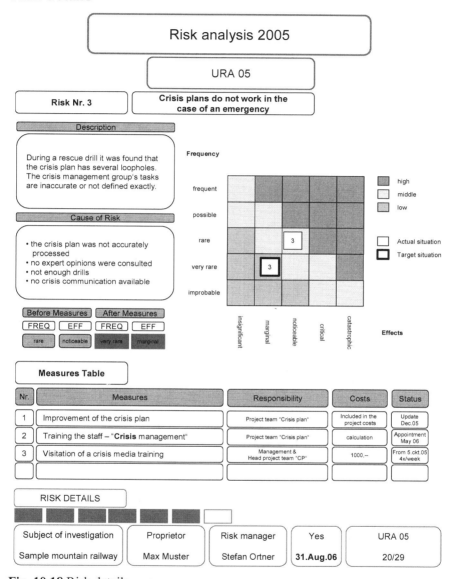

Fig. 10.18 Risk details

In the detailed report the individual process phases of risk management become apparent. Risk analysis is firstly responsible for providing identified risks with corresponding information. Secondly it assesses and illustrates the risks in the risk matrix (actual situation). Within the framework of risk control, measures are considered, newly valuated and depicted in a risk matrix. Risk monitoring controls the measures regarding competencies, etc. and the detailed report provides the information basis for risk monitoring.

Risk monitoring

Risk monitoring includes in addition to the information from the risk details, also information on observation and control. Thereby the release of the risk report (date and responsibility), the period of validity as well as the monitoring interval are stated.

In summary, the risk report provides a multiplicity of information and is an important instrument for the application of risk management in a company/organisation. The risk report is thus a further instrument, besides a risk management platform, that can be utilised for a holistic risk management approach.

10.5 Outlook

Risk management is understood as a holistic strategic instrument to analyze, control and monitor risks. The added value resulting from the management of risks lies in the risk sensitisation of the decision-makers, the employees, and the population and in the simultaneous enforcement of risk awareness through the structured and daily exposure to an organisation's risks.

The risk management platform and the risk report, as output-oriented instruments, can function as a gateway for crisis management in terms of a transmission of risk information. This comprehensive approach is necessary, to place measures and to be prepared in the case of an emergency. Today, providing thorough information is essential and required, in order to timely and actively implement the necessary steps in terms of an all-encompassing risk management.

References

Ammann W, 2003b Integrales Risikomanagement von Naturgefahren, 54. Geographentag, Jahrbuch 2003 DEF, Geogr. Institut Uni Bern

Ammann W, 2003d Die Entwicklung des Risikos infolge Naturgefahren und die Notwendigkeit eines integralen Risikomanagements, Tagungsbericht und wissenschaftliche Abhandlung, 54. Geographentag, Jahrbuch 2003, Eds. Werner Gamerith et al. 2003

Ammann W, Dannenmann S, Vulliet L, (2006) Risk 21: Coping with Risks due to natural hazards in the 21st Century, Taylor & Francis/Balkema, Leiden, The Netherlands

Back M et al. (2004) Einführung eines Risikomanagement-Systems bei Freudenberg. In: Risknews 01/04, pp 22-25

Brühwiler B (2003) Risk Management als Führungsaufgabe. Haupt, Bern

Buderath H, Amling T (2000) Das Interne Überwachungssystem als Teil des Risikomanagementsystems. In: Dörner D, Horvath P, Kagermann H (ed) (2000) Praxis des Risikomanagements. Schäffer Poeschel, Stuttgart; pp 127-152

Gleißner W (2001) Ratschläge für ein leistungsfähiges Risiko-Management – eine Checkliste. In: Gleißner W, Meier G (2001) Wertorientiertes Risiko-Management für Industrie und Handel. Gabler, Wiesbaden, pp 253-266

Hellbrück J, Fischer M (1999) Umweltpsychologie – ein Lehrbuch. Göttingen

Hinterhuber H (2004) Strategische Unternehmensführung, I. Strategisches Denken, 7, Auflage. Walter de Gruyter, Berlin

Kates RW, Hohenemser C, Kasperson J (1985) Perilous Progress: Managing the Hazards of Technology. Westview Press, Boulder USA

Kienholz et al. (1998) Begriffsdefinitionen zu den Themen: Geomorphologie, Naturgefahren, Forstwesen, Sicherheit, Risiko. Arbeitspapier, Bern

KPMG (1998) Integriertes Risikomanagement. Berlin

Malzahn D, Plapp T Eds., (2004) Disasters and Society – From Hazard Assessment to RiskReduction., Logos Verlag, Berlin

ONR 49000 (2004) Risikomanagement für Organisationen und Systeme – Begriffe und Grundlagen. Wien

PWC Deutsche Revision (1999) Unternehmensweites Risikomanagement. 2. Auflage, Wiesbaden

Romeike F (2003) Der Prozess des strategischen und operativen Risikomanagements. In: Romeike F, Finke R (2003) Erfolgsfaktor Risiko-Management. Gabler, Wiesbaden, pp 147-161

11 Laser scanning - a paradigm change in topographic data acquisition for natural hazard management

T. Geist, B. Höfle, M. Rutzinger, N. Pfeifer, J. Stötter

11.1 Introduction

Within the thematic area of *Databases and Modelling* a certain focus is placed on the effective acquisition and management of geo-data and the derivation of standardized products from this data, e.g. as input parameters in process simulation models. In most cases, in-situ data collection (e.g. run-off measurements) is state-of-the-art. As there is a growing demand for area-wide data collection, the utilization of remote sensing technology will gain ground in the future. Within alpS the project 'Determination of surface properties from laser scanning data' addresses these demands by incorporating certain aspects of remote sensing in natural hazard management. Remote sensing is the contactless collection of information about an object or process. This is done with electromagnetic waves and imaging methods. Remote sensing can be carried out from the earth's surface or from airborne and spaceborne platforms. For earth observation issues passive sensors are widely applied, which record the reflected radiation of natural energy sources (with the sun as the most important one). For many applications the method of choice is still aerial photography. Active sensors are more flexible as they have their own energy source. Laser scanning is such an active method.

In modern natural hazard management remote sensing data are used in manifold ways and are especially valuable in inaccessible terrain. A widespread application of remote sensing is the mapping and monitoring of area-wide impacts of natural hazards and the analysis of process disposition and triggering factors. Current research demands include the development of operational monitoring methods and the development of support tools based on automated analysis algorithms, both aiming to reduce the time gap between data acquisition, processing and data application, with real-time user-tailored data availability as the main goal for the future.

Laser scanning, a remote sensing method for the acquisition of topographic data, can meet these demands, allowing the calculation of high resolution and high-accurate digital elevation models and, additionally, providing information on characteristics and properties of the surface. This technology has been developed into an operational and reliable airborne method in recent years. Due to the availability of commercial off-the-shelf sensors and an increased awareness of the advantages of laser scanning by end-users, the use of airborne laser scanning data has grown rapidly and, consequently, the development of a wide variety of applications is under way. Therefore, applied research in the field of laser scanning is embedded in a dynamic and challenging frame. Fundamental knowledge about the technical accuracy of this method and the quality of produced digital elevation datasets has evolved in recent years. At the moment a paradigm change is taking place, with laser scanning replacing image-based photogrammetry as the standard method for acquiring topographic data (Kraus 2004).

One expression of this paradigm change is the acquisition of area-wide digital elevation models, on a regional or even on a national scale (e.g. the entire Netherlands or the federal state of Baden-Wuerttemberg, Germany). Recently, also in topographically complex terrain like the Alps, area-wide mapping campaigns were carried out, e.g. in Vorarlberg (Würländer et al. 2005), South Tyrol (Wack and Stelzl 2005) and Switzerland (Luethy and Stengele 2005). In Tyrol and other Austrian provinces the area-wide data acquisition has started or is in a planning and preparation phase. The responsible authorities' main motivation is the manifold and growing demand for high-resolution elevation data, e.g. for natural hazard management. Recent applications show that the acquisition of airborne laser scanning data is not only restricted to Europe and North America but is also applied in developing countries as an example from Honduras shows (Tamiru Haile and Rientjes 2005).

Applied research questions related to laser scanning focus on the optimization of data management and the development of tools for the user-driven extraction of information from this data. One research goal is the qualitative and quantitative assessment of surface properties (e.g. surface roughness) and the temporal change of these properties. More specifically the interest lies in:

- *improved data handling* - the conception of a management system for laser scanning data, which is flexible and easy to adapt to user requirements.
- *information extraction* - the development of a set of methods for a user specified analysis of laser scanning data, especially for the classification of surface properties and the quantification of temporal changes.
- *work flow integration* - strategies for utilizing laser scanning data for specific purposes of modern alpine natural hazard management.

In Section 11.2 and Section 11.3 the airborne laser scanning technology and the resulting data products are briefly explained, the state-of-art in utilizing LS data in natural hazard management is summarized in Section 11.4, current research and development in data management and data analysis are outlined in Section 11.5, while a short synthesis and outlook is formulated in Section 11.6.

11.2 Description of the technology

Laser scanning is an active remote sensing technology for directly measuring 3D coordinates of points on surfaces, including the terrain and objects thereupon (e.g. houses or trees). It is operated from airborne and terrestrial platforms. This section will provide a detailed description of airborne laser scanning (ALS), i.e. laser scanning from an airplane or helicopter. Nevertheless, applications with terrestrial laser scanning (TLS) are of growing importance for natural hazard management and related processes, and some statements in Section 11.4 refer to TLS.

Airborne laser scanning is also referred to as airborne LIDAR (LIght Detection And Ranging) or LADAR (LAser Detection And Ranging). First applications of laser altimetry were based upon the so-called laser profiling technology, designed to collect data following a virtual single line on the observed (recorded) surface.

An ALS system is a multi-sensor measurement system that incorporates the following time-synchronized components (Figure 11.1):

- The satellite global positioning system (GPS), which is used to determine the absolute position (x, y, z) of the sensor platform in a differential mode using ground reference stations.
- The inertial measurement unit (IMU) is used to determine the angular attitude of the platform (roll, pitch, and heading). The flight path of the platform is calculated from combined analysis of the GPS and IMU data.
- The laser scanner itself, consisting of the laser range finder, measuring the distance from the sensor on the airborne platform to a reflecting surface, and a beam deflection device, that deflects the laser beam perpendicular to the flight direction (±20° as common value). The laser range finder operates by measuring the two-way travel time required for a pulse of laser light (commonly in the infrared section of the electromagnetic spectrum) to travel to the location of reflection, and back to the receiver. The distance r can be computed from the travel time Δt by the

known speed of light c ($r = c \cdot \Delta t/2$). Current systems are capable of operating up to 5000 m above ground level. Range measurements provide pulse repetition rates up to 150 kHz. Some experimental systems utilize continuous wave lasers and examine phase differences between the transmitted and the reflected radiation.

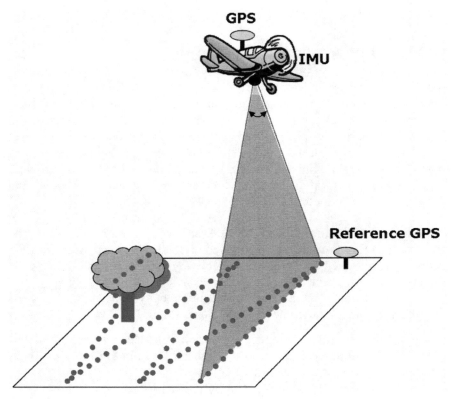

Fig. 11.1 Components of an ALS system

The laser beam has a specified divergence and consequently a certain diameter at the scanned surface, the so-called footprint, typically between 0.5 m to 1 m, depending on the flying altitude above ground (e.g. a beam divergence of 0.25 mrad causes a 0.25 m footprint diameter, when the system is 1000 m above ground level). Multiple reflecting surfaces may be found within one footprint. In those cases more than one *echo* of the emitted signal can be recorded, e.g. one from the first reflecting surface, which is ideally on top of the vegetation, and a second from the last reflecting surface, which may be the real terrain surface (Figure 11.2). The term *echo* is often referred to as reflection or pulse. Contemporary systems have the standard ability to record the first and the last echo of one single laser shot. There are a number of commercial ALS systems on the market with quite different technical properties (Baltsavias 1999c). The most recent scanners also provide the digitized full-waveform of the reflected laser beam (Wagner et al. 2004), which allows for the use of scanner independent echo detection algorithms (Persson et al. 2005). Due to the current data amount, for current operational use full-waveform data still has to be reduced to distinct echoes.

The primary product of ALS campaigns are numbers (time of measurement, x, y, z, intensity) for one or more reflections (echoes) of the emitted laser beam representing a *point cloud* of measurements in a global coordinate system. Most of the scanners save additional attributes for each point (e.g. intensity of the echo). As secondary products, raster models can be derived from this primary data (Section 11.3). The higher the density of point measurements the more accurate elevation models can be achieved (Kraus 2004).

ALS has the advantage (i) of penetrating vegetation and thus recording the ground surface also in wooded areas, (ii) of having a high degree of automation, ranging from data acquisition to digital elevation model generation, (iii) of having a high point density (several points/m^2) that allows a very detailed terrain description, (iv) of having a vertical accuracy of about ±10 cm, and (v) as an active system, of allowing data acquisition at night or over areas without texture (e.g. snow).

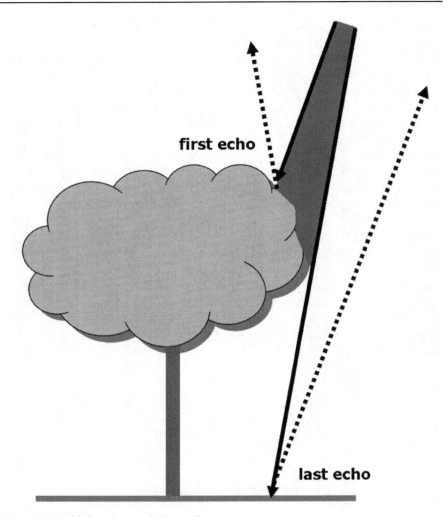

Fig. 11.2 Multiple echoes of a laser shot

The error budget for an ALS measurement is driven by the contributing error budgets from the core subsystems; the laser rangefinder, the GPS position solution and the IMU orientation solution. For a comprehensive discussion and detailed examples of how each system parameter contributes to the overall system accuracy see Baltsavias 1999a. The sum of the errors of the laser scanner and the precision of the flight path leads to a typical vertical and horizontal point accuracy of 0.1 m and 0.5 m, respectively, for a flight height of 1000 m. Accuracy can be improved, provided systematic errors are removed, and measurements are performed to flat, well-defined surfaces.

Principal applications have focused on urban areas and forestry studies since the detailed and accurate information can describe building shapes or canopy configuration and structure.

11.3 Description of data products

In general, laser scanning (LS) vendors deliver the original measurements, the so-called point cloud, but also further processed derivatives such as, for example, rasterized digital elevation models (DEMs). In addition to the topographical/geometrical information most LS systems record the strength of reflection (intensity) for each echo. The three-dimensional coordinates together with the intensity (x,y,z,i) are the root dataset for all further analyses. Basically, the algorithms for LS data processing can be divided into two groups: i) algorithms working on the point cloud directly and ii) algorithms using DEM(s) as input data (Figure 11.3). The most evident algorithms representing the first group are all procedures for generating raster datasets, i.e. converting the vector points into a regular grid, such as, for example, filtering and interpolation of Digital Terrain Models (DTMs). Both datasets, the point cloud and the DEMs, have individual advantages and disadvantages, which will subsequently be discussed.

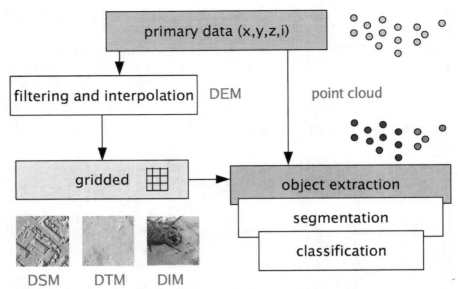

Fig. 11.3 Workflow of laser scanning data processing and classification

11.3.1 Laser point cloud

The original measurements contain the highest degree of information, as they (i) represent the original data delivered with least modifications, (ii) are three-dimensionally located and (iii) are additionally fixed in time (timestamp), which offers interesting possibilities, for example reconstructing the time flow of the scanning procedure or connecting the ground measurements with the corresponding position of the scanning system. This raw state of the data can also contain erroneous points (outliers), which have to be removed before further usage. Such outliers can be caused by redirection and re-reflection of laser shots or simply by temporary objects (e.g. birds). Together with the high degree of information a large data volume arises when every point measurement has to be stored. Due to the fixed scanning pattern of the laser scanning systems - the beam is usually redirected orthogonally to the flight direction - the laser beam cannot be pointed on particular objects directly (Brenner 2005). If a high accuracy of the shape of objects (e.g. buildings) is needed, a high point density (points/m^2) has to be chosen (Maas and Vosselman 1999). The unorganized spatial distribution of the point cloud complicates analyzes and leads to a need for data structures that improve the performance for the accessing and processing of the whole data set (Section 11.5.1).

To give an impression of the dimensions of the data volume the Austrian Federal State of Tyrol (12.648 km²) can serve as an example. If a moderate average point density of one point per square meter is assumed, then this results in about 13 billion single measurements for the whole area. Saving the coordinates and some additional attributes with altogether 50 Bytes per point will lead to a total of 590 GBytes data volume.

Nevertheless, sophisticated strategies have been implemented to reduce the flood of information to the essential parts. The *segmentation* of the point cloud connects and extracts homogeneous areas defined by a specific criterion of homogeneity - mostly homogeneous is defined by planar, connected areas (Section 11.5.2). Region growing algorithms and clustering methods for finding similarities (e.g. k-means) in the feature space are used most often. Further grouping and *classification* of such segments are, for example, used to reconstruct buildings (e.g. roofs, walls) and roads but also tree crowns (Filin and Pfeifer 2006, Morsdorf et al. 2003).

The laser point cloud is represented in the vector data model. Each feature, i.e. laser point measurement, is located in xyz space and can hold an unlimited number of attributes (e.g. time, intensity, echo number, classification). The timestamp is held as attribute because a single point cannot clearly be identified and connected to another point of a different epoch. For multitemporal analysis, spatially continuous data models are preferred (e.g. rasters, triangulations). On the one hand vector points are much easier to be transformed into another coordinate system and adequate datasets can easily be derived and generated (e.g. DEMs), but on the other hand a sparse number of algorithms for analyzing large point clouds are implemented in standard GIS and remote sensing software. Hence, until now the laser point cloud has been used particularly in practice for simple visualizations (e.g. cross-sections: Figure 11.9) and raster data calculation.

There a number of commercial ALS systems available on the market with quite different technical properties (Baltsavias 1999b), which leads to the end-users being confronted with a great variety of data formats. One attempt at defining a standard for ALS point cloud data can be seen in the LAS format (LAS specification 2005). The LAS format is a public binary file format for the interchange of ALS data between customers and ALS vendors, but also between software packages and between operating systems. Furthermore, the LAS format includes both meta data definitions and definitions for classified laser points (e.g. ground, non-ground, building or vegetation points).

11.3.2 Digital Elevation Models (DEM)

The most common types of DEMs generated from the point cloud are the Digital Surface Model (DSM) and the Digital Terrain Model (DTM). The DSM simply represents the upper hull of the Earth's surface including all natural and man-made objects. The DTM represents the bare earth, without any raised objects, which are by definition not part of the terrain. So far there is no universally valid definition of what belongs to terrain or not. For example, bridges are extensions of the bare earth, must not be fully raised but are man-made (Sithole and Vosselman 2006). Whether objects are by definition part of the bare earth or not depends on the application the DTM will be used for.

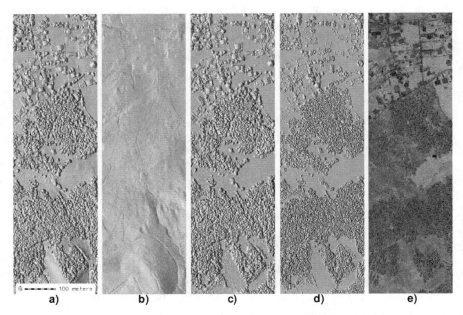

Fig. 11.4 Different types of digital models: a) DSM, b) DTM, c) nDSM, d) FLDM and e) Digital Intensity Model (DIM)

Having both a DSM and DTM the so-called normalized DSM (nDSM) can be calculated by simply subtracting the DTM from the DSM. That leads to a normalization of the absolute elevations to relative object heights. The nDSM is most often used for detecting and classifying objects without the disturbing influence of the changes in the underlying terrain, as for example building detection (Section 11.5.2) and generation of 3D city models (Kaartinen et al. 2005), as well as applications in forestry (Hyyppä et al. 2004). Other types of DEMs are for example models representing first or last reflections of laser shots, or a first and last echo difference model (FLDM) (Figure 11.4).

For the derivation of DTMs both algorithms filtering the point cloud directly and algorithms removing objects in a raster DEM (mainly of last reflections) exist (Sithole and Vosselman 2004). DTM filtering is combined with an interpolation step, i.e. filling holes caused by removing off-terrain points/cells. The percentage of filled and empty cells mainly depends on the point density, point distribution, the chosen target grid size, and the amount of filtered points/cells. In general, the higher the point density the more accurate elevation models can be achieved (Kraus et al. 2004, Karel et al. 2006). The interpolation of gridded DTMs results in a smoothed representation of the terrain. Briese (2004) states that with integrating explicitly modeled structure lines (breaklines) in the DTM generation process better results can be achieved. Including structure points, lines (breaklines) and areas as vectors in a raster DTM is called hybrid DTM (Kraus 2000). The raster data model is most frequently used for DEMs because its data structure is clearly defined and supported by many software packages. But overlaying the laser point cloud with a regular grid means a loss of information because the three-dimensional coordinates of one or more points are put into a two-dimensional matrix in which one interpolated value represents the elevation. Other data models well-suited for storing elevation information are triangulated irregular networks (TINs) and 3D volume rasters (Voxels) (Bartelme 2005, Kraus 2004, Mitas and Mitasova 2005, Neteler and Mitasova 2004).

11.3.3 Digital Intensity Models

In addition to spatial information, most LS systems record the received signal power with the coordinates, the so-called signal intensity (Figure 11.5). Hence, the intensity is already georeferenced. While traveling from airplane to target and back the emitted laser signal power is diminished by (i) spherical loss, (ii) target properties (e.g. reflectance, target area), (iii) topographical effects (shape) and (iv) atmospheric attenuation (Jelalian 1992, Rees 2001). If one wants to use the intensity to describe the target properties, all effects due to scan geometry and atmosphere have to be removed (Höfle and Pfeifer, 2007). The range and angle of incidence of a laser shot may rapidly vary within a small area, mainly due to surface elevation changes (e.g. high mountainous areas), which results in a heterogeneous, noisy representation of the intensity. Strategies for correcting laser scanning intensities for these influences and deriving a value proportional to the surface reflectance in the wavelength of the laser are presented by Coren and Sterzai (2006) and (Höfle and Pfeifer 2007).

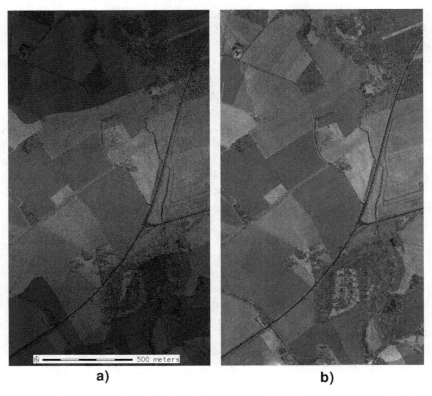

Fig. 11.5 Digital Intensity Models: a) uncorrected DIM representing original recorded intensity values and b) DIM corrected for influences of scan geometry, topography and atmosphere

The Digital Intensity Model (DIM) simply stands for an intensity image generated by rasterization of the original or corrected intensity values. The corrected Digital Intensity Model (cDIM) can be used as an additional data source for surface classification and object detection. DIMs have many advantages especially for the classification of surfaces with low texture in the visible light spectrum and with a high separability in the laser wavelength, e.g. on ice and snow surfaces (Lutz et al. 2003, Song et al. 2002).

11.4 Existing applications in natural hazard management

Mountainous regions are especially affected by geomorphodynamic processes causing damage, loss of property and human life. The mapping of the processes according to their surface properties and geomorphological structure and the analysis of their characteristics is the basis for risk and vulnerability analysis (Figure 11.6). The applicability of remote sensing methods for natural hazard assessment is predominantly governed by the following factors:

- *spatial resolution* - determines the degree of detail that can be detected from the data.
- *spatial coverage* - determines the area that can be included in the assessment procedure.
- *temporal resolution* or *revisit time* - has to be in agreement with the rate of hazard development or changes observed.

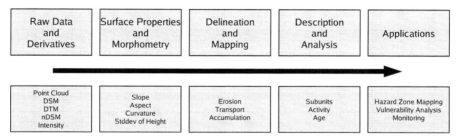

Fig. 11.6 From data processing to applications

ALS data deliver high quality topographic information in a spatial resolution that is unprecedented; the availability of data (spatial coverage) is steadily increasing. An active decision on the revisit time was made, for example, by the Austrian Federal State of Vorarlberg who acquired ALS data for several river catchments after the flood event in August 2005 in order to compare it with data that were acquired before the flood in the course of an area-wide data acquisition for the entire state (Section 11.1).

Elevation data are an important source for process models of different types (empirical, numerical and probabilistic). Most outputs of these models are highly sensitive to DTM characteristics such as resolution, level of detail, and vertical or horizontal errors.

Terrain elevation changes over time, i.e. vertical differences between repeated DTMs are indicators for geomorphodynamic processes. Thus, their detection is an important step in hazard assessment and disaster mapping. In general, changes in terrain elevation are derived by subtracting repeat DTMs. The accuracy of such-derived vertical changes is, in principle, on the order of the accuracy of the single DTMs that are used. If the DTMs represent independent measurements, the root mean square error (RMS error) of an individual elevation change can be estimated from the RMS errors of the DTMs involved.

The following sections summarize existing applications of using ALS data for different tasks in natural hazard management.

11.4.1 Rockfall

Rockfalls are spontaneous events in steep terrain where blocks are weakened by weathering. Falling rocks with high kinetic energy can cause damage to infrastructure, buildings, and endanger human life. Simulation models calculate the paths of falling rocks, the influence of protective measures, as well as the effect of the forest where the single tree is a barrier to reduce the kinetic energy of the falling rock (Dorren et al. 2005). High resolution DTMs are used to derive discontinuities, instability zones, and to model the tracks of falling rocks, which includes the calculation of the distribution of rock trajectories, kinetic energies by unit of mass, bounce height and stop points. Low point density in filtered DTMs derived from ALS data and the problem of modeling overhangs can be solved by combining ALS data with TLS surveys carried out for critical spots (Abellán et al. 2006, Janeras et al. 2004). Beside the supply of topographic information as model input especially TLS (for its flexible campaign upset) can be used for the monitoring of deformations and activity of rockfall areas within constant time steps (Scheikl et al. 2001). Methods and accuracies by comparing multitemporally acquired surfaces for volume determination and deformation monitoring are described in Tsakiri et al. 2006.

11.4.2 Landslides and debris flow deposits

Landslides are mass movements, which are triggered by geologically instable zones, heavy rainfalls leading to high soil water pressure, or earthquakes. The processes can be continuous creeping movements or are initiated spontaneously. The spatial and temporal distribution of landslides is of high interest because not only single buildings or infrastructure but whole settlement areas can be affected. In high resolution DTMs undisturbed terrain appears smoother than the mass movement area. Derived morphometric parameters from ALS DTMs are used to locate and map landslides. The process itself can be described and characterized in terms of spatial distribution and activity and age. Parameters used for landslide description are for example surface roughness coefficients, first and second order derivatives like slope, aspect and curvature. The relation between surface properties and activity of certain terrain parts makes it possible to distinguish kinematic units of mass movements (McKean and Roering 2004). Furthermore, active landslides can be distinguished from old, inactive ones by the analysis of surface roughness values. Current investigations show that active landslides are characterized by higher surface roughness (Glenn et al. 2006). A similar approach is used to investigate the spatial distribution of debris flow deposition on alluvial fans, which is necessary to understand the process itself. For selected fans an ALS DSM is investigated towards curvature and gradient properties in order to (i) detect deposit zones on an individual process level and (ii) to compare surface trends between different debris flow accumulation areas. To suppress noise caused by fine-scale surface forms like levees, lobes, debris dams, and channels formed from past events, a calculation window larger than these forms is used for parameter calculation (Staley et al. 2006). While in this work large scale structures, which have disturbing effects on the analysis of the general surface trends, are suppressed, the information of such small structures could be useful in classifying and characterizing erosion, transportation, and deposition areas of the entire process on a more detailed object level.

11.4.3 Hydrology (torrent activities and floods)

The last 15 years have been marked by severe flooding in many parts of Europe. For example, large parts of Central Europe were hit in August 2002 by two consecutive flood events, which caused significant damage to buildings and infrastructure. ALS has been evolved to an attractive technology for the acquisition of useful data for various river management tasks, e.g. floodplain vegetation classification for hydraulic modeling, the determination of soil volume, the determination of riverbed morphology (at low water levels) and measuring water levels and wave pattern parameters (Brügelmann and Bollweg 2004). DTMs form the basis for distributed hydrologic models as well as for two-dimensional hydraulic river flood models. Two-dimensional hydraulic surface flow models are mostly constrained by inadequate parameterization of topography and roughness coefficients, primarily due to insufficient or inaccurate data. DTMs and their derived parameters such as slope, aspect and drainage network form a fundamental input for the models mentioned. In addition to the topographical information, nDSMs have the advantage of offering the possibility to estimate object heights (vegetation, buildings). Detailed land cover maps can be derived in conjunction with the complementary information provided by high resolution color-infrared orthophotos.

Important tasks which can be supported by laser scanning data are the delineation of flood prone areas and the determination of landscape roughness as the standardized input parameter in two-dimensional river flood models.

Flood risk areas are modeled by delineating inundated land surfaces for different water levels. For this task, ALS data has become the preferred data source. Laser DTMs have a sufficient vertical accuracy for the modeling of design events. An additional requirement is the consideration of micro-topography effecting the flow routing. Certain structures like dams, ditches, levees, embankments and old channels are represented in laser DTMs. Grenzdörffer et al. (2002) compared ALS data with data from terrestrial surveying for the specific task of flood risk area delineation and found significant deviations in areas with a high surface roughness (e.g. bush, reed) seeing as the automatic ALS filtering methods reached their limits here. As the quality of the derived DTMs is often not completely sufficient for the modeling of inundation patterns in the case of flooding, approaches to extract hydraulically relevant breaklines have been a main focus in recent research efforts (Briese and Attwenger 2005). The determination of landscape roughness is of significant importance. In the case of flooding the flow of water in the floodplain of a river is influenced by the spatial distribution of forests, grasslands, agricultural fields and infrastructure like roads and houses. ALS data have a significant potential for assessing landscape roughness and vegetation structure (height, layering, spatial arrangement). This information is useful for determining relevant model parameters, such as the coefficient of friction.

For flood modeling the *Manning coefficient* of roughness is often used. Asselman et al. (2002) estimated hydraulic roughness of flood plain vegetation in the Netherlands while Smith et al. (2004) assessed the potential of using ALS data in analyzing the landscape for the estimation of roughness coefficients and compare ALS data with data derived from aerial photography/photogrammetry. They proposed automated techniques for a more objective estimation. Major advances are expected from upcoming full-waveform laser scanner systems (e.g. Wagner et al. 2004). The overall goal is a spatially distributed parameterization of friction as the standardized input parameter in two-dimensional river flood models. These model types are core elements in flood prediction systems.

Cobby et al. (2003) used ALS data for improving such models by decomposing a finite-element mesh to reflect floodplain vegetation features, such as hedges and trees having different frictional properties to their surroundings, and significant floodplain topographic features having high curvature values. The decomposition is achieved by using an image segmentation method that converts the ALS data into separate data sets of surface topography and vegetation height at each point. The derived vegetation height map is used to estimate a friction factor at each node, which results in a physically based, spatially distributed friction parameterization. Methodologies were developed to convert vegetation heights to friction coefficients (Mason et al. 2003). The use of the decomposed mesh also allows the prediction of velocity variations in the neighborhood of vegetation features such as hedges. These variations can consequently be used for predicting erosion and deposition patterns. Thoma et al. (2005) used multitemporal ALS data for riverbank erosion assessment by quantifying volume and mass changes. A study in Iceland (Smith et al. 2006) showed how multitemporal ALS data can support the estimation of sediment erosion and deposition in sander plains after a glacier outburst flood (jökulhlaup). French (2003) considered the application of ALS data for the provision of elevation data at accuracies and spatial densities in accordance with the current generation of high resolution hydraulic models. He specifically addressed the quality of the data via multi-scale calibration against surveyed sections and supplementary control points, and the use of image processing techniques for identifying regions of interest. He concluded that ALS provides topographic information at an accuracy and resolution close to the present limits of model representation. Charlton et al. (2003) discussed the derivation of representative cross-profiles of river channels.

The summarized advantages of ALS data for hydrodynamic modeling are as follows:

- adequate vertical accuracy
- adequate horizontal accuracy and data density
- potential for the derivation of breaklines
- potential for the derivation of land use classification and roughness coefficient
- increasing data availability

Most studies until now were constrained to small test sites. Hollaus et al. (2005) summarized the experiences to process ALS data for large mountainous regions, demonstrating the applicability for hydrological applications. Focusing on the exploitation of ALS for hydrological applications, the Christian Doppler laboratory *Spatial data from laser scanning and remote sensing* was founded at the Technical University of Vienna in 2003, whereby the extreme topographic environment of Austria is taken into consideration as a challenging boundary condition. One point of focus is also the development of new and advanced methods combining ALS with radar remote sensing and digital photogrammetry. In Germany, the federal state of Bavaria started the project *Floodscan* in 2006. The goal of the project is the optimization of ALS data processing for hydrodynamic modeling with the development of data thinning techniques.

11.4.4 Hazards related to glaciers and permafrost conditions

Hazards related to glaciers and permafrost conditions are of significance in densely populated, high mountain areas. On a global scale damages and mitigation costs related to disasters associated with this environment are on the order of several 100 million EUR as a long-term annual average sum (Kääb et al. 2005b). Glacier and permafrost hazards include floods triggered by glacial and periglacial processes, glacier fluctuations, glacier- and permafrost-related mass movements, and permafrost thaw-settlement and frost heave. In high mountain areas process interactions, chain reactions, as well as the present shift of hazard zones due to atmospheric warming call for the application of modern remote sensing techniques for hazard assessment. An overview of suitable air- and spaceborne remote sensing methods suitable for glacier and permafrost hazard assessment and disaster management is given by Kääb et al. (2005a) and Huggel (2004). They state that high resolution DTMs as derived from laser scanning represent one of the most important data sets for investigating high mountain processes. DTMs represent the core of any investigations of alpine hazards, because many relevant geomorphological processes are driven by the relief energy. Changes in terrain geometry (elevation, volume) can be measured by repeated laser scanning. Multitemporal data can additionally be used to derive surface velocities on glaciers (Bucher et al. 2006), and can be used likewise in permafrost areas. Due to the relatively rapid change of high mountain environments, hazard assessment shall be undertaken routinely and regularly. Laser scanning is particularly suited for monitoring purposes as is described by Geist et al. (2003) and Stötter (2007) for glacier monitoring.

11.4.5 Avalanches

Avalanche simulation models need highly accurate terrain information for a precise model output, which is the calculation of velocities, pressure distribution, and accumulation areas. Laser scanning DTMs show better model results compared to DTMs derived from other sources like photogrammetry stereo matching. Besides pure spatial resolution highly detailed representation of topography is an important factor for modeling the flow component of avalanches (Schmidt et al. 2005).

To make process and prediction models more reliable a better understanding of the snow cover stratigraphy and the snow distribution based on snow depth, wind impact and topography is needed. Snow depth distribution calculated from difference models between snow covered and snow free conditions contributes to the research of avalanche risk conditions (Deems et al. 2006).

11.4.6 Protection forest

Forests have a major protective function for different natural hazards. Site-protecting forests keep a certain area in stable conditions and prevent the occurrence of natural hazards. These forests help to avoid, for example, surface runoff occurrence (floods), they stabilize slopes, they protect from soil erosion (landslides, debris flows, torrent activity) and they have a potential impact on the snow cover stability avoiding snow drift and the generation of instable snow layers in potential avalanche starting zones. Protection forests are a barrier in case of an event and protect buildings, infrastructure and farm land from damages (e.g. rockfall and avalanches).

Protection forests must consist of heterogeneous patches of species mixture and have a high variability of age structure to guarantee continuous growing cycles with long-term stable conditions. Properties of the vertical and horizontal forest structure, like canopy closure rate, tree density, and canopy roughness derived from ALS data, provide information about the protective ability of a forest (Maier et al. 2006).

11.4.7 Object protection

Process models used in natural hazard management do not only need accurate topographic information but also information on the buildings, roads and public facilities to be protected. In addition, this information is needed for risk and vulnerability calculations. Several studies show that ALS data can be used to classify buildings and street networks (Kaartinen et al. 2005).

11.4.8 Conclusion

In summary, it has to be concluded that a variety of patchwork applications in line with the solution of specific tasks in natural hazard management exist. Many applications rely on manual interpretation and show a lack of automation in data processing. In the following section results of an applied research project at the *alpS - Center for Natural Hazard Management* are presented. The goal of the project is to overcome the limitations mentioned, whereby the focus is placed on a concept for an improved data management and the creation of a framework for feature extraction and classification.

11.5 Laser scanning data and products – a substantial input for natural hazard management

11.5.1 Data management: GIS and database embedding strategies

Laser Scanning delivers high quality and highly accurate vector and raster data models. In natural hazard management, there is a strong need for immediate and simple access to all LS data products and derivatives. The large data volume does not allow for the use of standard GIS software products, as these cannot handle billions of vector points and large high resolution DEMs. In the following section, a new concept for GIS and database embedding of country-wide LS datasets is explained.

Decision for Open Source software

The development and usage of *Open Source software* has a long tradition in the scientific world (OSI - Open Source Initiative 2006). Nowadays, Open Source solutions are also attracting more and more attention from administrative institutions and companies using the software for their operational sequences (Boulanger 2005). Large Open Source projects ensure a constant improvement through a constantly growing community, as for example GRASS GIS (GRASS Development Team 2007). The major advantages of using Open Source components are that many sophisticated programs already exist - developed and reviewed by a large community - and that a redistribution of modifications and enhancements leads to an even faster and better development of the Open Source project. Nonetheless, cost-saving solutions, such as freely available Open Source packages, are required in particular by institutions with a low budget. The developed information system *LISA (LiDAR Surface Analysis)* combines the advantages of Open Source software packages with the functionality of a spatial database management system (DBMS), a geographic information system (GIS) and statistical software packages. The open data format standards, especially for spatial data, allows for a simple data interchange but also direct access from many applications on one and the same data source (OGC 1999). Open standards (formats and interfaces) and fully transparent software solutions allow for a stronger cross linking of knowledge and data itself.

System architecture

LISA integrates data management, spatial analysis and processing (Section 11.5.2), as well as visualization tasks (Höfle et al. 2006). The most obvious requirements for such an information system - given by LISA - are (i) multiuser and client-server architecture to avoid redundant data storage (ii) full access to all LS datasets in highest quality (including point cloud) (iii) GIS and spatial statistics functionality and (iv) open and well-defined interfaces to other software packages (import and export). The system architecture design follows the key principles of using the best existing application for specific tasks, of filling the gaps where no existing solutions are available, and of optimizing the communication between the modularly built system components in terms of performance, data storage, usability, administration effort, data security, multiuser and cross-platform support. The idea is to use the advantages of every single component, for example, the raster capabilities of the GIS component. The implemented client-server architecture minimizes both data storage and administration efforts. Time intensive data transfer within LISA is avoided because every application of LISA can directly access the original laser points in the spatial database system and the raster derivatives in the GIS database. The data models for vector and raster datasets include the dimension *time*, which allows for multitemporal analyses between different epochs. Further processing, analysis and visualization of the LS data can be performed in both the original point cloud and the calculated raster datasets.

System components

The LISA system consists of two main components: the object relational database management system PostgreSQL (PostgreSQL Global Development Group 2006) with its spatial add-on PostGIS (Refractions Research 2006) and the geographic information system GRASS (GRASS Development Team 2007). The scripting language Python (Python Software Foundation 2006) was chosen to construct workflows, applications and interfaces. In addition to its rich pool of scientific libraries (Jones et al. 2001), Python supports a direct connection to PostgreSQL/PostGIS (Cain 2006), and to the GIS layers of GRASS (Warmerdam 2006). There are in fact no important limitations for the storage of huge amounts of data in a PostgreSQL database and GRASS GIS. The PostgreSQL database size is unlimited and the maximum table size is ca. 32 Terabyte. Organizing large rasters in tile sets (similar to a raster catalogue) allows for the management of country-wide high resolution raster datasets within GRASS. For the 3D visualization of point cloud subsets the Open Source program VTK (VTK 2006) is used.

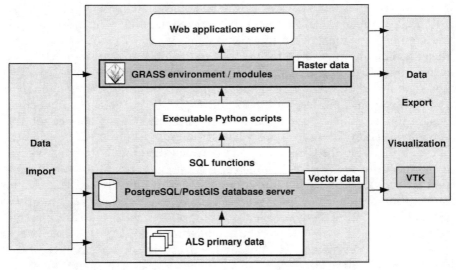

Fig. 11.7 LISA system components and workflow (Höfle et al. 2006)

Data management for point clouds

The original point cloud is stored within the spatial DBMS PostgreSQL/PostGIS, which provides user access management and multiuser capabilities. The point cloud vector data is therefore stored only once with assured access control and data availability. The spatial add-on PostGIS allows for the use of geometry objects defined by the OpenGIS *Simple Features Specification for SQL* (OGC 1999), as for example points, lines and polygons. Within the database environment the Structured Query Language (SQL) is used to apply geometric functions and spatial queries on the geometry objects. Geometry objects are internally stored as OpenGIS Well-Known Binary (WKB) geometries (OGC 1999, Refractions Research

2006). The WKB representation is used by external applications to recognize the content of a geometry object.

A hierarchical data model for table structures, relationships and constraints was built, which reflects the characteristics of a typical ALS campaign (Section 11.2). For example a typical ALS campaign consists of many overlapping flight strips. A single flight strip consists of millions of laser shots. And each laser shot can have none, one or more reflections. Additionally to the laser shots, the airplane positions are stored with a much lower resolution in time (about 250 laser shots temporally between two plane positions). With a linear interpolation approach over time the corresponding plane position can be calculated for each laser shot/echo. Figure 11.8 shows a simplified data model of ALS datasets.

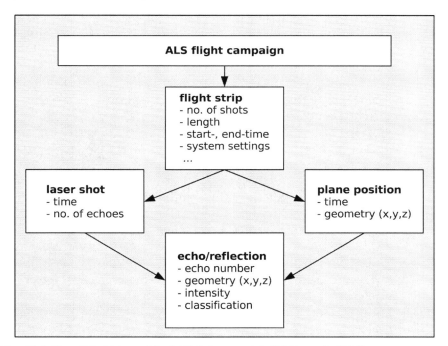

Fig. 11.8 Simplified hierarchical ALS data model (Höfle et al. 2006)

The chosen data model makes complex queries possible with any combination of spatial, temporal or attribute search criteria. To speed up queries on large databases, column indexing is necessary. A B-tree index on the time attribute/column drastically increases performance of multitemporal queries. Geometry columns are indexed by an R-Tree index implemented on top of a Generalized Search Tree (GiST), which is provided by

PostGIS (Bartunov and Sigaev 2006). Spatial queries consist of a prior bounding box query followed by a distance query. On the one hand indexing makes managing large geometry tables possible, but on the other hand much more disk space is needed to store the index tables. In comparison to a file based management system, the client-server DBMS guarantees data security and integrity for a running system while data manipulation (insert, update and delete entries) is performed. The transaction management of PostgreSQL and the defined foreign keys (constraints) between the data model relations (e.g. flight campaign, strips, laser shots) maintain data consistency.

Data management for raster datasets

Basically, raster data management in LISA is organized within GRASS GIS. On the clients side GRASS GIS is used for executing all GIS and LISA applications respectively, but also on the server side for providing raster basic data, such as DTMs, DSMs and orthophotos, to multiple concurrent users/clients. The *GRASS server* handles user requests (e.g. clipping and export of defined rasters) and user domains, as every user has their own rights and data spaces. Users with granted access can read and process the basic data in their own data space. The actual raster data management can either be installed on client or server side. For better performance and to avoid file size limits large raster datasets are tiled. A tile index vector holding the tile extents and meta data (e.g. timestamp, data type, bands) increases query possibilities and reduces access times on raster subsets.

Advantages of the central raster management are that the basic data is stored only once in the highest resolution, lower resolutions are generated on-the-fly. The full functionality of LISA and GRASS can be applied individually by each user to one and the same basic data because of the client-server principle. Well-defined interfaces additionally allow direct access from many different clients (Warmerdam 2006). Thus, for example, the UMN MapServer (UMN Map Server 2006) can directly visualize GRASS rasters.

Data processing and workflows

Workflow paths and applications for LISA are constructed following a hierarchical approach (Figure 11.7). First the LS primary data has to be imported into the spatial database. During the import procedure the relationships and constraints between the data model relations are built and tested. Once the point cloud is stored in the defined data structure further processing is straightforward due to the fixed interfaces. The spatial database is not only used for data storage and retrieval but also for data manipulation (e.g. coordinate transformation, plane position interpolation, and signal intensity correction). A workflow starting in the 3D point cloud makes use of SQL functions written in one of the procedural DB languages (e.g. PL/pgSQL, PL/Python) supported by PostgreSQL. Hence, most of the work load is already done by the SQL functions, so that the data is already preprocessed before being exported into the next higher level. The Python scripting language was chosen to construct the workflows, which are available as GRASS GIS commands/modules. Python can directly access PostgreSQL, run SQL statements and receive the resulting data for further processing. Python GRASS modules of LISA follow the standard of GRASS GIS commands and are therefore easy-to-use. The integration of LISA into GRASS allows full access to the existing GIS functionality. The hierarchical workflow offers a multitude of data accessing and data processing possibilities. Thus, for example, the user can work with the GIS commands of LISA within GRASS or through a web application server; he can directly access the point cloud with an appropriate program supporting PostGIS geometries (e.g. QGIS) or can directly read the raster data through the GDAL (= Geospatial Data Abstraction Library) interface.

Application example - point cloud cross-section

The point cloud cross-section application is a good example for interaction on all levels (Figure 11.9). Through a graphical user interface (GUI), either the GRASS command GUI or a web mapping application, the user interactively digitizes the vertices of the cross-section line and specifies the width of the cross-section, the laser point type (e.g. only last echo), the coloring mode (e.g. gray-scale intensity), and the output format (e.g. ASCII text file, graphic format). The GUI front end starts the GRASS command (a Python application), which further sends a SQL request to the spatial database. The database SQL function for cross-sections returns the selected laser points as query result records. In the next step the GRASS command generates the cross-section image, which is finally sent and visualized to the client.

334 Geist et al.

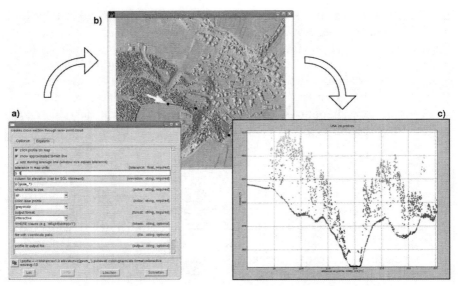

Fig. 11.9 LISA application example – interactive point cloud cross-sectioning within GRASS GIS: a) GRASS command GUI, b) GRASS monitor for interactive cross section digitizing and c) resulting cross-section image. The cross-section shows all laser points (first and last echoes) colored grayscale by intensity. The red line shows the additionally extracted terrain line

11.5.2 Data analysis

Using high resolution data means that traditional classification algorithms (e.g. clustering single pixels) lead to noisy classification results, which is known as the 'salt & pepper effect'. This problem is overcome using an object-based image analysis (OBIA) approach (Figure 11.10). In the first step single pixels are merged to segments representing a homogeneous area (Hay et al. 2003). Then object features are calculated for each segment, which are used as input for the classification procedure. Object features are statistical values of pixels forming one segment: (i) statistics related to segment geometry, (ii) topological information on neighboring segments and (iii) segments in an upper or lower hierarchy. This leads to a flexible classification tool with integrated methods for spatial analysis. It is possible to query neighborhood objects and to analyze the hierarchical connection between objects. The easy integration of different data types into the workflow (e.g. from optical sensors, ancillary data like cadastral information) is one of the strengths of the OBIA concept (Benz et al. 2004).

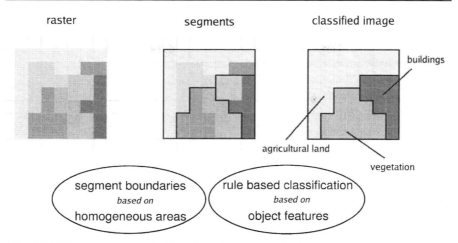

Fig. 11.10 Basic concept of object-based image analysis

The classification workflow subsequently described is primarily designed to classify ALS points and its rasterized derivatives. The used OBIA workflow is integrated in the LISA system and the GRASS GIS environment (Section 11.5.1). All classification procedures described here use ALS data only, as a consistent data set. Differences which occur when using additional data sources (aerial or satellite images or ancillary data) for classification are avoided using ALS data only. Ancillary data is only used to compare the classification results in the error assessment (Rutzinger et al. 2006a).

11.5.3 ALS land cover classification

The basic idea is to use an iterative OBIA workflow to classify rasterized ALS derivatives first and if an application requires more detailed information the original point cloud is used to derive objects on a finer scale. In the first step a vegetation mask is derived from the FLDM and a segmentation algorithm detecting 'raised' objects is applied. 'Raised' objects are distinguished from bare earth by jumping edges in the DSM. For these objects features like area, shape index, mean first/last echo difference or the standard deviation of heights are calculated. They are the input for a classification rule base. For every feature a value range and a weight are defined by the user in order to describe the object of interest best. Finally, the output is a map showing how well (based on the weights) a single object belongs to a specific class. The classification degree is converted to a layer representing a single class. This classification result can be compared with other classifications based on multitemporal data sets (change detection), with results of different parameter settings of the OBIA workflow (workflow calibration) or with ancillary data like a cadastral map (error assessment). If the comparison between two objects is not satisfying those objects are reclassified in a second iteration with new parameters to enhance the classification (Rutzinger et al. 2006b).

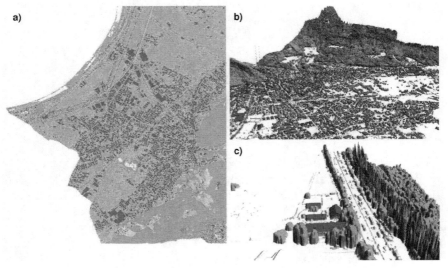

Fig. 11.11 a) Overview of the classified test site, b) 3D view with the GRASS visualization tool NVIZ, c) detailed 3D view

At the current stage this method is suitable to derive high vegetation and buildings (Figure 11.11) (Rutzinger et al. 2006b). After the classification parameters can be derived for each object. These are, for example, object heights and volume estimation derived from a nDSM or roughness values calculated from the DSM, which are the input for several applications in natural hazard management (Section 11.4).

11.5.4 Topographic analysis

On the one hand the derivation of surface parameters is an important input for process detection and monitoring (Section 11.4). On the other hand the extraction of breaklines is necessary for proper flood and hydraulic modeling. GRASS GIS provides several tools with which to calculate first (e.g. slope, aspect) and second order derivatives (e.g. curvature) from elevation models, which characterize terrain objects (Wood 1996, Horn 1981).

11 Laser scanning   337

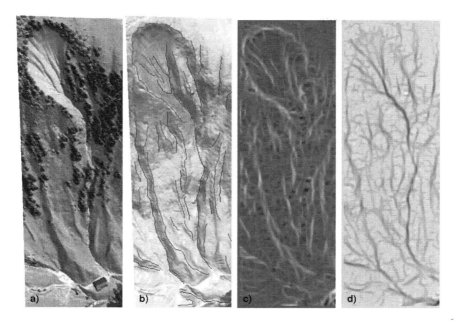

Fig. 11.12 a) 3D view DSM overlaid with an infrared orthophoto, b) 3D view DTM overlaid with classified ridges and erosion channels, c) maximum curvature of DTM, d) minimum curvature of DTM

The sample in Figure 11.12 shows the extraction of upper and lower edges, which are used for the classification of ridges and valleys. It shows a small torrent sub-catchment, which erodes shell shaped into the surrounding pasture. The representation of geomorphological structures depends on scale. The calculation window for the surface parameters must fit the size of the geomorphological structure in order to achieve meaningful results. If the window is too small, only substructures are derived and if the window is chosen too large only patterns in a superior scale are detected. The line objects in this case are extracted by the minimum and maximum curvature, describing the convexity and concavity of the surface respectively. The window for the curvature computation is 11 x 11 m. Figure 11.12 shows the maximum (d) and minimum curvature (c), which is used to derive upper and lower edges (b) respectively.

11.6 Opening new dimensions - future potentials of laser scanning in natural hazard risk management

In recent times both on the risk management side as well as on the laser scanning technology side considerable progress has been made.

On the one hand, coping with natural hazards has been further developed from mostly process-oriented approaches to new risk-based concepts (Stötter and Zischg, in print). This development results in a new demand to combine phenomena and processes of both the natural and the human environment. On the other hand, laser scanning technology has undergone a somewhat exponential development within a rather short time span, e.g. by the design of new sensors (from single echo to first echo/last echo to multi echo and, most recently, to full-waveform signal), the combination of ALS and TLS technologies or the potentials of multi channel laser scanners.

When trying to introduce risk-based models, one of the most interesting challenges is the scale of investigation. It seems a vital prerequisite for the combination of geodata describing the natural hazard and spatial information representing the damage potential to have a common spatial resolution or scale. As risk-based concepts increasingly focus on the potentially threatened object or subject, extremely detailed information is demanded, that allows for the delineation of risk for the individual object.

Existing process models/existing modeling concepts must be tested regarding their capacity for laser scanning data, as results from avalanche and flood modeling show. Chances are that a new generation of models or a modified data driven approach is required. The combination of airborne remote sensing methods and GIS have already led to very accurate digital elevation models (elevation accuracy better 0.5 m) for many flood endangered city areas. Hence, a predictive analysis for defined buildings is feasible. Such information supports decision making for individual citizens, authorities, emergency services or insurance companies.

As shown in this paper, laser scanning technology opens new and promising options for those risk management ideas. Nevertheless, there are also problems that arise. Most of the process models that are applied to simulating natural hazard processes and consequently the natural hazard situation are not yet designed on the spatial resolution of laser scanning data. As a possible consequence, a new generation of models is required, which can cope with high resolution and highly precise input data. Recent advances in computational resources, web-based data management, and improved scientific and engineering knowledge can be exploited to meet the requirements for the next generation models.

As an additional challenge, these models will have to include the multiple aspects of global change conditions, both on the natural and the human environmental side to meet the demands of sustainability. Through the option of multitemporal analyses, laser scanning data and the highlighted management and analysis system offer interesting and promising opportunities.

Finally, it can be stated that laser scanning provides many advantages for future concepts of risk management, but there are still many questions to be solved.

References

Abellán A, Vilaplana JM, Martínez J (2006) Application of a long-range Terrestrial Laser Scanner to a detailed rockfall study at Vall de Núria (Eastern Pyrenees, Spain). Engineering Geology 88 (3-4), pp 136-148

Asselman N, Middelkoop H, Ritzen M, Straatsma M (2002) Assessment of the hydraulic roughness of river flood plains using laser altimetry. IAHS publication 276, pp 381-388

Baltsavias EP (1999a) A comparison between photogrammetry and laser scanning. ISPRS Journal of Photogrammetry and Remote Sensing 54 (2-3), pp 83-94

Baltsavias EP (1999b) Airborne laser scanning: basic relations and formulas. ISPRS Journal of Photogrammetry and Remote Sensing 54 (2-3), pp 199-214

Baltsavias EP (1999c) Airborne laser scanning: existing systems and firms and other resources. ISPRS Journal of Photogrammetry and Remote Sensing 54 (2-3), pp 164-198

Bartelme N (2005) Geoinformatik - Modelle, Strukturen, Funktionen. Springer, Berlin Heidelberg New York

Bartunov O, Sigaev T (2006) GiST for PostgreSQL, http://www.sai.msu.su/_megera/postgres/gist/, last accessed 1. December 2006

Benz U, Hofmann P, Willhauck G, Lingenfelder I, Heynen M (2004) Multiresolution, object-oriented fuzzy analysis of remote sensing data for GIS-ready information. ISPRS Journal of Photogrammetry and Remote Sensing 58 (3-4), pp 239-258

Boulanger A (2005) Open-source versus proprietary software: Is one more reliable and secure than the other?, IBM Systems Journal 44 (2), pp 239-248

Brenner C (2005) Building reconstruction from images and laser scanning. International Journal of Applied Earth Observation and Geoinformation 6 (3-4), pp 187-198

Briese C (2004) Three-dimensional modelling of breaklines from airborne laser scanner data. International Archives of Photogrammetry, Remote Sensing and Spatial Information Sciences 35 (B3), pp 1097-1102

Briese C, Attwenger M (2005) Modellierung dreidimensionaler hydrologisch und hydraulisch relevanter Geländekanten aus hochauflösenden Laser-Scanner-

Daten. In: Bundesanstalt für Gewässerkunde (ed): Praxisorientierte und vielseitig nutzbare Fernerkundungseinsätze an der Elbe, pp 35-45

Brügelmann R, Bollweg A (2004) Laser altimetry for river management. International Archives of Photogrammetry, Remote Sensing and Spatial Information Sciences 35 (B2), pp 234-239

Bucher K, Geist T, Stötter J (2006) Ableitung der horizontalen Gletscherbewegung aus multitemporalen Laserscanning-Daten Fallbeispiel: Hintereisferner/ Ötztaler Alpen. In: Strobl J, Blaschke T, Griesebner G (eds) Angewandte Geoinformatik 2006 - Beiträge zum 18. AGIT-Symposium Salzburg, pp 277-286

Cain DJM (2006) PyGreSQL PostgreSQL module for Python, http://www.pygresql.org, last accessed 1. December 2006

Charlton M, Large A, Fuller I (2003) Application of airborne LiDAR in river environments: the River Coquet, Northumberland, UK. Earth surface processes and landforms 28 (3), pp 299-306

Coren F, Sterzai P (2006) Radiometric correction in laser scanning. International Journal of Remote Sensing 27 (15-16), pp 3097-3104

Cobby D, Mason D, Horrit M, Bates P (2003) Two-dimensional hydraulic flood modelling using a finite-element mesh decomposed according to vegetation and topographic features derived from airborne scanning laser altimetry. Hydrological Processes 17, pp 1979-2000

Deems JS, Fassnacht SR, Elder KJ (2006) Fractal distribution of snow depth from LiDAR data. Journal of Hydrometeorology 7 (2), pp 285-297

Dorren L, Berger F, Maier B (2005): Der Schutzwald als Steinschlagnetz. LWFaktuell 50, pp 25-27

French J (2003) Airborne LiDAR in support of geomorphological and hydraulic modelling. Earth surface processes and landforms 28 (3), pp 321-335

Filin S, Pfeifer N (2006) Segmentation of airborne laser scanning data using a slope adaptive neighborhood. ISPRS Journal of Photogrammetry and Remote Sensing 60 (2), pp 71-80

Glenn NF, Streutker DR, Chadwick DJ, Thackray GD, Dorsch SD (2006) Analysis of LiDAR-derived topographic information for characterizing and differentiating landslide morphology and activity. Geomorphology 73, pp 131-148

Geist T, Lutz E, Stötter J (2003) Airborne Laser Scanning Technology and its Potential for Applications in Glaciology. International Archives of Photogrammetry, Remote Sensing and Spatial Information Science 34 (3/W13), pp 101-106

Geist T, Stötter J (2007) Documentation of glacier surface elevation change with multi-temporal airborne laser scanner data - case study: Hintereisferner and Kesselwandferner, Tyrol, Austria. Zeitschrift für Gletscherkunde und Glazialgeologie 41, pp 77-106

GRASS Development Team (2007) Geographic Resources Analysis Support System (GRASS) Software. ITC-irst, Trento, Italy. http://grass.itc.it, last accessed 1. May 2007

Grenzdörffer G, Foy T, Bill R (2002) Laserscanning und andere Methoden zur Ausweisung potenziell gefährdeter Hochwasserbereiche der Unteren Warnow.

In: Strobl J, Blaschke T, Griesebner G (eds) Angewandte Geoinformatik 2002 - Beiträge zum 14. AGIT-Symposium Salzburg, pp 133-138
Hay GJ, Blaschke T, Marceau DJ, Bouchard A (2003) A comparison of three image-object methods for the multiscale analysis of landscape structure. ISPRS Journal of Photogrammetry and Remote Sensing 57 (5-6), pp 327-345
Höfle B, Rutzinger M, Geist T, Stötter J (2006) Using airborne laser scanning data in urban data management - set up of a flexible information system with open source components. In: Fendel E, Rumor M (eds) Proceedings of UDMS 2006: 25th Urban Data Management Symposium, Aalborg, Denmark: 7.11-7.23
Höfle B, Pfeifer N (2007) Correction of laser scanning intensity data: data and model-driven approaches. ISPRS Journal of Photogrammetry and Remote Sensing, in press
Hollaus M, Wagner W, Kraus K (2005) Airborne laser scanning and usefulness for hydrological models. Advances in Geosciences 5, pp 57-63
Horn BKP (1981) Hill Shading and the reflectance map, Proceedings of the IEEE 69 (1), pp 14-47
Huggel C (2004) Assessment of glacial hazards based on remote sensing and GIS modelling. Schriftenreihe Physische Geographie, Glaziologie und Geomorphodynamik 44, Zürich
Hyyppä J, Hyyppä H, Litkey P, Yu X, Haggrén H, Rönnholm P, Pyysalo U, Pitkänen J, Maltamo M (2004) Algorithms and methods of airborne laserscanning for forest measurements. International Archives of Photogrammetry, Remote Sensing and Spatial Information Sciences 36 (8/W2), pp 82-88
Janeras M, Navarro M, Arnó G, Ruiz A, Kornus W, Talaya J, Barberá M, López F (2004) Lidar applications to rock fall hazard assessment in vall de núria. Proceedings of the 4th ICA Mountain Cartography Workshop, pp 1-13
Jelalian AV, (1992) Laser Radar Systems. Artech House, Boston London.
Jones E, Oliphant T, Peterson P et al. (2001) SciPy: Open Source Scientific Tools for Python, http://www.scipy.org, last accessed 1. December 2006
Karel W, Pfeifer N, Briese C (2006) DTM quality assessment. International Archives of Photogrammetry, Remote Sensing and Spatial Information Sciences 36 (2), pp 7-12
Kaartinen H, Hyyppä J, Gülch E, Vosselman G, Hyyppä H, Matikainen L, Hofmann AD, Mäder U, Persson Å, Söderman U, Elmqvist M, Ruiz A, Dragoja M, Flamanc D, Maillet G, Kersten T, Carl J, Hau R, Wild E, Frederiksen L, Holmgaard J, Vester K (2005) Accuracy of 3D city models: EuroSDR comparison. International Archives of Photogrammetry, Remote Sensing and Spatial Information Sciences 36 (3/W19), pp 227-232
Kääb A, Huggel C, Fischer L, Guex S, Paul F, Roer I, Salzmann N, Schlaefli S, Schmutz K, Schneider D, Strozzi T, Weidemann Y (2005a) Remote sensing of glacier- and permafrost-related hazards in high mountains: an overview. Natural Hazards and Earth System Sciences 5, pp 527-554
Kääb A, Reynolds J, Haeberli W (2005b) Glacier and permafrost hazards in high mountains. In: Huber U, Bugmann H, Reasoner M (eds) Global change and mountain regions. Advances in global change research, pp 225-234

Kraus K (2000) Photogrammetrie, Band 3, Topographische Informationssysteme. Dümmler

Kraus K (2004) Photogrammetrie, Band 1, Geometrische Informationen aus Photographien und Laserscanneraufnahmen, De Gruyter Verlag, Berlin

Kraus K, Briese C, Attwenger M, Pfeifer N (2004) Quality measures for digital terrain models. International Archives of Photogrammetry, Remote Sensing and Spatial Information Sciences 35 (B2), pp 113-118

LAS Specification (2005): ASPRS LIDAR Data Exchange Format Standard Version 1.1, http://www.lasformat.org, last accessed 1. December 2006

Luethy J, Stengele R (2005) 3D mapping of Switzerland challenges and experiences. International Archives of Photogrammetry, Remote Sensing and Spatial Information Science 36 (3/W19), pp 42-47

Lutz E, Geist T, Stötter J (2003) Investigations of airborne laser scanning signal intensity on glacial surfaces - Utilizing comprehensive laser geometry modelling and orthophoto surface modeling (A case study: Svartisheibreen, Norway). International Archives of Photogrammetry, Remote Sensing and Spatial Information Sciences 34 (3/W13), pp 143-148

Maas HG, Vosselman G (1999) Two algorithms for extracting building models from raw laser altimetry data. ISPRS Journal of Photogrammetry and Remote Sensing 54 (2-3), pp 153-163

Maier B, Tiede D, Dorren L (2006) Assessing mountain forest structure using airborne laser scanning and landscape metrics. In: Lang S, Blaschke T, Schöpfer E (eds) 1st International Conference on Object-based Image Analysis (OBIA 2006), ISPRS XXXVI-4/C42, Salzburg

Mason D, Anderson G, Bradbury R, Cobby D, Davenport I, Vandepoll M, Wilson J (2003) Measurement of habitat predictor variables for organism-habitat models using remote sensing and image segmentation. International Journal of Remote Sensing 24 (12), pp 2515-2532

McKean J, Roering J (2004) Objective landslide detection and surface morphology mapping using high-resolution airborne laser altimetry. Geomorphology 57, pp 331-351

Mitas L, Mitasova H (2005) Spatial Interpolation. In: Longley P, Goodchild MF, Maguire DJ, Rhind DW (eds) Geographical Information Systems: Principles, Techniques, Management and Applications, second edition, Wiley, New Jersey

Morsdorf F, Meier E, Allgöwer B, Nüesch D (2003) Clustering in airborne laser scanning raw data for segmentation of single trees. International Archives of Photogrammetry, Remote Sensing and Spatial Information Sciences 34 (3/W13), pp 27-33

Neteler M, Mitasova H (2004) Open Source GIS: A GRASS GIS Approach. 2nd Edition. Kluwer Academic Publishers, Boston, Dordrecht

OGC Inc. - Open Geospatial Consortium Inc. (1999) OpenGIS Simple Features Specification for SQL, Revision 1.1, http://portal.opengeospatial.org/files/?artifact id=829, last accessed 1. December 2006

OSI - Open Source Initiative (2006) The Open Source Definition, http://www.opensource.org, last accessed 1. December 2006

Persson Å, Söderman U, Töpel J, Ahlberg S (2005) Visualization and analysis of full-waveform airborne laser scanner data. International Archives of Photogrammetry, Remote Sensing and Spatial Information Sciences 36 (3/W19), pp 103-108

Refractions Research Inc.(2006) PostGIS: Geographic Objects for PostgreSQL, PostGIS Manual, http://postgis.refractions.net/docs/, last accessed 1. December 2006

PostgreSQL Global Development Group (2006) PostgreSQL 8.1 Documentation, http://www.postgresql.org/docs/manuals/, last accessed 1. December 2006

Python Software Foundation (2006) Python programming language, http://www.python.org, last accessed 1. December 2006

Rees WG (2001): Physical Principles of Remote Sensing (Second Edition), Cambridge University Press, Cambridge

Rutzinger M, Höfle B, Geist Th, Stötter J (2006a) Object-based building detection based on airborne laser scanning data within GRASS GIS environment. In: Fendel E, Rumor M (eds) Proceedings of UDMS 2006: 25th Urban Data Management Symposium, Aalborg, Denmark: 7.37-7.48

Rutzinger M, Höfle B, Pfeifer N, Geist Th, Stötter J (2006b) Object-based analysis of airborne laser scanning data for natural hazard purposes using open source components. In: Lang S, Blaschke T, Schöpfer E (eds) 1st International Conference on Object-based Image Analysis (OBIA 2006), ISPRS XXXVI-4/C42, Salzburg

Scheikl M, Grafinger M, Poscher G (2001) Entwicklung und Einsatz eines automatischen Fernüberwachungssystems basierend auf einem Laserscanner (ALARM). In: Chesi G, Weinold T (eds) Internationale Geodätische Woche Obergurgl 2001, pp 205-214

Schmidt R, Heller A, Sailer R (2005) Vergleich von Laserscanning mit herkömmlichen Höhendaten in der dynamische Lawinensimulation mit SAMOS. In: Chesi G, Weinold T (eds) Internationale Geodätische Woche Obergurgl 2005, pp 131-140

Sithole G, Vosselman G (2004) Experimental comparison of filter algorithms for bare-Earth extraction from airborne laser scanning point clouds. ISPRS Journal of Photogrammetry and Remote Sensing 59 (1-2), pp 85-101

Sithole G, Vosselman G (2006) Bridge detection in airborne laser scanner data. ISPRS Journal of Photogrammetry and Remote Sensing 61 (1), pp 33-46

Smith L, Sheng Y, Magilligan F, Smith N, Gomez B, Mertes L, Krabill W, Garvin J (2006) Geomorphic impact and rapid subsequent recovery from the 1996 Skeikarársandur jökulhlaup, Iceland, measured with multi-year airborne lidar. Geomorphology 75, pp 65-75

Smith M, Asal FFF, Priestnall G (2004) The use of photogrammetry and Li- DAR for landscape roughness estimation in hydrodynamic studies. International Archives of Photogrammetry, Remote Sensing and Spatial Information Sciences 35 (B3), pp 714-719

Song JH, Han SH, Yu K, Kim, YI (2002) Assessing the possibility of land-cover classification using lidar intensity data. International Archives of Photogrammetry, Remote Sensing and Spatial Information Sciences 34 (3B), pp 259-262

Staley DM, Wasklewicz TA, Blaszczynski JS (2006) Surficial patterns of debris flow deposition on alluvial fans in Death Valley, CA using airborne laser swath mapping data. Geomorphology 74, pp 152-163

Stötter J, Zischg A (in print) Alpines Risikomanagement. In: Felgentreff C, Glade T (eds): Naturrisiken und Sozialkatastrophen

Stötter J, Weck-Hannemann H, Veulliet E (this volume) Global change - natural hazards: new challenges, new strategies

Tamiru Haile A, Rientjes T (2005) Effects of LiDAR DEM resolution in flood modelling: a model sensitivity study for the city of Tegucigalpa, Honduras. International Archives of Photogrammetry, Remote Sensing and Spatial Information Sciences 36 (3/W19), pp 168-173

Thoma D, Gupta S, Bauer M, Kirchoff C (2005) Airborne laser scanning for riverbank erosion assessment. Remote Sensing of Environment 95 (4), pp 493-501

Tsakiri M, Lichti D, Pfeifer N (2006) Terrestrial laser scanning for deformation monitoring. Proceedings of 12th FIG symposium on deformation measurement and 3rd IAG symposium on geodesy for geotechnical and structural engineering, Baden, Austria

UMN MapServer (2006) UMN MapServer Homepage, University of Minnesota, http://mapserver.gis.umn.edu/, last accessed 1. December 2006

VTK (2006) The Visualization Toolkit Homepage, http://www.vtk.org, last accessed 1. December 2006

Wack R, Stelzl H (2005) Laser DTM generation for South Tyrol and 3D visualisation. International Archives of Photogrammetry, Remote Sensing and Spatial Information Science 36 (3/W19), pp 48-53

Wagner W, Ullrich A, Melzer T, Briese C, Kraus K (2004) From single-pulse to full-waveform airborne laser scanners: potential and practical challenges. International Archives of Photogrammetry, Remote Sensing and Spatial Information Sciences 35 (B3), pp 201-206

Warmerdam F (2006): GDAL - Geospatial Data Abstraction Library. http://www.gdal.org, last accessed 1. December 2006

Wood JD (1996) The geomorphological characterisation of digital elevation models, PhD Thesis at University of Leicester, UK, http://www.soi.city.ac.uk/_jwo/phd, last accessed 1. December 2006

Würländer R, Rieger W, Drexel P, Briese C (2005) Landesweite Datenerhebung mit ALS technologische Herausforderungen und vielseitige GIS-Anwendungen. In: Strobl J, Blaschke T, Griesebner G (eds) Angewandte Geoinformatik 2005 - Beiträge zum 17. AGIT-Symposium Salzburg, pp 800-80

12 Improving Safety in Alpine Regions through a combination of GSM/GPRS with satellite communication, GIS, and robust positioning technology

S. Baumann, J. Czaja, W. Lechner

12.1 Introduction

The effective interaction of innovative navigation and communication technologies with intelligent GIS management systems have led to a number of successful applications in the domain of transport telematics (e.g. route guidance, fleet management and emergency call) in the last few years. Since low price Personal Digital Assistants (PDAs) and PDA-compatible GPS receivers are available, there is a potential to introduce new local and personalised mobile applications which require a vehicle independent combination of so-called core technologies like satellite positioning, mobile data communication, and GIS management systems. Hence, many processes in risk and catastrophe management can be optimized and the security of individuals (mountain rescuers, tourists, etc.) can be improved by this technology.

Requirements for these improvements are the availability of respectively prepared spatial data, sufficiently accurate and reliable localisation, as well as robust communication between the various persons and information sources involved. As mountain regions like the Alps have their own characteristics compared to other regions e.g. obstruction of GPS satellite signals, gaps within the GSM network, etc. the research project PANORAMA (Personalised ApplicatioNs based On ReliAble Positioning, Communication and GIS Management Systems in Alpine Regions) has identified possible problems related to the core technologies required and developed according ideas to solve them. The main objective of PANORAMA was to develop a modular platform which is based on Commercial off the Shelf (COTS) products and which can be adapted for various applications.

Based on a market and user requirements analysis, a concept for a demonstration system, a Mobile Mission Centre (MMC) for mountain rescue teams, has been developed and been used to evaluate the key components of the three core technologies in cooperation with selected users. The project with a duration of 21 months (August 2004 - April 2006) was carried out by alpS (Austria), Telematica (Germany), TeleConsult Austria (Austria), GPS GmbH (Germany) and the Institute of Navigation and Satellite Geodesy of the Technical University Graz (Austria).

12.2 User requirements and market analysis

12.2.1 User requirements

To evaluate the advantages of using the three core technologies, i.e. localisation, communication and GIS for mountain rescue operations, to identify the related requirements and to obtain an overview of ongoing and planned projects in these domains, extensive discussions were held with various stakeholders. The key findings derived from these discussions were:
- The use of GPS has increased in recent years, but GPS receivers are not part of the official equipment of mountain rescuers
- GPS reception in an Alpine environment suffers from signal blockage by mountains and vegetation
- The GSM availability in a remote Alpine environment is not sufficient for safety critical applications like mountain rescue. During large natural disasters the GSM network is often overloaded
- Digital maps for visualisation are available in suitable scales
- The knowledge of the position of the mobile mountain rescuers in regard to their previous tracks could significantly increase the situation awareness of local coordinators of the rescue operation
- The mobile terminals have to work in harsh environments (water proof, shock resistance, adapted to low temperatures, etc.).

12.2.2 Market overview

The market relevant to the PANORAMA project can be segmented into:
1. Localisation, tracking and navigation
2. Communication
3. Geographic Information Systems (GIS)

Between these stand-alone market segments are also specific cross-sectional markets providing e.g. combined navigation and communication or navigation and GIS products or services. In addition, another aspect covering all three segments has been taken into account: the market for ruggedised user terminals.

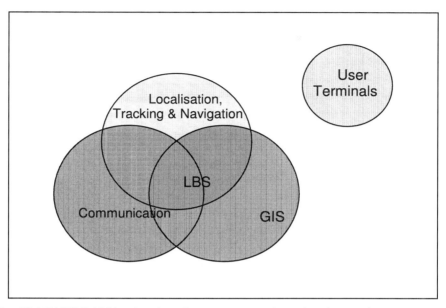

Fig. 12.1 PANORAMA related market segments and overlapping markets

Localisation, tracking and navigation

The market of so-called Personal Mobility applications is dominated by GPS/EGNOS handheld receivers, partly with integrated compass and/or barometric height sensor. Investigations have shown that e.g. 40 % of all mountain bikers use GPS receivers for Trans-Alp crossings.

Fig. 12.2 iQue 3600, GPSMap 60CS, EMTAC CruxII, Foretrex

Besides the above mentioned handheld receivers, various GPS Bluetooth receivers are available on the market, which can be connected to different hardware devices. The key advantages of Bluetooth receivers are their small size (cigarette packet size) and low weight (< 100g) and cableless connections. Both Bluetooth receivers and handhelds are in general robust enough for most outdoor activities. The life-time of the power supply depends strongly on the kind of application (tracking frequency), the environmental conditions (e.g. temperature) and the type of battery used. Recently "GPS-watches" have become available for outdoor activities.

In addition to satellite-only-based positioning technologies, so-called autonomous technologies could be beneficial for use in an Alpine environment, because they are completely independent from any infrastructure. The principle of autonomous positioning is based on an incremental determination of position which is made on the basis of a direction measurement and a distance measurement. This method is also called dead reckoning. Various concepts have been developed and tested in the previous years. Magnetometers and gyros have been used for direction detection, barometric height sensors for altitude determination and accelerometers for step detection (and subsequently distance determination).

The analysis of user requirements has shown that the accuracy provided by stand-alone GPS (and of course Galileo later on) will be sufficient for mountain rescue applications, but the availability and continuity of GNSS under Alpine conditions will not be sufficient for safety-related applications like mountain rescue. Combinations of GNSS with other (more robust) terrestrial navigation systems like Loran-C or positioning technologies based on mobile communication systems (e.g. Cell ID, Cell ID with Timing Advance, Angle-of-Arrival, Time-of-Arrival, Time-Difference-of-Arrival, etc.) will also not help to overcome this problem, because these systems are only partly available in the Alps. Therefore, the PANORAMA project focuses on the previously described GNSS/sensor approach.

Communication

Many transport telematics and tracking applications are based on GPS, GPRS or UMTS as communication links to transmit the position of the mobile users to a fixed centre. Due to the fact that the availability of these mobile communication systems is based on business considerations and therefore concentrates on populated areas and street networks, these systems will never cover rural areas completely and especially not remote areas. The following figure visualises the network-coverage of various Austrian mobile communication providers in Tyrol.

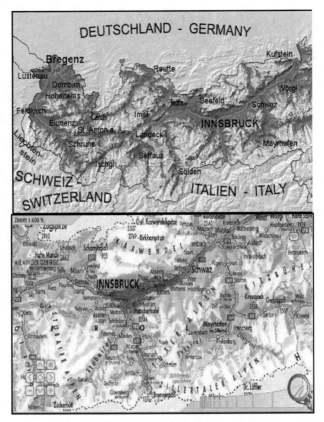

Fig. 12.3 Mobile communication network coverage in Tyrol (A1 top, 3 below)

Satellite communication (SatCom) offers the opportunity of transmitting position data from the mobile field teams to a rescue centre independent from terrestrial infrastructure. Within the framework of PANORAMA five SatCom systems have been analysed in detail: Globalstar, Iridium, Orbcomm, Inmarsat and Thuraya. Several SatCom-handheld devices for voice and/or data communication are available on the market for outdoor applications. The data rate varies from 2.4 to 144 kbps. Some SatCom-mobiles also support GSM, and some are quipped with an integrated GPS receiver. The communication fees have decreased significantly in the previous years and are today comparable to the costs for GSM calls between different countries. The main difference between the various SatCom systems mentioned above for applications in Alpine environment is the type of orbits flown by the satellites. Geostationary satellites orbiting quasi-stationary at a height of 36.000 km over the equator have an elevation angle below 35° for users in Austria, which means that the communication link will most probably be interrupted by using such a system in mountainous regions. Low Earth Orbit (LEO) satellites fly on an inclined polar orbit which means that from a user's perspective the satellites rise from the horizon up to the sky and down again. LEO satellites are "visible" for 20 minutes on average. Due to the high number of satellites used for a LEO SatCom constellation (between 40-60 satellites) a continuous communication connection can be assured.

Due to the fact that GSM is available in some parts of mountain rescuers' operation areas and LEO satellites have a very high availability in mountainous areas, a combined GSM/LEO approach has been selected for the PANORAMA-project to achieve an economic communication link with very high availability.

User terminals

Personal Digital Assistants (PDAs) are palm-sized mobile PCs with touch screens and (partly) keyboards. The terminologies Handheld PC or Pocket PC are used for the same type of devices. The integration of mobile communication interfaces into the PDAs has created the term Smartphone. In terms of computing performance, storage capability, resolution and quality of display PDAs have made significant progress in recent years. Multiple GPS receivers are available for PDA use too.

Fig. 12.4 Smartphones and PDAs (PalmOne Treo, Nokia 626, Pocket Loox)

Due to the size, similar performance and attractive prices compared to Pen computers or Tablet PCs, PDAs have become quite popular. Although so-called "protection cases" are available for the outdoor use of standard PDAs these devices are not suitable for mountain rescue applications. A market survey shown in the following table gives an overview of the (few) ruggedised PDAs, which fulfil the requirements for harsh environment.

Table 12.1 Ruggedised PDAs

Product	Type	Manufacturer
Reccon	PDA/Handheld	Trimble
CF-P1	Handheld PC	Panasonic
TimbaTec Pocket PC	PDA/Handheld	Latschbacher
FuturePad CE	PDA/Handheld	IBD
DA04M	PDA/Handheld	Roda Computer

The market analysis performed within PANORAMA regarding the three core technologies has shown that there are many products and services in each sector (localisation, communication and GIS), but currently there is no integrated solution available which combines all necessary elements and corresponds to the stringent requirements for safety and security related applications like mountain rescue.

12.3 Application scenario definition

12.3.1 Background

Mountain Search and Rescue (SAR) in the Alps is a prominent and challenging example of emergency response tasks. A significant number of search and rescue operations are carried out every year. In Austria 7.495 "Bergrettungsdienst" (mountain rescue) operations were carried out in 2002 (242 were search missions) and 17.229 members served a total of 40.720 mission hours.

Operating in small teams, rescue personnel have to be coordinated over large areas, often in rough terrain and adverse weather conditions. Additional assets like helicopters or reconnaissance planes are called in on demand. Operations control typically uses a mobile command post (car or van), that is driven close to the area. Due to the number of such tasks and complexity of operations, often with human lives at stake, there is a significant demand for assistance by technology. Key tasks of support are, on the one hand, position, situation and intent reporting of field teams to operations control. On the other hand, operations control needs to forward operational orders and assistance information to field teams, e.g. actual visual or thermal overhead imagery. Thus, there is an important need to exchange position related data between field teams and operations control.

12.3.2 Mobile Mission Centre scenario

For the demonstration of the PANORAMA system the application scenario of a Mobile Mission Centre was selected and defined in cooperation with the participating project partners and the Alpinesicherheit Tirol (asi) in Landeck, who supported the project team in questions regarding mountain SAR operation processes.

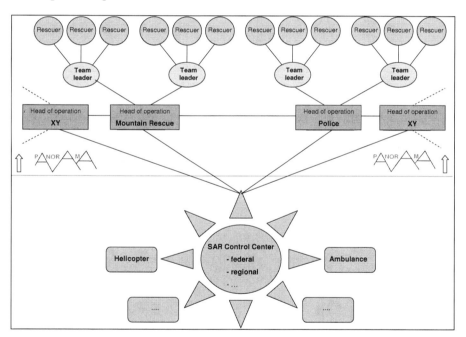

Fig. 12.5 Command hierarchy during an alpine SAR mission

Figure 12.5 shows a typical command hierarchy for search missions with rescuers on foot from different rescue forces. PANORAMA focused only on the following operation levels:
- The rescuer as a member of search team.

- The leader of a search team.
- The head of operation of a rescue organisation.

The scenario selected in PANORAMA can be described as follows:

Different SAR teams are tracked constantly by a mobile operations centre, located at the base of operation. The rescue forces are equipped with the PANORAMA Mobile Unit (PMU) and are navigating in the terrain with a topographic map displayed on their PDA. GPS indicates their actual position, which is displayed on the screen. Points of interests (POI) can be defined prior to the mission or sent to the PMU by the PANORAMA Central Unit (e.g. a GIS management system). Situational awareness is increased by displaying the other rescue teams' positions. Simultaneously, the coordinator of operations supervises the rescue forces by following their position and trajectories on the server application. By setting targets, sending POIs and by informing the rescue teams about dangers (weather, avalanches etc.) the mission is controlled. The functionality of the system shall be sufficient to provide enhanced situational awareness to all participating parties during SAR operations in mountainous areas. A recent analysis of large scale military operations showed that two seemingly contradicting critical issues arise, that can be applied also to SAR missions. On the one hand, up-to-date and accurate information on the tactical environment is needed for actors in the field. On the other hand, central operations management must not be overwhelmed by masses of data. At the same time, information flow between actors in the field can be crucial for mission success. Therefore, in the PANORAMA system, the information shall be available at the control centre and at every participating mobile team. The information shall be categorized, and each participating party shall have the option to select or deselect the display of information based on their actual need. Analyzing the functionality at the different operation levels leads to the results shown in the following table.

Table 12.2 Essential functionality of different operation levels

Level	User	Position	Mob-Com	User Terminal
1	Rescuer	X	X	(X)
2	Team leader	X	X	X
3	Head of operation		(X)	X

Several surveys with various organizations in SAR during the project showed that the concept of PANORAMA will be beneficial essentially in large scale SAR operations and regional disasters. It provides the rescue patrols and the mission centre an important overview on the mission. Furthermore, the rescue patrols benefit from the navigation information in often unknown areas. Communication is facilitated between several SAR organisations that are involved in the mission. Data exchange is highly profitable to transmit dangers (snow and mud avalanches, flood etc.), to guide patrols and to indicate neighbour patrols.

12.4 System architecture with innovative mobile modules

As already mentioned in the previous chapter and illustrated in the figure below, the PANORAMA system consists of two main components: the PANORAMA Central Unit (PCU) and the PANORAMA Mobile Unit (PMU). Each unit by itself includes several modules, all containing essential components of the core technologies, namely, positioning, communication and GIS/management systems, which could be combined using (standard) interfaces and protocols.

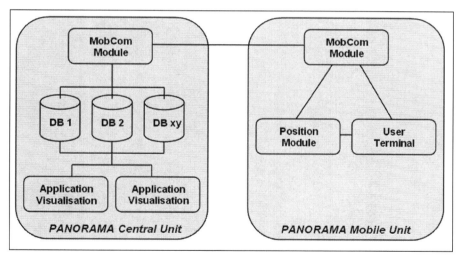

Fig. 12.6 Basis architecture PANORAMA

In detail the PCU consists of one or more central databases and process specific applications or visualisation software. For data exchange both units work with a communication module, which in the case of the PMU is supplemented with a positioning module and a (optional) user terminal (e.g. PDA). The mobile user equipment consists of different, physically independent modules. This characteristic permits the flexible future use of the PMU in other applications with different or even without certain navigation and communication devices. Beside the commercial products identified in the market analysis (e.g. GPS handhelds, PDA or smartphones) PANORAMA especially focused on the following innovative PMU modules:

a) Positioning-Module: Multi-Sensor Box (MSB)

The Multi-Sensor Box (MSB) is a development of TeleConsult Austria GmbH the project partner and provides position information either from a GNSS sensor or, in the case of satellite signal loss, from additional integrated sensors like accelerometers, gyros, magnetometers and barometers using different complex models (e.g. distance information from a step detection approach).

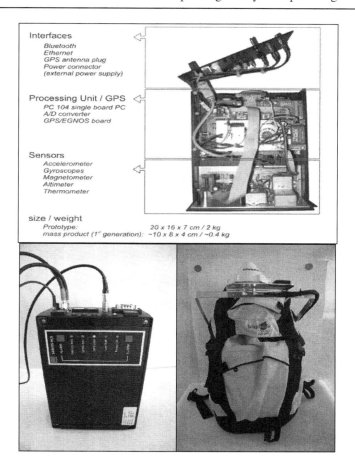

Fig. 12.7 Multi-Sensor Box, MCB and dedicated backpack

b) Communication Module: Multi-Communication Box (MCB)

The Multi-Communication Box (MCB) was developed by the Swiss-based teleCrossAlpina GmbH in Bern in close cooperation with the PANORAMA project team. The basic idea behind the MCB architecture is the seamless data communication routing over different wireless media and networks to provide complete coverage over a defined operations area. The selection of the means of communication is performed automatically by the equipment without any user interaction, according to pre-defined rules. Additionally, the MCB provides automated exchange of the actual positions at user-defined intervals using an integrated GPS/EGNOS receiver. The following COTS components have been included in the MCB:
- Satellite Modem: Qualcomm GSP 1620 (for Globalstar)
- GPRS Modem: Telit 862 (GMS/GPRS-Module)
- Bluetooth: D-Link DBT-900AP
- GPS: µblox GPS/EGNOS Receiver Board TIM LF
- PC104 Processor Board: Arcom Viper PXA255 CPU with Windows CE.net 4.2

12.5 Validation of system components

12.5.1 Simulation of SatCom availability

Within the PANORAMA project a visibility simulation for some selected geostationary communication satellites (Thuraya and Inmarsat) was conducted. This was done using the SRTM3-Digital Terrain Model (DTM) with a resolution of 90m and a DTM of the Austrian Surveying Administration (BEK) with a resolution of 30m. The ArcGIS 3D-Analyst and POV-Ray for Windows (v3.6) were used as software tools. The following figure shows the "shadowed areas" (blue areas) in the region of the Kitzbüheler Horn.

12 Improving Safety in Alpine Regions 359

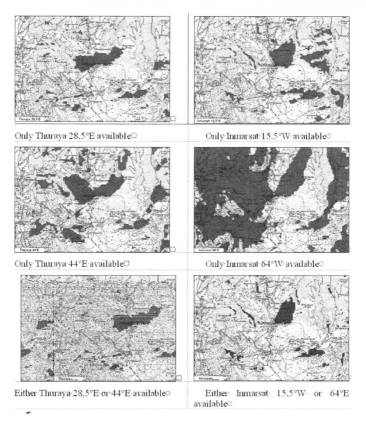

Fig. 12.8 Shadowing of geostationary satellites in the Alpine environment

The simulations show better results for the INMARSAT system, because the INMARSAT satellites are located west and east of the test area, whereas the two Thuraya satellites are both located in the east of the simulation area.

12.5.2 GSM/SatCom availability in the Alps

To validate the impact of mountainous topography on the availability of GSM and SatCom (LEO) the signal-power of both systems were recorded and analysed during multiple cross-country skiing outings in the Alps in the winter of 2005/2006. The signal-power was recorded by a proprietary software tool called "Phonydump". Both GSM- and SatCom-mobile phones document the strength of the received signal in an internal protocol using the so-called Communication Signal Quality (CSQ) value. This information can be exported, time stamped and recorded by "Phonydump" in free definable time intervals via an AT command. A parallel operated GPS receiver with data log capability was used to geo-reference the signal strength values via the common time stamp. As test equipment a Globalstar mobile phone, a GSM mobile phone, a Bluetooth GPS receiver and a laptop placed in a backpack were used (see figure below).

Fig. 12.9 PANORAMA GSM/SatCom test set-up

Several of the field-trials took place at the cross-country skiing course in the Eng valley in the border region between Bavaria and Tyrol. The availability of GSM respective SatCom is shown in figure 12.10.

The relationship between availability (green) and non-availability (red) is quite similar for both types of technology (60% / 40%). The figures show that the availability of the SatCom link is distributed more homogenously along the course, whereas GSM provides a good availability in the western part of the test course and a bad availability at the end of the Eng valley. It has to be noted that the measurements were performed under dynamic conditions, which means for the SatCom measurements that short outages caused by shadowing of the antenna e.g. by the head of the mobile user or vegetation near the track, occurred. In an operational mountain rescue scenario a connection could be established at these locations by a small adjustment to the antenna in many cases. The long-term SatCom-outages were caused by the Alpine topography. This is the reason why the long-term outages occurred at the same locations on the way back (with a different satellite constellation).

Fig. 12.10 Results of **GSM** Measurements in the Eng valley

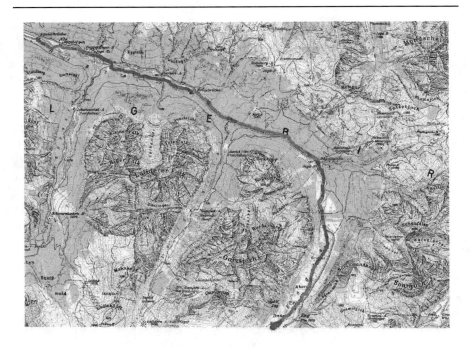

Fig. 12.11 Results of **SatCom** Measurements in the Eng valley

The different spatial distribution of availability/non-availability for GSM and SatCom clearly shows the advantage of a combined system. The following map shows the overlay of both measurements to visualise the benefits of an integrated GSM/SatCom system.

Fig. 12.12 Benefits of a combined GSM/SatCom system

12.5.3 Performance analysis of "low-cost" GPS-receivers

For the performance analysis "low cost" GPS receivers (Rx) were used. These receivers compute the navigation solution based on code-measurements. As a "truth-system" or reference a high end geodetic GPS receiver was used. The follwing hardware components were also used:
- Emtac Crux II BTGPS (SiRF Star IIe/LP, 12 Channel, L1, C/A-Code)
- Garmin GPS60CS (12 Channel, L1;C/A-Code)
- u-blox ANTARIS^{TM} (16 Channel, L1, C/A-Code)
- Ashtech Z-Xtreme^{TM} (12 Channel, L1/L2, Code & Phase)

To compare the different GPS receivers with each other, it must be ensured, that all receivers use identical satellite signals with the same strengths. For this purpose an antenna with a splitter was used. For the selection of the test points the following criteria was defined:
- Position with unobstructed satellite visibility ("open field")
- Position with signal attenuation/blockage due to vegetation

- Short distance between both positions and similar elevation
- Availability of precise reference coordinates

For the analysis, two test series each with 24 hour duration were carried out. NMEA data was recorded for all Rx and the GPGGA data sets were used for the evaluation. In the "open field" scenario no significant difference in the position determination between the different Rx could be identified. All Rx were able to consistently compute a position with good quality and no real weaknesses were found.

However, when testing began in the forest, the weaknesses under conditions of reduced satellite visibility and multipath could be clearly identified. In this environment the strength of the u-blox receiver became obvious and resulted in the excellent quality of the position and height determination. In contrast, the geodetic receiver Ashtech Z-Xtreme^{TM} shows significant weaknesses. This is due to the fact that this receiver is optimized for high precision surveying in open field environments with excellent signal availability. The result is a significantly reduced availability, but, if the Z-Xtreme^{TM} provides a position solution, this is very accurate. Worse results were delivered by the Emtac due to missing filter algorithms. The availability of the different receivers varies from 99.5% for the u-blox to 67.7% for the Z-Xtreme^{TM}. If high availability is a requirement, than navigation receivers should get preference. Additionally, it should be noted that all tests were done in a static mode. If dynamic conditions are added, the requirements on the GPS Rx increase even more, as varying conditions like multipath, signal obstructions and signal attenuation in general further decrease the performance of the position solution.

12 Improving Safety in Alpine Regions   365

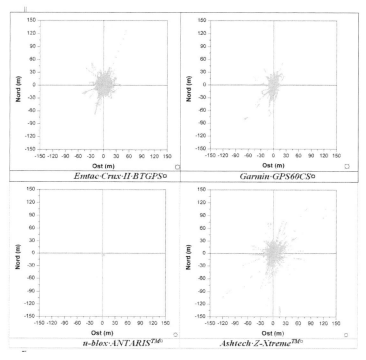

Fig. 12.13 Results of the GPS performance analysis the forest test environment

12.5.4 Multi-Sensor Box (MSB)

The figure on the left shows the MSB in highly vegetated terrain. The prototype together with the battery pack is in the backpack. Access to the position data is obtained through Bluetooth, either through a PDA or through the Multi-Communication Box (MCB).

The GPS antenna is placed on the head of the person to optimize the reception conditions. Placing the antenna on top of the backpack shows slightly reduced reception quality due to obstruction effects from the head and/or shoulders.

This could result in a reduced number of satellites and therefore on a less accurate position solution.

The MSB principle is based on dead reckoning. Within the PANORAMA project various analysis and testing was done to optimize step detection models and develop new algorithms. On the one side it was the aim to adapt the integrated positioning system to the selected application and on the other side to optimize the use of new methods. The following table shows the areas which were analysed in detail.

Table 13.3 Topics for the MSB analysis

Step detection:	- Accelerations
	- Data storage optimized correlation algorithms
Step-lengths models:	- PointResearch Model
	- Ladetto-Model
	- Weinberg-Model
	- Kaeppi-Model
	- Multi-Sensor-Model
Step-lengths calibration with GPS:	- Comparison of different models (plain/slope)
	- Comparison of different walking times
Magnetometer	- Calibration
	- Deviation
Barometric height determination	- Performance and influence factors

The following figure gives an example of a test done in Praxmar, Tyrol. The test person walked up and down the well paved road from Praxmar to Luesens and after that hiked along a snow-covered trail to Praxmar on a winding road.

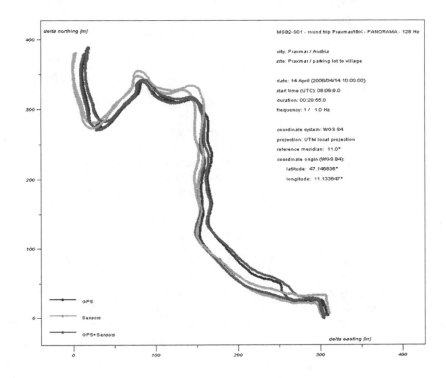

Fig. 12.14 Trajectory results with empirical step-length estimation of uphill 0.73 m and downhill 0.91 m

The results show that when using the optimized algorithms and calibration models, it is possible to determine positions with good accuracy in mountainous and pathless terrain even in cases where GPS is not available for longer periods. If GPS is available and used for calibration it must be made sure that the position quality is high, otherwise a mis-calibration could finally result in fatal position errors. Considering the results of the Praxmar tests, it can be estimated that even in cases of 30 minutes of GPS outages, an accuracy of 2-3% of the walking distance is achievable, the "smoother" and "regular" the walk is.

12.5.5 Multi-Communication Box (MCB)

The Multi-Communication Box (MCB) was developed by teleCross Alpina GmbH as part of the ESA project SARFOS (Search And Rescue Forward Operation Support System) based on the requirements of the

PANORAMA system and was tested in both projects. The results are summarized in the following section.

The core functionality of the MCB, which is the seamless change between GSM/GPRS and SatCom to distribute position and text information, was verified without any failure during test drives in the area of Linden (near Bad Toelz) in March 2006 and on a drive from Innsbruck to Munich on April 13, 2006. Both test areas are characterized by quite incomplete GSM/GPRS availability conditions. All position messages, which were transferred in two minute intervals, were received without any problem at the server.

As part of the SARFOS project another extensive trial took place in Thun, near Bern, in Switzerland on April 22, 2006. The test was performed in close cooperation with the Bernese Cantonal Mountain Rescue Commission (KBBK). In total there were 4 rescue teams (including tracking dogs) and the director of operations in action. The test scenario was a typical SAR mission in an area with very fragmented GSM/GPRS coverage. 3 out of 4 field teams were equipped with the MCB system. For comparative reasons the other team just used a "classical" VHF and SMS (text messaging) technique to communicate with the mission management. The capabilities of the MCB system were validated using qualitative and quantitative parameters.

Quantitative parameters:
- Change of GPRS/SatCom under different network situations
- Transmission of position information
- Transmission of text information
- Transmission of POIs
- Triggering of alarms
- Operation time of the batteries
- Stability of the overall system

Qualitative parameters:
- User friendliness
- Usefulness of the application (SAR operations)
- Proposals for improvement

During the whole demonstration (duration about 2.5 h) all position and text information was received at the server without any disturbance. The following proposals for improvement were received from the rescue teams and the director of operations:

- The Bluetooth connection between the MCB and the PDA showed some interference with the VHF communication. An alternative cable connection should be installed
- The installation of the SatCom antenna should be improved, e.g. using an extendable pole instead of a fixed connection to the backpack
- The weight of the MCB is acceptable, however, the size should be reduced for the next generation

12.6 Validation of the PANORAMA core system

12.6.1 System Overview

The PANORAMA core system consists of the sub-systems: Geo-data, Navigation / User Terminals, GIS with www- interface.

12.6.2 Geo-data

In principle all kinds of digital geographical data is supported by the specification of the PANORAMA system. Both vector- and raster-data can be imported and handled. During the validation campaigns the following raster-data was used:
1. Topographic Map 1:50.000 or 1:200.000 (AMAP 50 / 200)
    DAV Map „Innsbruck und Umgebung" 1:50.000.
2. DAV Map „Stubaier Alpen - Sellrain" 1:25.000
3. Coloured Orthophoto 1:5.000 (Source: TIRIS)

12.6.3 User terminal

Fig. 12.15 Recon, HP 6515, HP 4700

As quoted above ruggedized PDAs are required for mountain rescue applications. For cost reasons the validation of the PANORMA functionalities were performed with one ruggedized PDA (Recon) and two standard PDAs (HP 6515, HP 4700).

For visualisation purposes the Pocket PC application "GPSMap" of the company "CPU Control" was modified and used. The following functionalities were implemented:
- Display of current position, of self-defined (PMU) or external POIs such as position of people in distress, equipment depots, etc.
- Display, generation, reception and transmission of text messages

12.6.4 GIS with www-interface

The PANORAMA system could be interfaced with existing operational GIS/Management systems. To demonstrate this flexibility an interface with ESIS, a Tyrolean web-based system for the management of natural hazards, has been realised in cooperation with ASI-Tirol (Alpine Safety and Information Center). The ESIS operator now has immediate access to the position information of the mobile rescue teams in the filed, can visualise them on a digital map or photograph and exchange text messages with the team members.

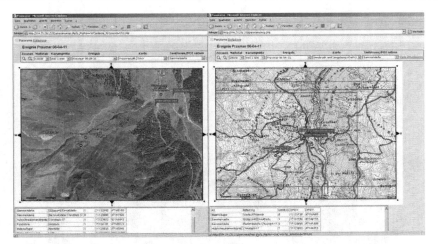

Fig. 12.16 www-interface PANORAMA Central Unit (raster/vector map)

12.6.5 Test campaign

The PANORAMA system was validated in multiple test campaigns in the Alpspitz area (Garmisch-Partenkirchen, Bavaria) and Praxmar (Tyrol). The test areas were selected based on the following criteria:
- High-alpine terrain with complex topography
- Expositions in all directions
- Vegetation canopy in the lower parts of the test area
- Typically GSM coverage
- Good access in winter season
- Facilities for equipment storage and online data evaluation during the test campaign

To validate the server-based GIS and the communication to the PANORAMA Mobile Units (PMUs), both GSM and MCB were tested. A laptop with GSM connection represented the PCU mission centre. A SAR scenario was selected for validation purposes. Figure 12.18 shows some impressions of the test campaign in Praxmar and some screenshots from the PMU graphical user interface.

12 Improving Safety in Alpine Regions 373

Fig. 12.17 Mobile Mission Center and Rescuer at validation campaign

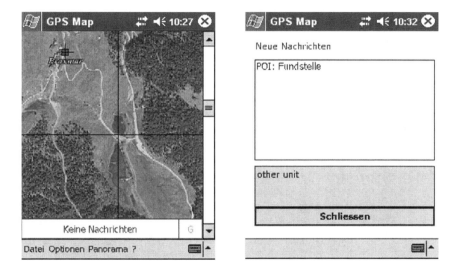

Fig. 12.18 Screenshots from the PANORAMA validation campaign in Praxmar

As a conclusion to the PANORAMA project, a public demonstration took place in Igls near Innsbruck, Austria on October 12, 2006. After the demonstration all participants agreed that the PANORAMA project was completed successfully and that the development of the system should be continued and brought into practise via long-term validation.

12.7 Conclusions

The primary difference between the requirements for Alpine users and those of the mass market, e.g. in urban areas, is the need of independence from local infrastructure. Therefore, the standard approach to establish an extended localisation infrastructure can only be realized to a quite limited extent in Alpine regions. Vice versa, systems, which are of great help for alpinists, can only be used with little success in urban areas, e.g. a magnetic compass will give some available information in the country side, but is quite significantly disturbed in an industrial environment or in cities.

The most popular positioning system for nearly all applications is the US Global Positioning System, GPS. Available since 1994, it allows a worldwide position determination with an accuracy of ±13 m horizontal and ±22 m vertical. Using augmentation systems like EGNOS (European Geostationary Navigation Overlay Service) it is even possible to achieve accuracies of about ±3 m. A prerequisite for getting a GPS based position is a direct line of sight to at least four satellites. In rough terrain like the Alps or in difficult environments like forests, the satellite signals might be obstructed, attenuated or corrupted. This has a strong influence on the availability and quality of the position solution. New technologies and high sensitive receivers have significantly improved the ability to track satellite signals even under difficult conditions, but the accuracy of the positioning results is quite often questionable. As soon as the European global satellite navigation system Galileo is operational, the user will have up to 60 satellites in total for position determination. In addition to this, it can be expected that GLONASS, the Russian system, will also revive so that in the end 3 independent satellite navigation systems will be available. Although Galileo and GLONASS will not be able to overcome the system-specific deficiencies of GPS (e.g. sensitivity to intentional or unintentional interference, line of sight requirements for signal reception), the additional satellites in combination with new combined receivers will satisfy more and more users even in environments where GPS is still critical today. But the question of reliability and the dependency on a pure satellite based solution will remain and can only be solved by combining satellites with autonomous sensors. This combination will allow a high quality and highly reliable positioning solution even in difficult terrain and under unfavourable signal reception conditions.

The manifold analyses as part of PANORAMA have shown that the approach to integrate a barometer, magnetometer and some step detection together with satellite navigation is applicable also in an Alpine environment. But for practical reasons a higher integration to reduce size, weight and power consumption is strongly required to fulfil the operational requirements. Technically, a system the size of a cigarette box seems to be feasible using today's technology. It can be expected that industry will provide such solutions as soon as the market evolves. First attempts are available and show promising results.

Integrated systems mostly use the complementary attributes of their components to maximize the overall availability. Redundant measurements and intelligent filters improve the system accuracy and reliability. This increases the overall performance and safety. However, the increasing complexity and the need for standardisation and certification for safety critical applications (e.g. SAR) require further effort.

In terms of communication the investigations, simulations and tests which were carried out as part of the PANORAMA project resulted in the following recommendations which should be taken into account to further optimize the future communication component of an operational PANORAMA system:

Simulations and field trials confirmed that LEO Communication satellites provide a significantly higher availability in alpine environments compared to systems using geostationary satellites. These are especially affected by topographic obstructions in valleys running in an East-West direction. Future systems therefore should use LEO based satellite communication services. During the tests Globalstar was used with good results, but in principle Iridium should also be capable of meeting the user requirements, although the reduced data rate and, at least currently, more expensive satellite modems must be considered. The weight of the MCB is acceptable but the size should be reduced; the current limiting factor is the size of the Globalstar modem.

The Bluetooth connection between MCB and PDA seems to sometimes be unstable and sensitive to interference from VHF. This will have to be further analysed and, an alternative cable connection should be provided as a back-up solution. The future system of choice should be able to work with and without PDA, as some users do not need the advanced communication capabilities of the handheld device.

As the TETRA infrastructure will be available in the next few years in Tyrol and other European regions, an integration of a TETRA modem as part of a modular design should be taken into consideration as an option to maximize the benefits to the users.

References

Baumann S, Czaja J, Wasle E, Hessing C (2005) PANORAMA D1: Anforderungs- und Marktanalyse. Version 1.0; 01/02/2005; projektinternes Dokument

Baumann S, Czaja J, Wasle E, Hessing C (2006) PANORAMA D2: Sachstand und Entwicklungspotential der Kerntechnologien und Technisches Konzept des Basissystems. Version 1.0; 01/02/2005; projektinternes Dokument

Baumann S, Czaja J, Wasle E, Hessing C (2006) PANORAMA D3: Realisierung des Basissystems. Version 0.9; 31/05/2006; projektinternes Dokument

Coordination Group on Access to Location Information for Emergency Services (CGALIES) final report v1.0, dated 18.02.2002

Deutscher Funknavigationsplan - Band 2, Bundesministerium für Verkehr. Bau- und Wohnungswesen, Forschungsvorhaben Nr. 96.0697/2001, 2004

Ladetto Q, Gabaglio V, van Seeters J (2002) Pedestrian navigation method and apparatus operative in dead reckoning mode. European Patent EP 1.253.404.A2. April 23, 2002

System and Policy Inventory – Development of the European Radionavigation Plan, Version 3.0, Helios Technology, 25.10.2004

GALA Definition and Sizing For The Safety and Life Market (Issue 3, 10/10/00), ESA 2000

GALILEI Required Performances for Local Application (Issue 1.1, 18/01/02), ESA 2002

Heinrichs G, Löhnert E, Mundle H (2004) GATE - The German Galileo Test & Development Environment for Receivers and User Applications. In the Proceedings of the 2nd ESA Workshop on Satellite Navigation User Equipment Technologies, NAVITEC 2004. 8-10 December 2004, ESA, ESTEC. WPP 239

Legat K, Abwerzger G, Weiss R, Wasle E (2001) GLORIA deliverable D4 - intermediate testing report. Information Society Technologies, Fifth EU Framework Programme, Contract no IST 20600

Legat K (2002) Indoor Navigation. Doctor Thesis. Department of Positioning and Navigation, Graz University of Technology. 2002

Retscher G (2002) Einsatz von Location Based Services (LBS) als Navigationshilfe: Integration in moderne Navigationssysteme. Österreichische Zeitschrift für Vermessung und Geoinformation. 90. Jahrgang 2002, ISSN 0029-9650 Heft 1/2002

ESA project SARFOS (2005) Technical Note on Interface Definition and Design for the PANORAMA project, teleCrossAlpina, 2005

ESA project SHADE (2004) Project Results. Special Handheld based Applications in Difficult Environment (FR-FD.6); Tender AO/1-4218/02/NL/GS Special GNSS Applications of the European Space Agency (ESA)

Caruso MJ (1997) Applications of Magnetoresistive Sensors in Navigation Systems. Solid State Electronics Center, Honeywell Inc., Plymouth, USA

Ladetto Q, Gabaglio V, van Seeters J (2002) Pedestrian navigation method and apparatus operative in dead reckoning mode. European Patent EP 1.253.404.A2. April 23, 2002

Käppi J, Syrjärinne J, Saarinen J (2001) MEMS-IMU Based Pedestrian Navigator for Handheld Devices. ION GPS 2001, Salt Lake City, USA, September 2001

Ladetto Q, Gabaglio V, Merminod B, Terrier P, Schutz Y (2000) Human Walking Analysis Assisted by DGPS. GNSS, Edinburgh, Scotland, May 2000

Ladetto Q, Gabaglio V, Merminod B (2001) Combining Gyroscopes, Magnetic Compass and GPS for Pedestrian Navigation. Proc. Int. Symposium on Kinematic Systems in Geodesy, Geomatics and Navigation (KIS 2001), pp 205-213

Product Validation Document for the Search And Rescue Forward Operation Support System (SARFOS), Version 4.0, 15.5.2006

Weinberg H Using the ADXL202 in Pedometer and Personal Navigation Applications. Application Notes AN-602, Analog Devices, 2002

13 Pros and cons of four years experience of alpS

E. Veulliet, H. Weck-Hannemann, J. Stötter

13.1 Introduction

Avalanches, debris flows, flooding, rock falls, landslides, and slow mass movements are examples of natural processes in mountain regions. As soon as these processes impact on spheres of human interest they become a threat, direct or indirect, and consequently are perceived as natural hazards. Ever since mankind began to settle and utilise the Alps, human environment has been affected by these processes. In areas where these natural processes and human activity overlap there is serious potential for human casualty and damage to buildings, property and infrastructure that may cause subsequent direct and indirect costs with far-reaching economic consequences. The increase of damages and ensuing costs arising from natural hazard processes has become arguably considerable over the last decades.

Since the end of the 19th century, a social transformation has been taking place through the industrial revolution and the growth of tourism, reaching from an extensively agriculture-based social system to one that focuses on industry and services. Together with a correlating change in demands on land use, this social transformation has so far been characterised by an ever expanding land requirement for housing, traffic, business and recreation. As a result, there is an increased probability for people and property, both mobile and immobile, to face risks. Alongside with undeniable shortages of public funds that meet with a visibly increasing reluctance on the part of a public that seemingly "accepts" risk, these factors indicate a clear need to rethink the analysis, evaluation and management strategies of natural hazard processes, especially when it comes to long-term and sustainable solutions. This consequently raises the questions "what can happen?", "what is acceptable?" and "what needs to be done?"

In order to meet new demands in the realm of management of natural hazards highlighted in Chapter 1, both inter- and transdisciplinary research and development models are required, which unite all institutions, public or private, under one roof. Only in this way, by utilising all available potentials for innovation, taking advantage of synergies and guaranteeing the smooth transfer of data and knowledge, an integral risk management program is possible.

Aware of these requirements, the *alpS – Centre for Natural Hazard Management* was founded in October 2002 as a part of the Kplus-Programme of the Austrian Government. Kplus-centres aim at conducting target-oriented and internationally competitive research and development work at a high standard in areas that are relevant to both industry and research.

It was the vision of the alpS research and development platform to bring together academic, commercial, public and non-public partner organisations. Furthermore, alpS wanted to both establish itself as an internationally recognised trademark and play a leading role in natural hazard management. The knowledge gained in scientific projects in the Alpine context can be transferred onto a global scale to subsequently increase the safety of individuals and society in mountain environments in general. Through practical research and development, damage from natural hazards can be more adequately limited to a socially and economically acceptable level in the long-term.

Through applied research and development, it is the mission of alpS to make a considerable and long-lasting contribution to the protection of human environment in mountain regions. Additionally it aims:

(i) at making a considerable contribution to the training of a highly qualified work force in the field of natural hazard and risk research, for business, science and public authorities;
(ii) at raising the level of innovation and competitiveness of the partner companies through tighter networking and smoother knowledge transfer between science and industry;
(iii) at making up-to-date scientific methods and knowledge accessible to the businesses and authorities involved;
(iv) at communicating scientific knowledge resulting from projects to a larger audience through publications, lectures and special events;
(v) at encouraging cooperation with other research institutes on a national and international level.

The activities of alpS, therefore, aim at protecting human beings, private and public properties and to preserve the general conditions of alpine environments in a sustainable way, taking climate and socio-economic change into consideration.

13.2 Main Objectives

The main goal of alpS is to deal with the intricacies of the cause-effect characteristics of natural hazard processes in an interdisciplinary environment. To guarantee this, one of the main strategies of the Centre is to develop interdisciplinary and highly qualified scientific human capital and to foster close ties in the long term.

In order to ensure the sustainable protection of human activities and interests, alpS pursues the following objectives:

i) Systematic compilation and evaluation of the current status: The systematic compilation and evaluation of the existing base data and the methodical approaches used so far is the fundamental prerequisite for achieving this stated goal.

ii) Development of a more efficient and effective way of natural hazard management: The present deficiencies in the assessment of the natural hazard situation and the planning of appropriate protective measures and strategies was identified to lie mainly in the use of non-standardised methods and data as well as the lack of cooperation between businesses, scientific research institutions and public administration. Based on a comprehensive compilation and evaluation of current scientific knowledge, alpS improves existing methods and develops new ways of recording and assessing the natural hazard situation as basis for adequate planning of protection measures and strategies. Thanks to the institutionalisation of multidisciplinary cooperation between businesses, public administration and research institutions under the umbrella of alpS, almost perfect conditions for improving, expanding and standardising methods can be offered.

iii) Implementing a paradigm change: alpS aims to implement paradigm change towards an interdisciplinary risk-based natural hazard management comprising:
  - the evaluation of natural hazards on data bases allowing reliable quantification of magnitude-frequency relations,
  - the development of preventative, knowledge-based and process-governing measures,
  - the application of expertise from business and economics as well as associated psychological measures.

iv) Development of natural hazard management strategies under changing conditions in the future: According to the most recent scenarios of global climate change, further warming will continue throughout the 21st century thus triggering drastic changes in mountain geospheres (hydrosphere, lithosphere and biosphere). Depending on the very different timescales of each sphere to react, both an increase in the magnitude of natural hazard processes and a higher frequency of floods, debris flows and mass movements are most likely. In addition to this, a probable change in man-made environments – expansion of areas utilised by people and a change in land use – will lead to increased risks in the future. So as to be prepared for this change and to be able to protect the limited space in mountain areas for fundamental demands of society in a sustainable way, one of the tasks of alpS, as part of its research and development program, is to devise possible scenarios and develop methods adapted to changed fundamental conditions.

13.3 Basic Approach

In order to achieve these goals, alpS endeavours to form sustainable partnerships and alliances with competent and financially strong business partners, on the one hand, and recognised academic institutions and research facilities on the other.

During the first four years the implementation of alpS was structured in three stages. In an initial stage, basic organisational structures were installed in order to guarantee a reliable operational system and the first stage functional operations around first generation projects were developed. In addition to the prompt realisation of these immediate projects, networks with business and scientific partners were set up. In October 2002, the Centre started with eight employees working in three research projects. These initial projects and additional seven projects were already defined during the application period. Within the first year the number of researchers increased to 25. Thereby a critical mass emerged generating a reputation recognised far beyond the boundaries of Tyrol.

The second half of the initial stage mainly consisted of consolidation activities, preparation work for future projects and acquisition of new business and scientific partners. In addition, the demand of public and private key-players (e.g. authorities, consulting engineers, scientific institutions) dealing with "Natural Hazard Management" in Austria, Switzerland, Germany and Southern Tyrol was assessed. This analysis provided the foundation for future strategic and scientific orientation. Throughout hundreds of personal interviews research fields with development potential for alpS were identified.

In general, new projects with partners from various sectors launched after a 6 to 12 months preparation phase. These second generation projects had to consider the following guidelines:
- The project needs to be value adding for the business partner.
- The project has to meet the scientific interests of the associated research institutions.
- The project should be of public interest.

- If the feasibility of the project is not already guaranteed at the beginning, evidence has to be provided within the first stage of the project (stop-or-go stage).
- The results of the project should be transferable to other applications and/or regions.
- The application and integration of up-to-date technology to natural hazard management has to be assured.
- The possibility to create synergies with existing alpS projects.
- Sustain the internationalisation of alpS.
- Increase and strengthen the alpS network.
- Support the positioning of natural hazard management as a cross-sectional subject.

This approach allowed compensating for the loss of an important business partner and the resulting cancellation of one project (five employees were laid off) and even starting an expansion period. During this phase the centre developed rapidly, managing 22 projects with about 50 employees and approximately 40 business partners. As the size of the administrative team stayed constant and rather decreased during the whole period, another phase of consolidation became necessary in the third year. Within this phase emphasis was put on increased linkages among the projects as well as the different working areas. Know-how created in working area C (socio-economic risk analyses) was considered particularly. The quality and design of this working area is hardly found elsewhere in Europe and clearly presents a competitive advantage of alpS.

13.4 Structure

At alpS, the aforementioned high demands are met through co-operation with competent commercial, scientific and administrative partners (Fig. 13.1), which in comparison to all other Kplus-centres makes alpS unique.

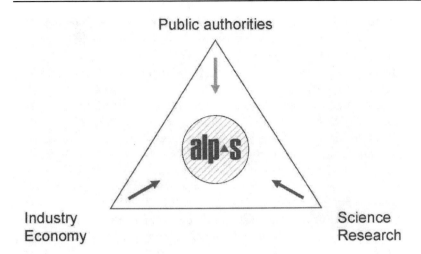

Fig. 13.1 alpS in a triangle connecting public authorities, science and business

alpS deals with issues of high social and political importance. In the process, alpS does not only combine industry with research – in keeping with Kplus principles – but it also incorporates officials and authorities in almost all its projects. Occupying the central position in the triangle of stakeholders allows for credible objectivity and neutrality. Thus alpS is increasingly developing into a consensus platform, which is a fundamental prerequisite for an integrated approach.

The inclusion of public authorities in the realm of natural hazard management is essential since the pertinent administrative and legal expertise in most countries is found almost exclusively at these institutions. Furthermore, due to the long term existence of these institutions, they offer a great deal of know-how and possess important base data.

The overall acceptance public authorities grant research findings resulting from alpS projects has significant relevance for business partners too. This crucial acceptance is guaranteed by involving public authorities in the process of selection of projects, by orientating the projects technically, by implementing them, and additionally this is assured by the representation of public authorities in the board of alpS.

13.5 Research Program

Cooperation of competent commercial, scientific and administrative partners took place in three interactive Areas (Fig 13.2)
- A – Databases and Modelling,
- B – Hazard Mitigation and Protection Measures,
- C – Socio-economic Risk Analysis,

as well as through interdisciplinary projects in order to make a valuable contribution to the long term protection of living conditions in mountain areas.

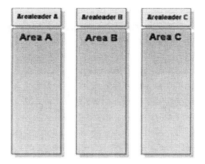

Fig. 13.2 Areas during the years 1-4

- **Area A**

Area A is focused on methods of data acquisition and processing, investigation into the magnitude and frequency of processes and the evaluation of existing models, respectively the development of new models. As Area A deals with databases and modelling, which are fundamental to all other activities (Areas B and C), the projects aim at further improving the discrepancy and deficiences between the databases and models required for process- and risk-orientated decision-making on the one hand and the actual situation on the other hand. Therefore, the following objectives are addressed:

A0 Systematic compilation and evaluation of available data on alpine natural hazards and current methods and models
A1 Further and new development and evaluation of methods for standardised data management
A2 Further development and application of models at various scales
A3 Contribution to and introduction of a paradigm change in the underlying assumptions of hazard assessment

A4   Development of models for the acquisition of spatial data according to global change scenarios

- **Area B**

The data bases and modelling results from Area A establish an important basis for Area B, which encompasses the complete spectrum of active and passive protective measures. The aim of this Area is to evaluate existing approaches for protective measures regarding construction and forestry, development planning and organisation as well as to develop innovative new approaches.

Since Area B deals with the whole range of hazard mitigation and protection measures in a preventive matter, further studies on i) alternations in the current state of research in mitigation strategies and measurement planning and ii) deduced possible future needs in these issues are addressed. The actions aim at providing best-practice procedures to deal with natural hazards in alpine areas. This approach fits to the idea of integrative risk management. The objectives of Area B can be summarised as follows:

B0   Systematic compilation of mitigation strategies and evaluation with respect to the risk concept
B1   Development and evaluation of process-specific measures
B2   Implementation of risk-reducing measures
B3   Implementation of a platform for the construction in alpine areas
B4   Enhancement of the integrative risk management facing global change processes

- **Area C**

Both the systematic examination of databases and the modelling by Area A and the assessment of the development of protective strategies by Area B are supplemented by investigation into the related socio-economic aspects carried out by Area C. In this Area, issues regarding the perception of natural hazard-induced risks, the financial estimation and social acceptance thereof as well as the conveyance of strategies for disaster prevention and management in particular are of primary importance.

Socio-economic hazard research considers risks to human health, social welfare and the ecosystem which result from natural hazard processes. Area C focuses on socio-economic aspects of natural hazard processes and related measures in mountain regions, i.e. the overall process of risk perception, risk analysis, risk assessment, risk management and risk communication. In particular, social sciences and economics are asked to contribute to the questions „what happens or what is possible to happen?" (risk analysis), „what may happen or is acceptable to happen?" (risk assessment) and „what has to be done?" (risk management).

And in effect, the contributions are significant and valuable: (i) in risk analysis, the quantification of the damage potential (in monetary terms) is called for; (ii) risk assessment asks for acceptable risks evaluated from the point of view of society or individual firms; and (iii) in risk management, the adequate measures have to be identified for implementation in the political process or on an individual or firm basis, respectively.

In order to identify strategies which are best suited, sustainable and acceptable for society, the emphasis is lead on further improving the tools for risk analysis, risk assessment and integral risk management. Consequently, the following objectives are addressed:

C0   Systematic compilation and evaluation of socio-economic models
C1   Socio-economic assessment of potential damages caused by natural hazards under alternative protection measures
C2   Development of efficient strategies for risk and disaster management
C3   Development of adequate strategies for decision making relating to natural hazards
C4   Socio-economic assessment of potential damages and risk management strategies according to global change scenarios

13.6 Lessons learned

The research plan for the second period (years 5-7) put strong emphasis on the experiences of the first funding period.

Since the multidisciplinary character of alpS, including the consideration of different methodologies and procedures of different scientific disciplines, substantiates the claim for the uniqueness of the Centre, priority in this second period is given on a strengthening of interdisciplinary linkages between the three research areas. The extension of this approach by specific projects is the new over-all goal, including the development of respective theories.

To stay abreast of these changes, alpS decided to adapt its internal structure and organisation.

13.6.1 Integrative Organisation

The established model of the three columns (Areas A, B, and C) is supplemented by a new horizontal organisation, facing the multidisciplinary character of the Centre (Fig. 13.3). The integrative approach of dealing with alpine natural hazards necessitates a comparative perspective on technical and organisational measures under consideration of societal requirements, precaution principles, and recovery strategies.

The special focus on societal needs is expressed by the new structure of the Centre: The natural processes studied in Area A have to be evaluated from a societal perspective, and alternative protection measures resulting from Area B have to be assessed with respect to the postulates of precaution and recovery, as well as with respect to possible consequences for the society. Thus, the new organisational structure of the Centre mirrors the societal relevance of the disciplinary areas. Additionally, since as much projects as possible were situated in the intersection between the areas, there was a need for a new horizontal orientation of alpS.

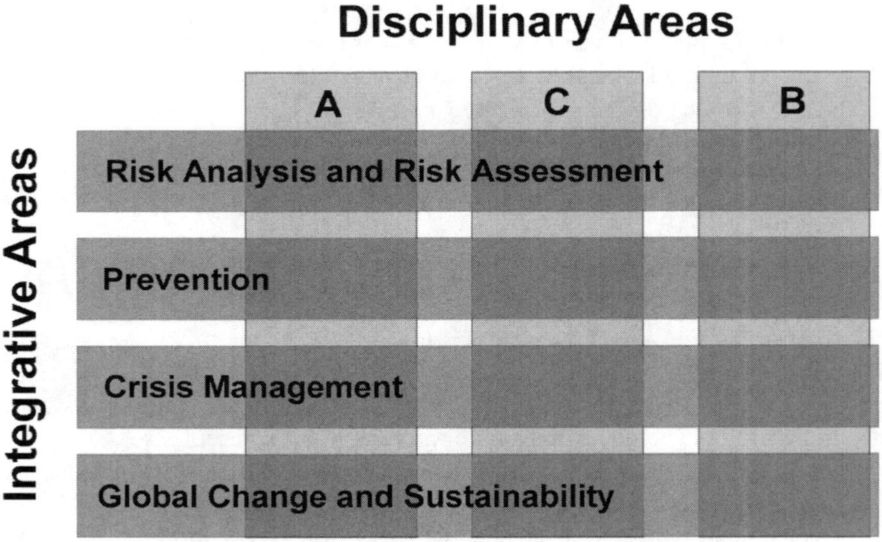

Fig. 13.3 Structural transition of alpS from the three column model to an organisational structure based on a more integrative concept

This new orientation is predominantly based on the idea of integrative risk management complemented by a fourth area focussing on global change and sustainability:
- Risk analysis and risk assessment: As a modification to the traditional risk circle, aspects of regeneration are included in this field, which further comprises risk analysis and assessment.
- Prevention: This field of activity includes all methods of (active and passive) measurement, planning and preparedness.
- Crisis management: Crisis management is an area dealing with coping strategies, recovery issues, and communication policies and practice.

- Global change and sustainability: For to achieve the overall goal of sustainable protection of mountain areas affected by natural hazards, the consideration of natural and man-made elements facing global change processes is of primary interest.

The intersection of the three traditional areas A, B, and C with the structure described above results in a matrix expressing both, the integrative and multidisciplinary procedure intended in almost all projects and the continuing specialised scientific approach.

13.6.2 Challenges, Development and Adaptation

Without doubt, block data and modelling issues dealt with in Area A desperately called for the introduction of modern technologies of data collection, processing, graphic display, and modelling. Thus, advancing developments in IT simultaneously allows the recording of environment-related data to be carried faster and more precise as well as in a more cost-efficient way. This research manner subsequently produces larger amounts of data that are only useful when combined with up-to-date databases and the application of geo-information systems. Hence geo-information systems established in Area A are used in all alpS projects– a manner which has led to improvement of hardware and updating of software resources as well as continuous learning of up-to-date know-how. Therefore, broader trends in geo-informatics and data management are also reflected in higher hardware and software costs of Area A.

Despite all these advanced technical opportunities, full documentation and description of the environment appears useless without scientific knowledge on the underlying natural processes. Therefore, a sound theoretical education and the practical experience of project workers in the real natural environment are still essential. Dynamics of natural processes are accelerated and intensified while leaving their stamp on both nature and the human environment – a fact largely shared by the public and within society. By and large, these dynamics and changes directly influence the data required. Thus data-collection needs to be conducted quicker and more precise within shorter intervals while at the same time being arranged in suitable formats in order to simplify its further processing and analysis.

The constantly growing demand of society for protection from natural hazards represents a further challenge to the implementation of appropriate coping-measures - a trend that gains intensity in the light of restricted public resources. Furthermore, increasing dynamics towards threatening natural processes paired with a constantly growing and changing damage potential as well as with a continuing strain on limited land resources for housing heavily underlines the pressing need for more research and development in this field. Innovative, marketable and resource-responsive solutions have to be developed, if coping with future demands is an aim. In this case, the sustainability of these measures plays an important role, too. The principle of sustainable measures needs to constantly reside right at the centre of all undertakings. Thus the demands for data modelling, future protection measures, products and procedures for risk mitigation are influenced considerably by climate change and changing environmental conditions. Set aside a higher frequency and magnitude of extreme events, comprehensive and permanent protection is clearly at hands and has to be taken for granted by all stakeholders.

The complex economic connections regarding restrictions in tourism, for instance, lead to a regionally high construction density. In such a sphere, temporary measures are considered important since they have inevitably proven more favourable with regard to the sustainability aspired.

Due to constantly reoccurring natural disasters, research and development activities in natural hazard management have been intensified all over the world – a fact that incorporates increased competition as well. Therefore, short development times are even more crucial, especially regarding the realisation of new procedures and products. Comparable dangers and danger scenarios world-wide directly lead to globalisation effects in this segment of the market. It is assumed that solutions which have proven satisfactorily in the Alps can be applied in an adapted form to other mountain regions around the world. While investment primarily focused on the construction of new protection structures, large percentages of the expenditure will most likely flow into the renovation and preservation of facilities that already exist within coming years. For this purpose, products, procedures and standards suitable for the future will be developed and put on the market. Protective structures have so far been erected with the idea of having a technical effectiveness of about 50 years and planned and built according to legally defined specifications. Due to the changing dynamics of natural processes, the assessment philosophy of the planning of protective measures is regarded inappropriate for future challenges. Consequently, applying risk-oriented planning and the realization of protective structures raises new questions, not only in alpine areas. Not less importantly, corresponding legal and socio-economic aspects subsequently fostered developments into this direction. Therefore, socio-economic forms of risk analysis such as cost-benefit analysis, cost-effectiveness analysis, or multi-criteria analysis, are increasingly necessary for the selection and implementation of appropriate coping-measures.

Nowadays we recognize and accept that socio-economic risk analysis continuously plays an important role in natural hazard management.

Through a dwindling supply of resources combined with increasing demands for a high safety standard, the economic and social aspect is becoming significantly more relevant. The natural disasters of recent years led to enormous economic losses and thus considerably influenced the national budgets of all countries concerned. Hence a paradigm shift in the realm of natural hazard management was clearly at hands for several years. Although goals are far from met, a lack of public funding and future threat scenarios directly leads towards an increased acceptance of socio-economic and risk-based viewpoints.

The communication of risk and threat, an indispensable component in the management of all hazards and risks, proved successful and thus constitutes another way to cope with danger in an optimal way. To this extent, communication is an inherent component of risk analysis, risk prevention and disaster management. Risk communication supports the transfer of technical information about danger and risk potentials, the recognition and the minimization of evaluation discrepancies and the avoidance of rising conflict in the event of disagreements over the subject.

Communicating with the public is crucial. The communication of risk aims at providing people with a clear picture of the hazard and risk situation, at the same time involving them in the assessment of risk and taking their opinions on risk management into consideration. Its special significance for the management of natural hazards arises from the following points:
- For the organisations responsible for the protection against natural disasters a clear focus on reoccurrence-rates of natural catastrophes inevitably produces communication challenges. At this stage, however, it appears necessary to clarify liability and avoid any harm to reputation. These challenges are hard to cope with in situations of lacking communication skills.
- The development of the civil society also affects risk governance structures of natural hazard management such as non-governmental organisations that increasingly operate side-by-side with the state. Given the increased demand for community participation, the decisions made by authorities are not always accepted right away. Thus a greater need for communication emerges.
- The influence of the media on democracy requires all decision makers to legitimise their decisions via communication. Hence it becomes necessary to develop and employ instruments to observe and influence public opinion. Risk communication is thus made an important part of politics, being able to influence political opinions, particularly risk awareness. Whether disagreements over risk assessment lead to public controver-

sies, largely depends at last on the way in which it is communicated, if at all.
- The shortage of public funds in relation to natural hazards also leads to the increasing use of the subsidiary principle. Hereby it is asked for to take further personal precautions for protection against natural hazards. Moreover, wide-ranging preventative measures against any extreme event are not always sensible. Therefore it becomes necessary to discuss risk acceptance, to negotiate the distribution of state funds as well as to consider the risk perceptions of those concerned. Without risk communication this can not be achieved.

13.7 Future Challenges

For the future, alpS has to cope with various challenges:

- **Expanding the National and International Network**

The alpS - Centre for Natural Hazard Management has established itself as a widely recognised research and development platform in the community of alpine natural hazard management on national and on international level. alpS strives to strengthen this position beyond the Kplus funding period.

Meanwhile, the findings resulting from alpS projects are considered as state of the art and part of the aspired global excellence. Thus, alpS could be regarded as generally accepted in the scientific community and approved by public authorities.

- **Increase of Unique Selling Proposition**

The comparison of the research and development portfolio, the specific projects and associated results with the international state of the art shows that alpS has two fundamental unique selling propositions. The first one is attributed to the innovative and consequent cooperation culture, integrating all public and private stakeholder interests, which is a basic principle of alpS and is acknowledged and complimented by (inter-)national groups and organisations. The consistent transformation is regarded as an attractive model, both, on the level of projects (e.g. principle of unanimity) and on the level of the centre (e.g. structure of the board). The culture of cooperation will be preserved and strengthened in the future.

The second unique selling proposition is the emphasis on including a socio-economic approach to risk analysis and risk assessment in natural hazard management. As in other spheres, an economic consideration (including an economic analysis of social and political processes and an economic analysis of law) turns out to be important and relevant also in dealing with natural hazards. Facing sustainability as interdisciplinary goal, the economic consideration has to be considered as equal to ecological and societal concerns.

With a team of about ten scientists, alpS was able to develop this research area and to become a key player in this field, a position that alpS is intended to extend in the next years.

- **Further development towards a "Centre for Climate Change Adaptation Technologies"**

Together with the University of Innsbruck, the Austrian Academy of Science, the University of Natural Resources and Life Sciences Vienna, the University of Applied Sciences Vorarlberg and the European Academy Bolzano the Centre alpS is going to build up the new research centre "alpS – Centre for Climate Change Adaptation Technologies". Given its geographic location, Austria is predestined to take the leadership role in mountain research and development issues. This centre will fulfil this mandate and guarantee international visibility in the field of climate change adaptation technologies in mountain regions.

It is the objective of the alpS – Centre for Climate Change Adaptation Technologies to provide sound understanding of the regional effects of global climate change. Moreover, the aim is to analyse the resulting constraints and opportunities in key resource areas, and to develop solutions in the technological and socio-economic realm as answer to present and future challenges. Hence, an inter- and transdisciplinary network of stakeholders from science, business and politics will be installed for the development of sustainable methods, technologies and strategies that will enable societies in mountain regions to adapt to climate change.

The centre will focus entirely on adaptation measures, i.e. strategies that respond to impacts of global climate change on regional and local human-environment systems. These adaptation strategies will be context specific, can be re- or proactive, are of socio-economic or of technological nature, and have to increase adaptive capacity. To comply with sustainable regional development goals, adaptation strategies require careful assessment and are to be devised in concert by private and public enterprises, individual households and communities, research, industry and government. The principles of risk management are applied to ensure that negative impacts can be constrained, while opportunities can be taken.

The centre will be organised in five research areas. Within the areas Land, Water and Energy adaptation technologies are to be developed for three essential spheres of human life: housing, provision and recreation. In the Scenarios area detailed regional and local understanding of the impacts of climate change and economic development on sustainability goals will be developed. In the Tools area a portfolio of instruments facilitating the implementation of adaptation technologies will be compiled. While key investigations will take place in the Alps, transferability of methods and concepts will be tested in mountains regions worldwide. Successful adaptation will ensure regional sustainability in mountain regions and will play a key role in the sustainable development of societies in adjacent lowlands.

List of reviewers

Dr. James Bathurst Newcastle University, Institute for Research on Environment and Sustainability (IRES), Newcastle upon Tyne, UK

Prof. Dr. Thomas Glade University of Vienna, Department of Geography and Regional Research, Vienna, Austria

Prof. Dr. Bernd Hansjürgens Centre for Environmental Research UFZ, Department of Economics, Leipzig, Germany

Ass.-Prof. Mag. Dr. Peter Mandl University of Klagenfurt, Institute of Geography and Regional Research, Klagenfurt, Austria

Dr. Jan Christof Otto University of Bonn, Geomorphological and Enviromental Research Group, Bonn, Germany; *now:* University of Salzburg, Research Group "Geomorphology and Environmental Systems", Salzburg, Austria

Dr. Oliver Sass University of Augsburg, Institute for Geography, Chair of Physical Geography and Quantitative Methods, Augsburg, Germany; *now*: University of Cologne, Department of Geography, Cologne, Germany

Prof. Dr. Walter S. Schwaiger MBA Vienna University of Technology, Institute of Management Science, Vienna, Austria

Dipl.-Päd. Holger Schütz Research Centre Jülich, Program Group Humans, Environment, Technology, Jülich, Germany

Prof. Dr. Klaus Spremann University of St. Gallen, HSG Swiss Institute of Banking and Finance, St. Gallen, Switzerland

Prof. Dr. Jean-Jacques Wagner University of Geneva, CERG – Centre d'Etudes des risques géologiques, Section Geosciences and Environment, Geneva, Switzerland

Dr. Jürgen Weichselgartner University of Basel, Institute for Coastal Research, GKSS Research Center, Geesthacht, Germany

Prof. Dr. Karl Friedrich Wetzel University of Munich, Institute of Geography, Munich, Germany; *now*: University of Augsburg, Institute for Geography, Chair of Physical Geography and Quantitative Methods, Augsburg, Germany

List of contributors

Asztalos J. alpS – Centre for Natural Hazard Management, Innsbruck, Austria; *now:* Oberösterreichische Landesregierung, Linz, Austria

Baumann S. alpS – Centre for Natural Hazard Management, Innsbruck, Austria; *now:* IABG Industrieanlagen-Betriebsgesellschaft mbH, Ottobrunn, Germany

Brandner R. University of Innsbruck, Institute of Geology and Palaeontology, Innsbruck, Austria

Brückl E. Vienna University of Technology, Institute of Geodesy and Geophysics, Vienna, Austria

Chwatal W. alpS – Centre for Natural Hazard Management, Innsbruck, Austria; Vienna University of Technology, Institute of Geodesy and Geophysics, Vienna, Austria

Czaja J. alpS – Centre for Natural Hazard Management, Innsbruck, Austria; *now:* Eureka Navigation Solutions AG, Munich, Germany

Eder S. ILF Consulting Engineers GmbH, Rum, Austria

Eitzinger C. alpS – Centre for Natural Hazard Management, Innsbruck, Austria; *now:* UMIT – University for Health Sciences, Medical Informatics and Technology, Hall in Tirol, Austria

Fellin W. University of Innsbruck, Institute of Infrastructure, Innsbruck, Austria

Gamper C.D. alpS – Centre for Natural Hazard Management, Innsbruck, Austria; University of Innsbruck, Institute of Public Finance, Innsbruck, Austria; *now:* The World Bank, Washington, USA

Geist T. alpS – Centre for Natural Hazard Management, Innsbruck, Austria; *now:* FFG – Österreichische Forschungsförderungsgesellschaft mbH, Vienna, Austria

Geitner C. alpS – Centre for Natural Hazard Management, Innsbruck, Austria; University of Innsbruck, Institute of Geography, Innsbruck, Austria; *now:* Austrian Academy of Sciences, Mountain Research: Man and Environment, Innsbruck, Austria

Gruber M. alpS – Centre for Natural Hazard Management, Innsbruck, Austria; University of Innsbruck, Department of Banking and Finance, Innsbruck, Austria

Hegg C. Swiss Federal Research Institute WSL, Birmensdorf, Switzerland

Höfle B. alpS – Centre for Natural Hazard Management, Innsbruck, Austria; *now:* Vienna University of Technology, Institute of Photogrammetry and Remote Sensing (I.P.F.), Vienna, Austria

Jenewein S. alpS – Centre for Natural Hazard Management, Innsbruck, Austria; *now:* i.n.n. ingenieurgesellschaft für naturraum-management GmbH & Co KG, Innsbruck, Austria

Kirnbauer R. Vienna University of Technology, Institute for Hydraulic and Water Resources Engineering, Vienna, Austria

Kirschner H. alpS – Centre for Natural Hazard Management, Innsbruck, Austria; *now:* GEOCONSULT Consulting Engineers, Wals, Austria

Lammel J. alpS – Centre for Natural Hazard Management, Innsbruck, Austria

Lechner W. Telematica e.K., Dietramszell-Linden, Germany

Leiter A. alpS – Centre for Natural Hazard Management, Innsbruck, Austria; *now:* University of Innsbruck, Department of Economics, Innsbruck, Austria

Leonhardt G. alpS – Centre for Natural Hazard Management, Innsbruck, Austria

Meißl G. University of Innsbruck, Institute of Geography, Innsbruck, Austria

Mergili M. alpS – Centre for Natural Hazard Management, Innsbruck, Austria; University of Innsbruck, Institute of Geography, Innsbruck, Austria; *now:* Austrian Academy of Sciences, Mountain Research: Man and Environment, Innsbruck, Austria

Mertl S. alpS – Centre for Natural Hazard Management, Innsbruck, Austria; Vienna University of Technology, Institute of Geodesy and Geophysics, Vienna, Austria

Moran A.P. alpS – Centre for Natural Hazard Management, Innsbruck, Austria

Oberparleiter C. alpS – Centre for Natural Hazard Management, Innsbruck, Austria; *now:* Hydrographisches Amt, Autonome Provinz Bozen, Südtirol, Bolzano, Italy

Ortner S. alpS – Centre for Natural Hazard Management, Innsbruck, Austria

Pfeifer N. alpS – Centre for Natural Hazard Management, Innsbruck, Austria; *now:* Vienna University of Technology, Institute of Photogrammetry and Remote Sensing (I.P.F.), Vienna, Austria

Pöckl M. alpS – Centre for Natural Hazard Management, Innsbruck, Austria

Poscher G. ILF Consulting Engineers GmbH, Rum, Austria; *now:* p+w - Baugrund+Wasser GEO, Hall in Tirol, Austria

Prager C. alpS – Centre for Natural Hazard Management, Innsbruck, Austria; ILF Consulting Engineers GmbH, Rum, Austria

Raschky P.A. alpS – Centre for Natural Hazard Management, Innsbruck, Austria; University of Innsbruck, Institute of Public Finance, Innsbruck, Austria

Renk D. alpS – Centre for Natural Hazard Management, Innsbruck, Austria; University of Innsbruck, Institute of Infrastructure, Innsbruck, Austria

Rickenmann D. University of Natural Resources and Applied Life Sciences, Institute of Mountain Risk Engineering (IAN), Vienna, Austria; Swiss Federal Research Institute WSL, Birmensdorf, Switzerland

Rinderer M. alpS – Centre for Natural Hazard Management, Innsbruck, Austria

Rutzinger M. alpS – Centre for Natural Hazard Management, Innsbruck, Austria; *now:* ITC (International Institute for Geo-Information Science and

Earth Observation; Department of Earth Observation Science), Enschede, The Netherlands

Schneider-Muntau B. alpS – Centre for Natural Hazard Management, Innsbruck, Austria; *now:* University of Innsbruck, Institute of Infrastructure, Innsbruck, Austria

Schöberl F. University of Innsbruck, Institute of Geography, Innsbruck, Austria

Schönlaub H. TIWAG - Tiroler Wasserkraft AG, Innsbruck, Austria

Senfter S. alpS – Centre for Natural Hazard Management, Innsbruck, Austria; *now:* REVITAL ZT GmbH, Nußdorf-Debant, Austria

Stötter J. University of Innsbruck, Institute of Geography, Innsbruck, Austria

Tentschert E. Vienna University of Technology, Institute for Engineering Geology, Vienna, Austria

Thöni M. alpS – Centre for Natural Hazard Management, Innsbruck, Austria; *now:* UMIT- University for Health Sciences, Medical Informatics and Technology, Institute for Medical Law, Human Resources and Health Politics, Hall in Tirol, Austria

Veulliet E. alpS – Centre for Natural Hazard Management, Innsbruck, Austria

Weck-Hannemann H. University of Innsbruck, Institute of Public Finance, Innsbruck, Austria

Wiedemann P.M. Research Centre Jülich, Program Group Humans, Environment, Technology, Jülich, Germany

Wiesner R. alpS – Centre for Natural Hazard Management, Innsbruck, Austria

Zangerl C. alpS – Centre for Natural Hazard Management, Innsbruck, Austria

List of partners

Austrian Academy of Sciences Mountain Research: Man and Environment, Technikerstr. 21a, 6020 Innsbruck, Austria

Bergbahn AG Kitzbühel, Hahnenkammstr. 1a, 6370 Kitzbühel, Austria

GPS – Gesellschaft für professionelle Satellitennavigation mbH, Lochhamer Schlag 5a, 82166 Gräfelfing, Germany

Graz University of Technology, Institute of Navigation and Satellite Geodesy, Steyrergasse 30/II, 8010 Graz, Austria

GRID-IT Gesellschaft für angewandte Geoinformatik mbH, Technikerstr. 21a, 6020 Innsbruck, Austria

Hypo Tirol Bank AG, Hypo Passage 2, 6020 Innsbruck, Austria

i.n.n. ingenieurgesellschaft für naturraum-management GmbH & Co KG, Grabenweg 3a, 6020 Innsbruck, Austria

IIG – Innsbrucker Immobilien GmbH & CoKEG, Roßaugasse 4, 6020 Innsbruck, Austria

ILF Consulting Engineers GmbH, Feldkreuzstr. 3, 6063 Rum bei Innsbruck, Austria

KGV (Präventionsstiftung der Kantonalen Gebäudeversicherung) Prevention Foundation, Bundesgasse 20, 3011 Bern, Switzerland

Klenkhart & Partner Consulting ZT GmbH, Dörrstr. 85, 6020 Innsbruck, Austria

p+w - Baugrund+Wasser GEO ZT GmbH, Salvatorgasse 2, 6060 Hall in Tirol, Austria

Passer & Partner Civil Engineering GmbH, Andechsstr. 65, 6020 Innsbruck, Austria

Research Centre Jülich, Program Group Humans, Environment, Technology (INB-MUT), 52425 Jülich, Germany

Schlick2000 Schizentrum AG, Dorfzentrum, 6166 Fulpmes/Stubaital, Austria

Swiss Federal Research Institute WSL, Zürcherstr. 111, 8903 Birmensdorf, Switzerland

TeleConsult Austria GmbH, Schwarzbauerweg 3, 8043 Graz, Austria

Telematica e.K., Baiernrainer Weg 6, 83623 Dietramszell-Linden, Germany

TILAK – Tiroler Krankenanstalten GmbH, Anichstr. 35, 6020 Innsbruck, Austria

Tirol Werbung, Maria-Theresien-Str. 55, 6020 Innsbruck, Austria

Tiscover AG, Maria-Theresien-Str. 55-57, 6020 Innsbruck, Austria

TIWAG – Tiroler Wasserkraft AG, Eduard-Wallnöfer-Platz 2, 6020 Innsbruck, Austria

TopScan Gesellschaft zur Erfassung topographischer Information mbH, Düsterbergstr. 5, 48432 Rheine, Germany

University of British Columbia, Department of Earth and Ocean Sciences, 6339 Stores Road, Vancouver, British Columbia, V6T 1Z4, Canada

University of Innsbruck, Department for Strategic Management, Marketing and Tourism - Strategic Management and Leadership, Universitätsstr. 15, 6020 Innsbruck, Austria

University of Innsbruck, Institute of Banking and Finance, Universitätsstr. 15, 6020 Innsbruck, Austria

University of Innsbruck, Institute of Basic Civil Engineering - Technical Mathematics, Technikerstr. 13, 6020 Innsbruck, Austria

University of Innsbruck, Institute of Basic Civil Engineering, Surveying and Geoinformation Unit, Technikerstr. 13a, 6020 Innsbruck, Austria

University of Innsbruck, Institute of Geography, Innrain 52, 6020 Innsbruck, Austria

University of Innsbruck, Institute of Geology and Palaeontology, Innrain 52, 6020 Innsbruck, Austria

University of Innsbruck, Institute of Infrastructure, Division of Geotechnical and Tunnel Engineering, Technikerstr. 13, 6020 Innsbruck, Austria

University of Innsbruck, Institute of Infrastructure, Unit of Hydraulic Engineering (IWI), Technikerstr. 13, 6020 Innsbruck, Austria

University of Innsbruck, Institute of Meteorology and Geophysics (IMGI), Innrain 52, 6020 Innsbruck, Austria

University of Innsbruck, Institute of Psychology, Innrain 52, 6020 Innsbruck, Austria

University of Innsbruck, Institute of Public Finance, Universitätsstr. 15, 6020 Innsbruck, Austria

University of Innsbruck, Institute of Public Law, Innrain 80/82, 6020 Innsbruck, Austria

University of Natural Resources and Applied Life Sciences, Institute of Mountain Risk Engineering (IAN), Peter Jordanstr. 82, 1090 Vienna, Austria

University of Technology Munich, Chair of Geodesy, Arcisstr. 21, Munich, Germany

Vienna University of Technology, Institute for Engineering Geology, Karlsplatz 13//203, 1040 Vienna, Austria

Vienna University of Technology, Institute of Geodesy and Geophysics, Gußhausstr. 27-29//128, 1040 Vienna, Austria

Vienna University of Technology, Institute of Photogrammetry and Remote Sensing, Gusshausstr. 27-29//122, 1040 Vienna, Austria

Vienna University of Technology, Water Resources Engineering Department, Institute for Hydraulic and Water Resources Engineering, Karlsplatz 13//222, 1040 Vienna, Austria